DATE			

THE GENERAL HISTORY OF ASTRONOMY

Volume 2

Planetary astronomy from the Renaissance to the rise of astrophysics

Part A: Tycho Brahe to Newton

THE GENERAL HISTORY OF ASTRONOMY

General editor: Michael Hoskin, University of Cambridge

Volume 2

Planetary astronomy from the Renaissance to the rise of astrophysics
Part A: Tycho Brahe to Newton

EDITED BY

RENÉ TATON

Centre Alexandre Koyré, Paris

and

CURTIS WILSON

St John's College, Annapolis, Maryland

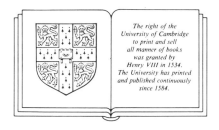

The right of the
University of Cambridge
to print and sell
all manner of books
was granted by
Henry VIII in 1534.
The University has printed
and published continuously
since 1584.

CAMBRIDGE UNIVERSITY PRESS

Cambridge

New York   Port Chester

Melbourne   Sydney

Published by the Press Syndicate of the University of Cambridge
The Pitt Building, Trumpington Street, Cambridge CB2 1RP
40 West 20th Street, New York NY 10011, USA
10 Stamford Road, Oakleigh, Melbourne 3166, Australia

First published 1989

Printed in Great Britain at the University Press, Cambridge

British Library cataloguing in publication data
The General history of astronomy.
Vol. 2: Planetary astronomy from the
Renaissance to the rise of astrophysics.
Pt A: Tycho Brahe to Newton
1. Astronomy to 1950
I. Taton, René II. Wilson, Curtis
520'.9

Library of Congress cataloguing in publication data
Planetary astronomy from the Renaissance to the rise of astrophysics
edited by René Taton and Curtis Wilson.
p.   cm. – (The General history of astronomy; v. 2)
Includes bibliographies.
Contents: pt A. Tycho Brahe to Newton.
ISBN 0 521 24254 1 (pt A)
1. Astronomy – History. 2. Astrophysics – History. I. Taton,
René. II. Wilson, Curtis. III. Series.
QB15.G38 1984 vol. 2
520'.9 s–dc19
[520'.9'03]   88-25817

ISBN 0 521 24254 1

CONTENTS

Preface vii   *Acknowledgements* x

THE CONTENTS OF PART B OF VOLUME 2 ARE:

PREFACE

Volume 2 of *The General History of Astronomy* deals with the history of the descriptive and theoretical astronomy of the solar system, from the late sixteenth century to the end of the nineteenth century.

In the European tradition from the time of Plato to the sixteenth century, theoretical astronomy viewed its task as the reduction of the apparent celestial movements to combinations of uniform circular motions. This formulation was still axiomatic for Copernicus; indeed, Ptolemy's violation of the axiom in his *Almagest* was an important stimulus leading Copernicus to undertake his renovation of astronomy. To many astronomers working in the later sixteenth century, after the publication of Copernicus's *De revolutionibus* (1543), the major achievement of this work was that it had freed astronomy from such violations of principle; in contrast, the associated rearrangement of circles that put the Sun at the centre and attributed motion to the Earth was, in the view of many, an absurd error. Copernicus's work, both in its adherence to the long-held axiom of uniform circular motion and in its organization, was thoroughly traditional; and it is thus fitting that Volume 1 of *The General History of Astronomy*, which is devoted to the history of ancient and medieval astronomy, should conclude with it. But, as all the world knows, the Copernican rearrangement contained the seeds of a further transformation.

The story of that transformation begins with the work of the Danish astronomer Tycho Brahe, which is also the starting-point for Volume 2. In 1563 young Tycho, then a student at Leipzig, was shocked to discover that the accepted astronomical tables of the day (the Alfonsine based on Ptolemaic theory, and the Prutenic based on Copernican theory) were, both of them, days out in predicting the conjunction of Jupiter and Saturn that took place on 24 August of that year. Tycho's response, over the remaining years of the century, was to give to the accumulation of accurate astronomical observations a priority and importance it had never had before. The store of observations he accumulated in the process became the empirical basis for Kepler's justly titled *Astronomia nova* (*New Astronomy*), which made possible a new level of accuracy in the prediction of planetary motions.

Inextricably bound up in Kepler's theory was a new celestial physics, founded on the Copernican vision of the solar system. It was an attempt to account for the observed motions by means of hypothesized quasi-magnetic forces and structures. Kepler did not and could not derive his 'laws' of planetary motion – elliptical orbit and areal rule – from observations alone; on the contrary, his analysis of Tycho's observations was directed by his own highly conjectural celestial physics. Consequently, his contemporaries and successors could not but find the hypothetical foundations dubious. Yet the new predicative accuracy that Kepler achieved was too remarkable to be ignored. Astronomers were ineluctably faced with a challenge: to achieve results as successful as Kepler's, but with more convincing physical foundations. It is not easy to imagine what the history would have been if Kepler's *Astronomia nova* had never appeared to pose this problem. At least we can say: exceedingly different.

The seventeenth century was a period of complex transition in thought, belief, and knowledge of the world; delineation of the aspects of the transition that bear on astronomy is a major concern in Part A of Volume 2. From 1610 onward, telescopic observations of the heavens revealed new facts that had to be incorporated in the system of the world, whether this was conceived to be geocentric or heliocentric. Studies of refraction and parallax showed the Earth to be some 18 times further from the Sun than had previously been believed, and so nearly 6000 times smaller in relation to the volume of the Sun. New studies of motion and force, of falling bodies and the impacts of bodies, were proposed by Galileo, Descartes, and others, a major motive being to show that the heliocentric system was not incompatible with physical principle. The Keplerian planetary tables were tested repeatedly, and continued to prove superior to earlier tables.

An impressively persuasive replacement for Kepler's celestial physics was at length provided in 1687 by Isaac Newton's *Principia*. Part A of Volume 2 concludes with the story of its emergence, and a summary of the astronomical results that Newton succeeded in deriving from his principle of universal gravitation.

Part B, after a section devoted to the gradual acceptance of the Newtonian doctrine during the first half of the eighteenth century, takes up the history of the efforts, from the 1740s onward, to deduce detailed mathematical consequences from universal gravitation. It is the story of what the eighteenth century called 'physical astronomy', but what Laplace in 1799 renamed '*mécanique céleste*', celestial mechanics. The challenge of deducing the consequences triggered the development of new forms of mathematics. Of notable importance here were the trigonometric series, first introduced by Euler, but with later contributions from d'Alembert, Clairaut, Lagrange, and Laplace; a new mechanics of the rotation of rigid bodies, to which d'Alembert and Euler were the chief initial contributors; and the method of variation of the constants of integration in the solution of differential equations, developed by Euler, Lagrange, and Laplace. The story also involves the introduction of statistical calculations into astronomy: at first, with a less than satisfactory outcome, by Euler; then more successfully by Tobias Mayer, whose example was followed by Laplace. With Laplace and Gauss the use of statistical procedures in bringing multiple data to bear on the determination of astronomical constants became *de rigueur*.

Through the nineteenth century the grand theoretical questions remained those that Laplace had forcefully posed and prematurely claimed to answer: the question of the stability of the solar system, and the question of the adequacy of universal gravitation to account for the observed motions of its constituent bodies. By the end of the century Simon Newcomb and his collaborators would achieve a precision in their prediction of planetary motions measured in seconds or fractions of a second of arc. Yet some inconsistencies remained, among them an anomalous motion of the node of Venus and an anomalous motion of the perihelion of Mercury. The first of these was a statistical artifact; the second would become evidence for a new theory of gravitation, which would supersede that of Newtonian physics.

The scope of Volume 2 does not include general relativity. Nor does it embrace astrophysics, which came into being in the 1850s with the development by Kirchhoff and Bunsen of spectral analysis. Both topics belong to Volume 4, *Astrophysics and Twentieth-century Astronomy to 1950*. Similarly excluded from the scope of Volume 2 are the topics of stellar astronomy and cosmology, and astronomical instruments, institutions and education, from the Renaissance to the beginnings of astrophysics: these topics constitute the subject-matter of Volume 3. At certain points the concerns of Volumes 2 and 3 overlap. Tycho's sighting instruments were highly relevant to his observational achievement; and from the 1660s and 1670s onward, the pendulum clock, telescopic sights, and the filar micrometer were similarly relevant to the programmes for the improvement of lunar and planetary tables that were adopted by the newly founded Paris Observatory and Greenwich Observatory. Again, Bradley's discoveries of the aberration of light and nutation were prerequisites for the attainment of seconds-of-arc accuracy in the prediction of planetary and lunar positions. For these and similar topics concerned with the stars and with observational instruments and their institutional context the reader must be referred to Volume 3.

Our aim in Volume 2, as in the other volumes of the series, has been to throw light on the development of astronomy as an inventive human activity. We have sought to view the questions and problems of the astronomers of a given time in the very way in which those astronomers saw them, without regard for what has later come to be accepted as 'correct'. We have not attempted an encyclopaedic completeness of coverage; rather, our goal has been to provide an intelligible account of the major endeavours through which astronomy has evolved.

A word is in order with respect to the division of tasks between the two editors of Volume 2, René Taton and myself. Professor Taton resigned his editorship in 1983. By this time he had drawn up the general plan of the volume, and had engaged authors to write rather more than half of the sections envisaged. The engaging of authors for the remaining sections has been my responsibility, and I have also undertaken some reorganization of materials, reducing the original tripartite scheme to a two-part plan, and introducing into Part B a number of new sections which focus on the application of theory to observation.

Finally, I have the grateful duty of acknowledging the extensive guidance and generous assistance provided a neophyte editor by the General Editor of the series, Michael Hoskin. He has again and again given generously of his time, knowledge, and thought to the solution of the problems and difficulties, whether major or minor, encountered in the assembling and editing of this volume. Our undertaking has been, in every aspect, a joint endeavour.

Annapolis, Maryland   Curtis Wilson

ACKNOWLEDGEMENTS

Albert Van Helden, Chapters 6 and 7, is grateful for comments received from Owen Gingerich and Olaf Pedersen.

Ewen Whitaker, Chapter 8, thanks the following individuals and institutions for kindly supplying, gratis, negatives of lunar images and maps: O. Van de Vyver, SJ, Henri Michel (Société Belge d'Astronomie and the Cabinet des Estampes, Bibliothèque Nationale, Brussels), A. Orte (Observatorio de Marina, San Fernando), the Astronomer Royal for Scotland (Royal Observatory, Edinburgh), and F. Warren Roberts (Sid W. Richardson History of Science Collection, Humanities Research Center, University of Texas at Austin). He also thanks the following for placing their collections at his disposal: the Earl of Egrement and Leconfield, the Royal Greenwich Observatory, the Royal Astronomical Society, and l'Observatoire de Paris. Finally, he thanks Godfrey Sill for translating the entire text and quotations from Van Langren's map, excerpts from which were used in the text, and Micheline van Biesbroeck Wilson for obtaining data, while visiting Paris, regarding the whereabouts of the large la Hire drawing of the Moon.

S. Débarbat and Curtis Wilson, Chapter 9, are grateful for the contributions and discussions at the Table Ronde du CNRS held in Paris in June 1976 on "Roemer et la vitesse de la lumière".

Curtis Wilson, Chapters 10 and 13, is grateful to the Program for History and Philosophy of Science of the US National Science Foundation for support of his research on the history of lunar and planetary astronomy in the seventeenth century after Kepler. An earlier version of the first part of Chapter 13 was published in *The Great Ideas Today* for 1985 (Encyclopaedia Britannica, 1985) under the title "Newton's path to the *Principia*". To D.T. Whiteside, for extensive and painstaking criticisms and suggestions regarding both chapters, his indebtedness is immense. I. Bernard Cohen and R.S. Westfall read a near-final version of Chapter 13 and so made possible the elimination of a number of erroneous or misleading formulations. For the interpretations finally adopted, however, the author alone is responsible.

PART I

Tycho, Gilbert and Kepler

1

Tycho Brahe

VICTOR E. THOREN

For as long as histories of astronomy have been written, heliocentrism has been regarded as the hallmark of modern astronomy. In accordance with this tradition, Nicholas Copernicus (1473–1543), as the effective originator of heliocentric doctrine, has been hailed as the founder of modern astronomy. In fact, however, except for the motion of the Earth, the revolutionary element in Copernicus's work is very small: in most respects his *De revolutionibus* (1543) follows Ptolemy's *Almagest* so closely that he can equally well be regarded (see Volume 1) as the last great practitioner of ancient astronomy. On this view, it was the seventy-year period following Copernicus's death in 1543 that actually saw the transition to modern astronomy. And insofar as any such development can be attributed to the influence of one person, that transition was wrought by the ideas and efforts of the Danish astronomer Tycho Brahe (1546–1601).

Only a few facts are necessary to establish the background of Tycho's career as an astronomer. The first is that he was born into the small oligarchy of families that had controlled political, economic, and social power in Denmark for at least 200 years: had he lacked the status conferred by birth into this privileged class, any work he might have done as an astronomer would have been carried out on a completely different basis. The second is that, except for the rather bizarre circumstances of being raised as a fosterson by his father's brother, Tycho simply would not have been given enough freedom to develop predilections for astronomy: his four younger un-fostered brothers were esquired into castles in their early teens and knighted before they were twenty. Third, Tycho's education in astronomy was essentially a private affair (as it was for any sixteenth-century astronomer), and a very extended one. Long after he could have begun to make his mark in society, this oldest son and oldest grandson of Councillors of the Realm was still drifting from one university circle to another. Only in his twenty-fifth year did Tycho take up his residence in Denmark. And then it was not at court, but at his maternal uncle's provincial Abbey of Herrevad – where he soon generated more controversy by forming a permanent liaison with a commoner.

Tycho's influence and reputation stems from achievements that fall into three quite distinct categories. The one that was most important during Tycho's life and for the fifty years following it was cosmological in character. It was initiated in 1572 by the appearance of what has come to be called Tycho's nova (now classified as a supernova), and raised to the dimension of a crusade by the appearance of the even more spectacular comet of 1577. During the ensuing decade Tycho composed lengthy monographs on each phenomenon. These would form the great bulk of his life's literary output, and would include the discovery which he himself undoubtedly regarded as the outstanding achievement of his career – the so-called Tychonic system of the world. Although Tycho naturally sought to endow his cosmological writings with the authority so justly due to observations made with his best instruments, the New Star and comet were both observed with instruments that were primitive by Tycho's standards. The mature instruments – as well as all of the other remarkable facilities associated with Tycho's name – were products of the second strand of Tycho's career, and appeared only during the decade following 1576, when Tycho was granted the island of Hven and provided with annual stipends to underwrite his work there. By 1585 he had established the modern prototype of the scientific research institute (Figure 1.1), featuring housing, instrument-mak-

ing works, instruments, observatories; a collection of artisans, students, unpaid assistants, and salaried co-workers to staff his manifold activities; and even papermaking and printing shops to publish his results. Only with the completion of these facilities could he proceed to the third facet of his work – technical studies of the celestial motions that would culminate in nothing less than the complete renovation of astronomical science.

Tycho's cosmological activities

The event that signalled the formal beginning of Tycho's career as an astronomer occurred on 11 November 1572, a month before his twenty-sixth birthday. As he was returning from his alchemical laboratory that evening for supper, he noticed an unfamiliar star-like object in the sky – one that was not only clearly alien to the constellation in which it appeared, but was also brighter than any star or planet he had ever seen. Its appearance seemed to deny that it could be a comet, and a few nights of observation revealed that it was not only too stationary to be a comet, but too stationary even to be associated with the realm below the Moon to which comets were supposed to be restricted.

By 1572 the Aristotelian dichotomy between the earthly sphere of *change*, and the heavenly spheres beyond the Moon, where things were assumed to retain forever a pristine perfection unimaginable to anyone living on Earth, had been accepted by philosophers from Greek, Moslem, and Christian cultures over a period of nineteen centuries. The primary reason for the longevity of Aristotle's cosmology, of course, was the fact that, in general, it fitted the evidence. With respect to comets it was seriously erroneous, but even there it was considerably less vulnerable than might be expected. The accepted nature of comets as atmospheric phenomena placed them outside the domain of astronomy; and their motions must have produced enough phantom parallaxes to mislead or discourage the few sceptics who sought to investigate them prior to Tycho's day. Nor, indeed, was the situation greatly different in 1572. For most of Tycho's contemporaries, change was not only theoretically but, by virtue of the weight of scienti-

fic terminology, almost *logically* restricted to the world below the Moon. Most of them were so compelled by the logic of Aristotelian cosmology that it did not occur to them to doubt that there would be parallax in a body so manifestly 'generated', and thus so obviously associated with the terrestrial sphere of change. Many others actually obtained observational results through which they satisfied themselves that the new star was indeed below the Moon where it was theoretically expected to be, while some insisted on calling it a comet even though they could find no measurable parallax for it. Even men who were willing to accept observations that placed the 'star' above the Moon could remain mired in the rest of Aristotle's doctrine: the Landgrave of Hesse-Kassel, John Dee (1527–1608), and Thomas Digges (*c.* 1546–95) assumed that any change of brightness of the 'star' must be purely apparent, and therefore the reflection of a change in its height. But despite the fact that even the very terms available for discourse on the subject implied commitments to the Aristotelian world view, Tycho succeeded in detaching himself sufficiently from his preconceptions not only to test the object for parallax, but to believe the negative results he obtained.

By the beginning of 1573, Tycho had decided that the object was indeed nothing less than a new star, and had written a manuscript that attempted to rationalize its origin and interpret its astrological significance. Early in the summer these reflections, along with a description of the progressive decline of the star (to the brightness of Jupiter in December, and down to that of a second magnitude star by May), and computations of the lunar eclipse predicted for 8 December 1573, appeared as a small book published in Copenhagen.

From the standpoint of sixteenth-century thought, Tycho's nova (Figure 1.2) had implications in principle as radical as those of heliocentrism. The net of Aristotelianism encompassed everything: once it started to unravel, the whole world would fall apart. But in fact nothing so serious happened in 1572, and what little perturbation of opinion there was owed little to Tycho's obscure volume. Only in conjunction with the

1.1. Tycho's observatory on Hven, Uraniborg, viewed from the east. The observatory building itself is at the centre, and it is surrounded by ornamental gardens and a protective wall. At the southern corner is the printing house, and at the northern the quarters for the servants.

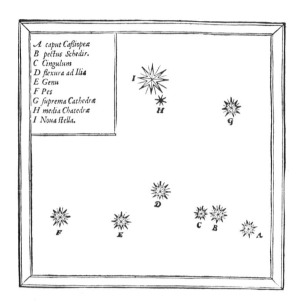

1.2. Tycho's nova (*I*) shown in relation to nearby stars of Cassiopeia.

comet of 1577 would the New Star, and Tycho, begin to have some impact.

Already in his publication on the New Star, Tycho suggested implicitly that comets might also turn out to contradict the Aristotelian view of the cosmos. His first chance to test this hunch proved every bit as interesting as he anticipated. The comet of 1577 was as bright as Venus and attended by a tail 22° long. It attracted so much attention as to seem almost made to order for Tycho's purposes. Of course, it guaranteed an audience for everyone else's opinion, too; and Tycho's experience with the literature on the New Star gave him a very good idea as to how closed-minded and muddle-headed those opinions would be, and how difficult it would be to prevail against them among a public who tended to share these intellectual handicaps. Accordingly, he seems to have conceived from the beginning a book considerably more ambitious than his ineffective work on the New Star – one that would stand out from its competitors in rigour, in detail, in sheer weight, if nothing else. And so he presented his findings in a detail that was unprecedented in the history of astronomy. Starting from observed distances between the comet and various reference stars whose coordinates he re-determined himself, Tycho painstakingly determined successive positions for virtually every one of the

thirty non-cloudy nights of the comet's duration. He then rationalized these positions into a trajectory which he defined in both ecliptic and equatorial coordinates, by determining its intersection with and inclination to both the ecliptic and the equator. While his results are not especially interesting, his methods certainly are. For, by making – and reporting – seven trials of each element, rather than the single determination that any of his predecessors would have offered, Tycho inaugurated the modern scientific practice of using redundant data and admitting scatter in his results.

After ninety pages of preparatory computations the actual task of proving that the comet "ran its course high above the sphere of the Moon in the ether itself" was almost anticlimactic. Several of Tycho's determinations showed no parallax at all. But for reasons of candour and partly of hunch, Tycho chose for his most detailed presentation a determination that showed the comet to have been some four times as far away as the Moon, and generalized his various results to estimate a minimum distance for the comet of about 300 Earth-radii (five times as far as the Moon).

While Tycho was labouring over such aspects of his analysis as re-determining the coordinates of his reference stars, his contemporaries rushed to comment on the comet. By the time he was struggling to prove such things as his contention that through some kind of optical illusion, the tail of this comet appeared to be directed away from Venus rather than away from the Sun (Figure 1.3), well over a hundred publications had descended on the reading public of Europe. Relatively few of them were concerned with the questions that interested Tycho, and fewer still handled them with any kind of competence. If Tycho was to prevail against these odds, he knew he would have to make a special effort. Accordingly, he struck on the stratagem that was probably to be responsible for the impact both of his book on the comet and of the great volume he subsequently published on the New Star – that of appending to his own analysis, critical reviews of the results of his competitors. Aggregating to a length exceeding the 200 pages of his own text, these reviews involved point-by-point analyses of the efforts of eight authors, and curt but reasoned dismissals of the results of eleven others. The principal goal, of course, was to expose the errors of those who had derived a measurable par-

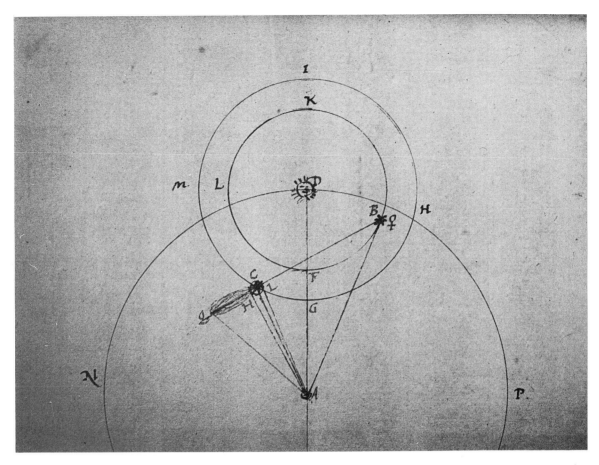

1.3. Tycho's placement in 1585 of the comet of 1577 in a geoheliocentric framework. Earlier Tycho had followed Apian and others in supposing that the tail pointed away from the Sun, but by 1585 his calculations were showing that the tail was directed opposite to Venus. Later still, as indicated in the text, he concluded that this effect was an illusion.

allax. But even those (such as Michael Mästlin) who had obtained results showing the comet to be above the Moon found their instruments, their observing procedures, and even their computations subjected to rigorous examination. And in the case of the Landgrave of Hesse-Kassel, who had neglected to derive any conclusions from his observations, Tycho converted the raw data into a minor treatise on the detection of parallax.

Tycho's arguments were finally circulated from his private press in 1588 as *De mundi aetherei recentioribus phaenomenis*. He contemplated but never completed another volume generalizing his result by means of observations of several succeeding comets. In fact, however, little room remained for intellectually respectable dispute concerning the astronomical – as opposed to meteorological –

status of comets; and after the volume was officially published in 1603 and again in 1610, the issue was essentially settled.

The Tychonic system

Closely associated with Tycho's challenge to Aristotelian cosmology was a parallel challenge to the world systems of Ptolemy and Copernicus. As early as 1574, in guest lectures given at the University of Copenhagen, Tycho had expressed the ideal of avoiding both the mathematical absurdity of Ptolemy's equant and the physical absurdity of Copernicus's moving Earth. Nor was the ideal original with him. Already in the first generation after Copernicus, Erasmus Reinhold (1511–53) and Gemma Frisius (1508–55) had made statements that suggest that they understood fully the

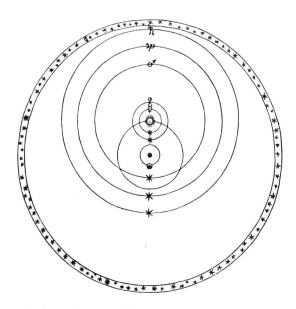

1.4. The Tychonic world system. The Earth is at rest at the centre, encircled by the stars. The five planets orbit around the Sun, while the Sun and the Moon orbit around the Earth.

geometrical possibility of utilizing Copernican models for a geostatic cosmology. By 1578, Tycho's readings and ruminations on the subject had led him as far as the so-called Capellan system, with the inferior planets in motion around a revolving Sun: and by 1584 he had envisaged applying the conception to the superior planets as well. But in an age in which practically everyone believed that some kind of solid sphere carried each of the planets on its appointed rounds through the heavens, the resulting intersection of Mars's orbit with the Sun's (Figure 1.4) at first prevented Tycho from taking the scheme seriously. He was finally induced to so by the implications of the comet of 1577.

When Tycho first saw the comet, it was moving rapidly away from the Sun in the evening sky. After the first week the elongation increased less quickly, until, after reaching a maximum of almost 60°, it actually began to diminish again. During all this time the comet had progressively faded, suggesting that it was moving away from the Earth. These data virtually cried out for a (retrograde) orbit circling the Sun, and that is how Tycho accounted for them already in 1578 (Figure 1.5). This geometry could be used to deduce distances for the

comet – distances (from the Earth) that ranged from a small fraction of the 1150 Earth-radii attributed to the Sun's 'altitude', to well above the Sun, when it finally disappeared. Only some years later, however, does Tycho seem to have realized that if the comet had thus gone right through the Ptolemaic spheres of Venus and Mercury, there could be no solid spheres, and no reason why the orbit of Mars could not intersect the orbit of the Sun.

The Tychonic system was published in 1588, as a hastily inserted (eighth) chapter of *De mundi*. There can be little doubt that Tycho regarded it as his most significant achievement, and in the short term it surely was. As a geometrical equivalent of the Copernican system, it was capable of representing every aspect of the astronomical phenomena without demanding allegiance to a moving Earth, for which there would be no proof until much later. During the seventeenth century, therefore, after telescopic discovery of the phases of Venus and Mars rendered Ptolemaic cosmology untenable, professional favour tended to shift either to Tycho's scheme or to the so-called semi-Tychonic version of his disciple Longomontanus (1562–1647) and others involving a rotating Earth (see Chapter 3).

Tycho's conception of the rest of the physical world was, likewise, a sensible compound of tradition and innovation. For the most part, he subscribed to the account of its workings synthesized by Aristotle and adapted by medieval Christendom. But his studies of the (non-)parallaxes of comets led him to doubt either that there was any kind of sphere of fire just below the Moon (from which comets and meteors were supposed to be generated), or that the universe beyond the Moon was partitioned off by solid spheres into individual domains for each of the planets. For him, rather, the Aristotelian sphere of air extended essentially up to the Moon, thinning gradually from a thick layer filled with impurities (which produced refraction) up to a completely free aether-filled universe beyond the Moon. The planetary portion of this universe, of course, was essentially Copernican, even as far as scale, since Tycho retained through his lifetime the 3′ parallax adopted by Copernicus from the ancients. But the stellar portion remained Ptolemaic. Although Tycho could find no parallax for the stars, he had no doubt that they were situated just beyond Saturn. No small part of his unwillingness to consider Copernicanism stemmed from his inability to imagine that God would have created a universe containing as much wasted space as was implied by the calculated Copernican distance to Saturn combined with the absence of stellar parallax. In addition, stars whose (pre-telescopic) angular diameters (1′ for third magnitude) implied quite reasonable sizes (four times the Earth's diameter for first magnitude) at a distance just beyond Saturn's, had to be 700 times as large, or comparable to the size of the Earth's *orbit*, if their parallax were as little as 1′. Tycho did not believe that all the stars were necessarily at precisely the same distance, however. He conceived of them as ordered in a thin shell, fixed into constellations and alignments that were still identical to descriptions given in Antiquity, but probably rotating axially to produce the phenomenon of scintillation.

Tycho's astrological views

So involved in his cosmology as to be virtually part of it was Tycho's attitude toward astrology. That the universe was an articulated unit created by God as a home for man was the foundation of Tycho's metaphysics. Again, it was a traditional theme strongly influenced by certain recent ideas due to Paracelsus (1493–1541) which linked man to the physical world even more closely than did Aristotle's doctrines. The notion that all aspects of the natural world were mutually interconnected provided Tycho with life-long motivation to elucidate the consequences of heavenly events and the effects of earthly substances on human beings. Experiments in medical alchemy earned him early recognition as an adept, and enabled him to dispense medications whose effects were highly valued and whose recipes were circulated long after his death. His investigations into the effects of celestial influences were less rewarding. Empirical attempts to relate weather to the cyclings of the planets induced him to record daily conditions for fifteen years, and circulate (under the names of two of his disciples) publications on weather forecasting in 1586 and 1591. Studies of the horoscopes of famous men proved even less instructive, and during his lifetime Tycho steadily lost confidence in the accuracy of 'personal' forecasts. The belief that the free will of an individual could counter astral influences assumed progressively greater prominence in his prognostications, and by 1587 was provoking Tycho to express serious reservations about the

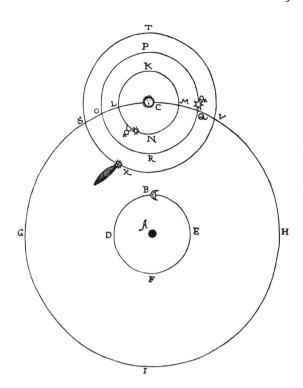

1.5. The comet of 1577 in relation to the inner bodies of the Tychonic system. Mercury, Venus, and the comet orbit the Sun, while the Sun and the Moon orbit the Earth.

whole horoscope enterprise. Yet Tycho could never quite shake off his feeling that in a created universe, the constituent parts would have to harmonize sufficiently to make some kind of astrological prediction possible. So, while the gulf between the intellectual allure of astrology and its operational value provided such mixed reaction that Tycho often managed both to denounce and defend it in virtually the same breath, his denunciations usually touched either the practice of the art or the state of the art, rather than the essence of the art itself. At bottom, he seems to have remained convinced that the weak link in the system was the science on which it was founded – astronomy.

The renovation of astronomy

According to Tycho, the conviction that nothing less than a renovation from the ground up would solve the problems of technical astronomy was a product of his student days. Only with his enfiefment on the island of Hven at the age of thirty, however, was he able to begin serious progress

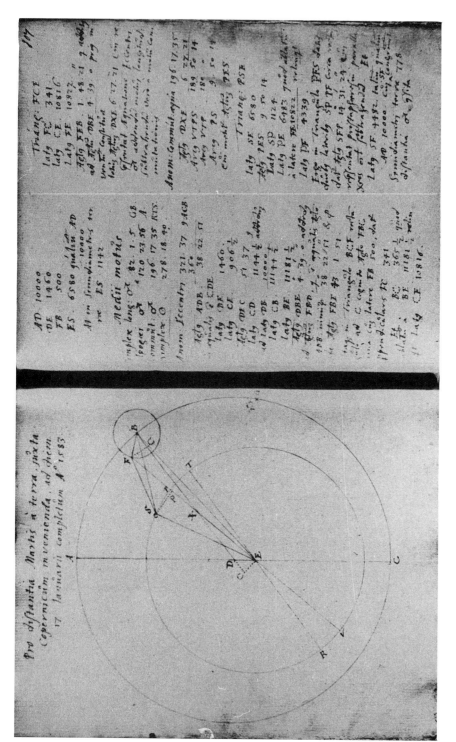

1.6. A diagram for determining, on the basis of the Copernican system, the distance of Mars from the Earth on 17 January 1583, with calculations in the hand of one of Tycho's assistants. The Earth–Mars distance (SF) is shown to be less than the Earth–Sun distance (SE); the same would be true in Tycho's geoheliocentric system.

According to Kepler (*Astronomia nova*, p. 64), Tycho often affirmed that from the observations of 1583 he had found the parallax of Mars to be greater than that of the Sun. But using Tycho's observational data Kepler found the parallax of Mars to be indetectable. He conjectured that Tycho's assistants had merely determined the Earth–Mars distance on the assumption of the Copernican system, as in the above diagram, and that Tycho had taken their result as a genuine empirical determination of parallax. But Kepler adds that he does not know whether this mistaken trust in his assistants' calculation was the sole basis for Tycho's pronouncement on the parallax of Mars.

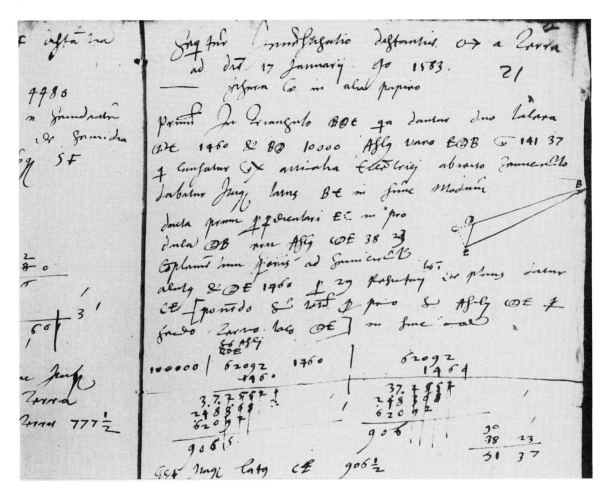

1.7. In Tycho's own hand, calculations corresponding to the diagram in Figure 1.6. Thus, as J.L.E. Dreyer showed, Tycho cannot have been ignorant of the basis of the calculation performed by his assistants, and Kepler's conjecture is wrong. In his *Progymnasmata* (I, 414), Tycho states that with a very exact instrument, measuring to parts of a minute, he has diligently and precisely measured the parallaxes of Mars when in opposition, and that in a more convenient place he will set forth his results – a promise that was never fulfilled. The parallax of Mars when in opposition to the Sun can rise to about 23″, but for the solar parallax Tycho always retained the ancient value of 3′; hence his pronouncements concerning the Martian parallax must be judged to be without foundation.

towards this goal. And then it was nearly five more years (late 1580) before he moved into his still-unfinished home/observatory (Uraniborg), and began to see instruments issue from his shop. As the size and number of these instruments overflowed the space available in Uraniborg, he built a separate underground observatory named Stjerneborg in 1584. Still, the results were worth the wait. Prior to the establishment of his own shop, he had to depend on instruments designed by and commissioned from commercial artisans. They were good enough to ensure that his observations of the New Star and the comet were as accurate as any ever made in Europe – perhaps to within 4′, rather than the 10′–15′ which was the best most earlier astronomers had been able to achieve. (There are some exceptions: the mean relative error in Bernhard Walther's observations, made about the end of the fifteenth century, is about 5′, once the error in the position of the reference body is removed.) From his own facilities, however, he turned out a succession of instruments capable of measuring accurately to less than 1' of arc. The sights, divisions, designs, and modes of construction which made this precision possible are described in Volume 3 of this work. It will suffice here to say that Tycho had an unprecedented concern for accuracy and a unique ability to detect and eliminate flaws in either observational procedure or mechanical design; and that he also had access to extraordinary resources and to craft traditions whose techniques had blossomed conspicuously during his lifetime. The result was an arsenal of perhaps ten first-line (but variously specialized) instruments, and about the same number of less successful ones which would still have been the pride and joy of almost any astronomer before 1700. With them he would achieve observational results that remained unsurpassed and only rarely matched long after the advent of the telescope should have rendered his open-sight instruments obsolete. More significantly, he created almost single-handedly the empirical ethic on which modern astronomy is founded.

That the establishment of an entirely new standard for astronomical instrumentation was enormously expensive in money and time goes without saying. What Tycho tended to emphasize was the pecuniary cost, which was, indeed, enormous. He also alluded frequently to the price he paid in professional time, although probably not even he ever fully realized just how much time he invested in founding his shop, designing the instruments, supervising their production, setting them up for effective service, and monitoring their continual performance – all under the inefficiency factor inherent in any protracted, intermittent task. Compounding this factor was the delay imposed by the nature of his construction programme – the fact that the prospect of increasingly better instrumentation compelled Tycho to postpone virtually every aspect of his work until he finally completed his mature quadrants around 1585. This, in turn, led to the ultimate cost – loss of recognition by posterity for his accomplishments in the third, most technical, mode of his activity.

Although the accumulation of observations, *per se*, constituted only a minor fraction of Tycho's work, whether gauged in effort or results, Tycho's reputation lies as exclusively in the category of 'pure observer' as anyone's in the history of astronomy. The reason is not hard to divine: it is the notoriety of his role as provider of the data for the brilliant researches of Johannes Kepler (1571–1630) on the laws of planetary motion. No one has ever suggested either that that role was unimportant, or that Tycho himself would have achieved the same results with his data. Yet it is ironic, to say the least, that Tycho's reputation should derive primarily from his work on the one aspect of his renovation of astronomy that he failed to complete. For, in fact, he was one of the great theorists in the history of the discipline: one of the two major cosmologists between Copernicus and Newton, the first serious analyst of observational technique, and the author of studies of the Sun, Moon, and stars that rival in importance the theoretical achievements of Ptolemy, Copernicus and Kepler.

The solar theory

If there is any area in which the nature of Tycho's efforts conforms to the popular image of his work, it is the solar theory. From 1578 until well into the 1590s, he recorded meridian altitudes for more than a hundred noons a year, usually with several different instruments. Of course, this provided many more data than could be used rigorously. It would be more than two hundred years before C.F. Gauss would show astronomers how to utilize excess data to maximum advantage. Tycho therefore had to make an *a priori* assumption concerning the nature of the solar orbit. It is unlikely that he ever considered the possibility that it might be anything but the circle eccentric to the Earth that had been used by all his predecessors. And when he reported his results formally to fellow-professionals he would have to use the traditional three items of data to position the circle. But through most of the research phase of his work he proceeded much more empirically than had his predecessors. He began, probably in 1580, by making comparisons between the longitudes derived from his meridian altitudes over the years, and those predicted by the Alfonsine and Copernican ('Prutenic') tables. These comparisons showed that the maximum correction provided by Ptolemaic theory (2° 23') was too large, but that the one adopted by Copernicus (1° 51') was too small. Already by the autumn of 1580 Tycho had decided to try a correction of about 2°, and by the end of 1581 he had computed trial tables of his own using that value and a

1.8. Tycho's great globe of the heavens. The body of the globe, which was over $1\frac{1}{2}$ m in diameter, was made of wood, with parchment added to the surface where necessary to improve its sphericity. The whole was then covered in brass, and carefully engraved.

similarly estimated value for the longitude of apogee (97½°). During 1582 and 1583 Tycho checked this theory against observations for over a hundred places of the Sun. These 'second approximation' comparisons, preserved in Tycho's logs, display a pattern of error from which Tycho certainly could have obtained a very good revision of his elements. Around the end of 1583, however, he decided to see what kind of results he could get from a geometrical positioning of the orbit. This required knowing very precisely the amount of time taken by the Sun to traverse two sections of its orbit. For Hipparchus (and Ptolemy) one of these sections had been an entire season, which entailed fixing the time of a solstice. But since solstices were notoriously difficult to establish precisely, and since the geometry could be done just as well with a partial-season section, Copernicus had followed several Islamic astronomers in working with a half-season. Such consideration for the pragmatics of observation had a strong appeal for Tycho. He managed to push it one step further by switching to summer octants, where the Sun's higher altitude rendered parallax and refraction much less problematical. By early 1584, Tycho had what was essentially his mature solar theory. Four years of further empirical comparison culminated in final adjustments (and printed tables computed) to an apogee of 95½° and an eccentricity of 0.0358, effectively equivalent to a post-Keplerian, bisected eccentricity of 0.0179.

Tycho's solar theory included two features that had never before been considered in that context. The first, solar parallax, was totally mischievous. By correcting his solar altitudes according to the time-honoured horizon value of nearly 3′ (the one parameter of the ancients he seems never to have thought to check), he introduced trigonometrically into his comparison longitude errors mounting upward from 7′ as the altitude of his observations dropped below 20°. Moreover, by introducing these parallax corrections into the observations of winter solstice that he used to determine the obliquity of the ecliptic, Tycho arrived at an excessive figure (23° 31½′) that fed back through his trigonometric conversion formulae to magnify the error to 11′ at altitudes of 20°. Much of the seventeenth-century progress toward essentially the modern parameters for the solar theory would stem from a gradual reduction of the pernicious parallax value to something approximating its true 8″.8 magnitude. Fortunately, even Tycho managed to mitigate its worst effects through the second novel feature of his theory, refraction.

Tycho's recognition of the effects of refraction resulted from his efforts to determine the latitude of his observatory. Of course, this latitude was of fundamental importance, so he began attempts at it in his first years on Hven, long before he had either the time or the instruments for more complicated ventures. Upper and lower culminations of the Pole Star quickly converged (1578) on a value around 55°53′. Meridian altitudes of the Sun at summer and winter solstice were less consistent. Not until the summer solstice of 1583, when an extensive series of computations indicated a latitude below 55°50½′, does Tycho seem to have realized that his determinations from the Sun had been consistently lower than those made from the stars. By the following winter solstice, Tycho was primed for extensive observations, expecting not only to find a problem, but to explain it as the consequence of an artificially high meridian altitude due to atmospheric refraction.

During the later 1580s, Tycho undertook extensive investigations of the effects of refraction. Refraction as such was not a new discovery – it had been used in Antiquity to explain the simultaneous visibility of the Sun and the partially eclipsed Moon on opposite horizons, and had been discussed by both Moslem and Latin writers in the Middle Ages. Ptolemy had even attempted to quantify its effects on light passing from air to water. Tycho's, however, were the first serious attempts to measure atmospheric refraction. He conducted them at summer solstices, when the Sun's declination could be regarded as essentially constant through a half-day's movement from horizon to meridian. In this situation, any observed change in the declinations of the rising or setting Sun could be attributed to refraction, and tabulated accordingly. Even under such simplified conditions, however, it was an extremely difficult task. Although optical theory had been one of the most thoroughly treated subjects of medieval science, it could offer little real help. Tycho's conception of the cause of refraction (vapours and other impurities in the air) suggested that the effects might vary not only from season to season, but also from place to place and from one weather condition to the next. The outcome of his investigations was a table that Tycho thought could provide only approximate values, and which, even at that, had some strange features. Most peculiar to the modern eye is its provision for no refraction at all for altitudes greater than 45°. And what is more remarkable still is that those are the altitudes where Tycho's table was technically best. For there, where the true values of refraction are all less than 1′, Tycho was in fact correct in his judgement that his table was accurate within 1′. Below 45°, the table necessarily incorporated Tycho's 'corrections' for solar parallax. Because he determined his values empirically, from the difference between true and observed declinations, and his 'true' declinations contained a correction for a 3′ solar parallax, Tycho's table of refractions is systematically too large by roughly the size of his parallax corrections.

While Tycho's table was not aesthetically perfect from a modern standpoint, it was empirically sound when applied with its theoretical matrix. Because it was, for Tycho, a table of *solar* refractions, he never used it in situations where he was not also using the parallax corrections and (slightly exaggerated) solar equations that were both the cause of and the compensation for the imperfections of the table. Starlight he had already found to suffer less refraction than sunlight did (because it did not require any parallax corrections, of course), and he would eventually construct a third table for moonlight as well. Above all, he used his tables sparingly. Convinced that refraction was inherently variable, he not only consistently conducted his observations so as to minimize its effects, but wrote enough about the problem to ensure that his successors would do the same.

Although Tycho's observations spanned some 35 years, most of them (and the best) were made between 1581 and 1597. During this period Tycho averaged some eighty-five night-time observing sessions annually. Most of them were in the early evening hours, and over four-fifths of them occurred during the dark months of September to March. In general, the observations were not carried beyond the immediate task of registering the sighting. For much of the time Tycho was as interested in checking the performance of a new instrument or verifying the orientation of an old one as he was in recording the position in question. And, in any case, since he always made multiple sightings

with several instruments, only a few selected observations would ever be used for theoretical purposes. Furthermore, the great bulk of the observations were made long before Tycho had either opportunity or intention of occupying himself with theory; for during the 1580s, Tycho was composing the 800 pages of argument he published on the comet (1588) and the New Star (1602). However, when Tycho managed to complete his solar theory and his studies of refraction, he decided, for reasons that are not entirely logical, that these results should be included in his magnum opus on the New Star; and the same considerations apparently dictated the inclusion of his star catalogue in the volume.

The star catalogue

Like virtually every other undertaking of Tycho's, the star catalogue was a ten-year project. He had begun working towards it in 1581, when the large instruments began to emerge from his workshop on Hven. Two attempts to use clocks failed to provide even the accuracy he was sure he could get without them, so he early, albeit reluctantly, abandoned that labour-saving device. But otherwise his instruments did everything he asked of them. By using Venus as an intermediary to tie his stellar longitudes to the Sun, he slowly and carefully established the reference system for his catalogue. By 1588 he had Alpha Arietis located within 15″ of its true longitude, and twenty other fundamental stars placed almost as accurately. From there on the project was carried to fruition relatively quickly, but on a much lower level of accuracy. Most of it was carried out during 1589–91. The general pattern was to observe somewhat more than half of a constellation in one pass, and then follow up with the rest of the constellation – usually with some duplication – on a later occasion. Each sweep involved two measurements for each star: declinations (or meridian altitudes) one night, and distances from some convenient bright star (but frequently not one of the twenty-one fundamental ones) on another night. The vast majority of measurements were taken twice, and provided readings that agreed within a minute. These then had to be evaluated for a 'best' result, converted trigonometrically to ecliptic coordinates (epoch 1600), and entered into the catalogue.

In view of the fact that the star catalogue is for

many people the ultimate symbol of Tycho's image as an observer, it is ironic that it falls below the standards Tycho maintained for his other activities. Being prepared from far fewer observations than any of Tycho's planetary work, and with calculations that must have been checked much less zealously than, say, Tycho's computations on the comet, the catalogue left the best qualified appraisor of it (Tycho's eminent biographer J.L.E. Dreyer) manifestly disappointed. Some 6% of its final 777 positions have errors in one or both coordinates that can only have arisen from 'handling' problems of one kind or another. And, while the brightest stars were generally placed with the minute-of-arc accuracy Tycho expected to achieve in every aspect of his work, the fainter stars (for which the slits on his sights had to be widened, and the sharpness of their alignment reduced) were considerably less well located. It is tempting to think that this comparative neglect was because Tycho himself seems to have had little to do with the later stages of the cataloguing. But, actually, observing of any kind was something he did so infrequently in later years, that he took to noting in the log those observations in some aspect of which he had participated. However, Tycho generally kept a tight rein on his assistants, and could certainly see that the stars were being charted only twice instead of the numerous times that were routine for any planetary position. Even the annotations in his handwriting that show that he was supervising many of the concurrent planetary observations are lacking among the stellar observations. It is therefore difficult to reach any conclusion other than that Tycho himself was simply less interested in the mechanics of cataloguing than in the manifold other activities in which he was engaged at the time – that he regarded his star catalogue as primarily a means to an end, and one in which he rapidly lost interest once the zodiacal stars marking the paths of the planets were recorded.

The lunar theory

Tycho began the research phase of his work on the planetary theories in 1590. Although his first efforts were far from conclusive, they were sufficiently promising by 1592 to induce him to lay plans for publishing the results. At the time, however, planning was all Tycho could do. The legal and political difficulties that would eventually result in significant cuts in his patronage were beginning to make serious inroads on his work time, and he was able to do little more than prepare his correspondence for publication (1596), and write up the thin volume (*Mechanica*, 1598) describing the instruments, facilities and other achievements produced during his twenty-one years at Hven. Thus, even though the twenty-four-page discussion of the motion of the Moon was all that stood in the way of publishing his (otherwise already printed) 800-page book on the New Star, Tycho settled down to serious work on the lunar theory only more or less accidentally.

Although Tycho did not begin systematic observation of the Moon until 1581, and even then averaged only about thirty evenings of attention to it annually, he never missed an opportunity to attend an eclipse. On Hven, this could mean turning out the whole staff to assemble three independent crews of observers. Since such an event required some advance preparation, it was important to know when the eclipse would begin. To cope with the fact that ephemeris predictions could easily mis-time first contact by an hour either way, Tycho decided to try to predict the event himself, by observing the Moon's position a night or two beforehand, and then estimating its arrival at opposition from the theoretical velocity of the Moon during the interval. Initially, the idea brought only trouble. When used for the eclipse of 30 December 1590, it left Tycho missing the first hour of the eclipse completely, and registering his perplexity in his log. In 1594, something similar seems to have happened again. At any rate, after the lunar eclipse of 19 October 1594, Tycho evidently developed the suspicion that the Moon was speeding up as it went into opposition, and deduced that it must, accordingly slow down somewhere else in the orbit – probably in the quadrants. Moreover, he seems to have perceived that the most noticeable effects of this variation in velocity would be *displacements* at the octants, halfway between syzygies and quadratures; for during the next several weeks, eight months after his last previous observations of the Moon, Tycho checked its progress through the various points critical to (and only to) the inequality which has since been known by the name Tycho gave it, the 'variation'.

Justly inspired by his discovery of the first new

astronomical phenomenon to be recognized since Ptolemy's era, and one that removed perhaps three-quarters of the error formerly inhering in predictions of the Moon's longitudes, Tycho turned his attention to its latitudes. He had already devoted some considerable effort to them earlier – not because he was interested in them *per se*, but because they were observationally bound up with the Moon's parallaxes. From the time of Johannes Regiomontanus (1436–76) at least, it had been recognized that Ptolemy's lunar theory provided a completely inadequate treatment of the Moon's distances from Earth and, hence, its parallaxes. Operationally, the issue was not particularly important: the theory was reasonably accurate for eclipses, and nobody in recorded history had concerned himself seriously with non-eclipse positions. But as a subscriber to the conviction that astronomical theories ought to reflect reality, Tycho was very interested in seeing whether Copernicus's obviously superior representation of the Moon's distances was actually correct. In such a context, it would not be easy to recognize that an observational discrepancy was due to the theoretical latitude rather than the theoretical parallax – particularly given that Ptolemy and Copernicus used identical theories for the latitudes. But Tycho succeeded in this, deciding that his problems stemmed from a previously undetected secular change in the Moon's orbital inclination. For eight years after 1587, he computed lunar latitudes from an inclination of $5\frac{1}{4}°$ rather than the $5°$ used by everyone since Ptolemy.

What Tycho contemplated in 1595 was a precise determination of the inclination, under optimum conditions. At that time, the lunar orbit in its eighteen-year retrograde shift around the zodiac had reached the point where its northern limit was in Cancer, and thus at the highest altitude – and lowest susceptibility to problems from parallax and refraction – at which he could observe it. It was a discrepancy of $5'$ that caught his eye. In the course of trying to discover how his observation and expectation could disagree so greatly, he finally realized that the critical difference between his and Ptolemy's determinations had been not the epoch of the observation, but the phase of the Moon – that the lunar orbit must be inclined by $5°$ when the Moon is in syzygy (where, because of the eclipse possibilities, virtually all previous attention had

been concentrated), and $5\frac{1}{4}°$ when it is in quadrature.

In Tycho's era, there was only one way to conceive of such a situation: imagine the pole of the lunar orbit nutating in a circle during a period of half the synodic month, so that the pole (P in Figure 1.9) would be the traditional $5°$ from the pole of the ecliptic (EP_S) in eclipse situations, and $5\frac{1}{4}°$ (EP_q) away at both quadratures (where $2a = 180°$). It would also provide a smooth continuous variation of the inclination (i) between those extremes, easily computed from a circular function

$$i = 5° + 7\frac{1}{2}'(1 - \cos 2a) = 5° + 15' \sin^2 a$$

in the best astronomical tradition.

Within two months, Tycho saw that his model implied a further phenomenon. If the pole of the lunar orbit (P) really gyrated around a circle, the line defining the limits and nodes of the lunar orbit, EPL', would oscillate back and forth around L. It was the same type of arrangement as Copernicus had used to connect and account for the changing obliquity of the ecliptic and the supposed oscillation of the equinoxes. This, in turn, would produce a second correction to the existing theory of latitudes – one that Tycho was able to confirm quickly from past observations.

Esoteric as the 'variation' and, especially, the perturbations in latitude were, they were not the end of Tycho's contributions to the lunar theory. Some time before 1598 he completed the discovery of the Moon's major inequalities by adding a correction for the annual equation, somehow estimating its $11'$ magnitude accurately, even though he erred by $24°$ in establishing the timing of its yearly phases. By 1598, also, he had succeeded in integrating all of these modifications into a model that was already the most complicated of the planetary theories. While he was in the very process of printing an exposition of his theory, however, one last consideration forced a complete re-shuffling of the model. The problem was the range in the Moon's distances from Earth. Although the Copernican model onto which Tycho had grafted his new inequalities involved only about half the variation implied by its Ptolemaic counterpart, it still provided almost twice the variation observationally detectable in the Moon's angular diameters. In order to compress the range, Tycho (actually, his disciple Longomontanus) discarded the single

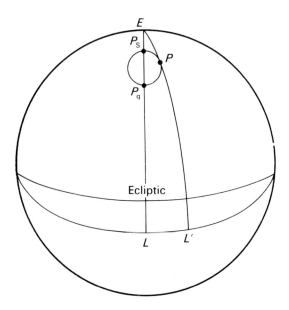

1.9. Tycho's method of determining the inclination of
the Moon.

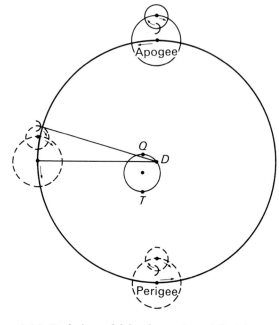

1.10. Tycho's model for the motion of the Moon.

epicycle used by Ptolemy and Copernicus for the first (elliptic) inequality, and substituted a double epicycle of the kind used by Copernicus for the planetary theories. Because he required more compression than Copernicus (and he) had used for the planetary theories, however, Tycho gave the small epicycle a greater effect by apportioning one-third of the eccentricity to it, instead of only one-fourth. The larger epicycle, assigned the two-thirds of the eccentricity, was conceived to move anti-clockwise from apogee (see Figure 1.10) in the anomalistic period of the Moon, while the centre of the second epicycle moved clockwise in the same period, so as to be at distances $R + \frac{2e}{3}$ at apogee, and $R - \frac{2e}{3}$ at perigee. On the second epicycle, the Moon was conceived to travel anti-clockwise at twice the (anomalistic) rate of the first two motions. As in the planetary theory, it was referenced so as to be at pericentre of the small epicycle when the large epicycle was in the line of apsides, and farthest from centre when it was in the mean distances. The subtractive effect in the line of apsides provided an apogee distance of $R + \frac{2e}{3} - \frac{e}{3}$ and a perigee distance of $R - \frac{2e}{3} + \frac{e}{3}$, and hence radius vectors that varied by only one-third of the traditional amount. The additive effect in the mean distances ($\frac{2e}{3} + \frac{e}{3}$ at either side), on the other hand, resulted in corrections

that were (everywhere) essentially equal to those provided by the traditional single epicycle eccentricity e.

Although Tycho's model was the first seriously to attempt to account rigorously for the distances of a 'planet', it was something less than completely successful. For the chief source of distortion of the Moon's distances had always been the mechanism used to account for the second inequality in the motion of the Moon, discovered by Ptolemy. And Tycho's representation of it – the circle at the centre of the deferent – was merely an inverted form of the one Copernicus had used. This provided a large and completely spurious variation tied to the phases of the Moon, which had nothing to do with the true anomalistic variations arising from the elliptical shape of the Moon's orbit. These latter had to be accounted for quite rigorously for eclipse computations. And while Tycho's representation of the first (anomalistic) inequality provided marginally better angular diameters than did the traditional single epicycle mechanism, its range, $\pm 1'$, was almost as far from reality as the $\pm 3'$ implied by Ptolemy's and Copernicus's models. Like them, therefore, Tycho had to fall back on a pragmatic $\pm 2'$ for the calculation of eclipses. For solar eclipses, moreover, he found still another empirical adjustment to be necessary. Although Ptolemy had adopted diameters for the Sun and Moon that theoretically ruled out the possibility of an annular eclipse, Tycho found, on the contrary, that solar eclipses tended to have a noticeably exaggerated annular appearance. Measurements of the angular diameters of New Moons showed that the corona effect was only apparent. But even if it were merely an optical illusion, it had to be accounted for if astronomers were to predict appearances successfully; so, for solar eclipse computations, Tycho advocated a pragmatic diminution of the Moon's diameter by one-fifth.

With two epicycles used for the first inequality, there was little alternative to putting the second inequality (which has since been dubbed the 'evection') back in the centre where Ptolemy had placed it. This was accomplished by making the centre of the deferent (*D*) revolve on circle *TQ* in half the synodic month; so that in syzygy it was always at *T* (the Earth) and providing no correction at all, while in quadrature it was at *Q*, providing its maximum effect. The 'variation' was then accounted for geometrically with a little circle carrying the centre of the large epicycle. But since the trigonometry of the model was already much more complicated than that for any previous theory, the effect of the circle was reduced to a simple libration of $40\frac{1}{2}'$ sin $2a$. The annual equation, having no very obvious place to fit into the model, was simply omitted. Only the fact that it was partially re-inserted by the extraordinary expedient of simply cancelling the portion of the equation of time that depends on the mean anomaly of the Sun, allowed a posterity ignorant of the historical development of Tycho's lunar theory to associate his name with the discovery of the annual equation.

Concurrently with the recasting of the lunar theory, Tycho finally embarked on serious work on the planets. It was a task he had taken up early in the decade, but laid aside when preliminary attempts indicated that it would involve something more than merely using precise observations to make the traditional determinations of apogee, eccentricity, inclination, and node. We must not suppose that for Tycho – or, indeed, anyone but Kepler – 'something more' would ever have included abandoning the circle. One way or another, Tycho would have rationalized his data in terms of the Copernican double-epicycle model, just as his dis-

ciple, Longomontanus, eventually did for his *Astronomia Danica* of 1622. In fact, the knowledge that Longomontanus had already managed to account for Mars's observed longitudes within 2' cannot have been a negligible factor in Kepler's notorious unwillingness to settle for 8' disagreements in his own work (see Chapter 5). Astronomy was well served, therefore, by the permanently unsettled conditions under which Tycho laboured during the last four and a half years of his life.

The end of Uraniborg

Concerning the specific events and underlying causes of Tycho's departure from Hven in 1597, much has been written and more probably remains to be discovered. The essence of the situation, however, is that Tycho's work had been endowed by King Frederick of Denmark and maintained after his death by a regency in which Tycho's brothers and friends played a dominant role; but with the transfer of power to the new king (eighteen-year-old Christian IV) Tycho quickly lost his special status. Stung by the loss of support and lack of appreciation for his work, and influenced also, probably, by indications that passing his estate to the off-spring of his morganatic marriage would not be a routine undertaking, he elected to emigrate. In every objective sense the decision turned out well. He found patronage at a princely stipend from the emperor of the Holy Roman Empire, Rudolph II. But the transition would occupy two valuable years of time, and even then would never provide the stability and security Tycho had taken for granted during twenty-one years on Hven. It cost him access to his best instruments for well over three years, and it took him the remaining year of his life merely to get his instruments re-orientated at Prague. In the move, he also lost access to the other institutional facilities on Hven – the assistants, the print-shop, the engraver, and the paper mill that over the years had rendered possible the printing of six books. This sealed the fate of such ventures as his projected volumes of comet-studies and miscellaneous correspondence. It meant that even the lunar theory remained unprinted at his death on 13 October 1601, so that the long-delayed volume on the New Star had to be overseen in its final stages by Kepler, his assistant and successor in Prague. With that volume, however, Tycho realized a considerable portion of his life's ambi-

1.11. The tomb effigy of Tycho in Tyn Church, Prague.

tion. The star catalogue, solar theory, and lunar theory included in it justified a revision of the title to *Astronomiae instauratae progymnasmata* (*First Exercises in a Restored Astronomy*), (1602). If the restoration of the planetary theories remained a dream, it would not be for long. And the man who fulfilled it – Kepler – would see so clearly that Tycho had satisfied his desire not to have lived in vain, that he would publish the long-awaited results (*Rudolphine Tables*, 1627) under the name of "that Phoenix of Astronomers Tycho Brahe".

Further reading

J.L.E. Dreyer, *Tycho Brahe: A Picture of Scientific Life and Work in the Sixteenth Century* (Edinburgh, 1890; reprinted New York, 1963)

C.D. Hellman, *The Comet of 1577: Its Place in the History of Astronomy* (2nd edn, New York, 1971)

Christine Jones Schofield, *Tychonic and Semi-Tychonic World Systems* (New York, 1981)

Y. Maeyama, The historical development of solar theories in the late sixteenth and seventeenth centuries, *Vistas in Astronomy*, vol. 16 (1974), 35–60

V.E. Thoren, An early instance of deductive discovery: Tycho Brahe's lunar theory, *Isis*, vol. 58 (1967), 19–36

V.E. Thoren, Tycho Brahe's discovery of the variation, *Centaurus*, vol. 12 (1967), 151–66

V.E. Thoren, New light on Tycho's instruments, *Journal for the History of Astronomy*, vol. 4 (1973), 25–45

V.E. Thoren, The comet of 1577 and Tycho Brahe's system of the world, *Archives internationales d'histoire des sciences*, vol. 29 (1979), 53–67

2

The contemporaries of Tycho Brahe

RICHARD A. JARRELL

The history of astronomy in the first half of the sixteenth century was dominated by Copernicus's formulation of the heliocentric system and by its early reception. The first half of the seventeenth century saw the emergence of celestial physics, largely the creation of Kepler and Galileo. The later sixteenth century was a period of consolidation and transition: consolidation of the mathematical techniques of Copernicus and transition from the purely mathematical account of planetary motions to a wider discussion of the actual nature of the universe. Tycho Brahe (1546–1601) holds centre stage in this era, as much because of the sheer force of his personality as for his ingenuity and systematic observational methods. Nonetheless, the European astronomical community was alive with clever and ambitious men, whose contributions helped pave the way for the full reception of the new cosmology and the ultimate overthrow of traditional natural philosophy. Indeed, it is this ambition and love of controversy that helps account for the sheer size of Tycho's books on the New Star of 1572 and the comet of 1577; his contemporaries recognized his abilities but few would concede to him intellectual leadership of the community.

Much of the transition from the Ptolemaic geocentric theory, linked with Aristotelian physical concepts, to the Copernican system, linked with a new celestial physics, occurred during the last third of the sixteenth century. To a surprising extent, this transition was the product of German-speaking astronomers and those foreigners educated by or in contact with them. Tycho, although a Dane, was as much a part of German astronomy as were the Scots Duncan Liddel and John Craig or the Czech Tadeáš Hájek (Hagecius). Indeed, we find very little critical discussion of Copernicus outside German-speaking countries. In England, for example, Copernicus was known and mentioned by Thomas Digges, John Dee, John Feild and Thomas Blundevile, among others, but none contributed to the primary debate. French, Scandinavian, Spanish, Italian and eastern European astronomers had little to offer until the next century.

The shift from geocentric to heliocentric perspective was not a unitary process but took place in stages. Three may be identified: first, the growing disillusionment with Aristotelian natural philosophy; second, the reception of Copernican technical astronomy; and third, the move to a realist view of heliocentricism. All three strands would culminate in the work of Johannes Kepler (see Chapter 5) towards the end of Tycho's life. The unification of Ptolemaic technical astronomy with a natural philosophy that was inspired by Aristotle, if not always truly Aristotelian, was a legacy of the Middle Ages. Two events within five years of each other raised grave doubts concerning this synthesis in the minds of several astronomers. The first, the star of 1572 (see Figure 2.1), simply did not fit into the accepted view of celestial events. The star (a supernova) appeared where none was observed before, eventually fading from view, a clear example of coming-to-be and passing-away, and so presumably a phenomenon of the sublunary realm of corruption and decay. Yet, the object was stellar in appearance, called in many accounts *stella nova*. Were the object to lie beneath the lunar sphere, then it could be classified as 'meteorological', even if its behaviour was curious. A number of astronomers, including Digges, Michael Mästlin, Paul Fabricius, Cornelius Gemma, Paul Hainzel and Tycho, employing Regiomontanus's method of determining diurnal parallax (as proposed in his *Problemata XVI*, published posthumously in 1531), found no measurable parallax, suggesting that the New Star was indeed in the heavens. Some observers, such as Hainzel and Caspar Peucer, although

2.1. Tycho's nova of 1572, as seen from Prague. To indicate that the nova was visible in the daytime, the anonymous engraver has shown the Sun setting (with Venus to its right!).

finding no parallax, refused to believe the object was supra-lunar. Still others, like Andreas Nolthius – not a particularly good observer – found a large parallactic displacement. However, the Landgrave of Hesse-Kassel, Wilhelm IV, who was equipped with the best instruments in Europe, believed that the star could not have a parallax greater than 3′. Even apparently crude determinations by Digges and Mästlin, who both held a thread at arm's length to connect pairs of stars at right angles to each other, revealed no displacement. Thus, a wide range of instrumentation and observing abilities led to the same conclusion, that the New Star posed a substantial challenge to the traditional terrestrial–celestial dichotomy.

The absence of parallax was not evidence for the Copernican theory, but two of the observers of the nova used their results to argue in its favour. Mästlin saw the linkage between Ptolemaic astronomy and Aristotelian philosophy as inseparable; if there was a flaw in one, it followed that the other must be seriously questioned. Digges held a similar opinion and suggested that a test might be made for the heliocentric theory by observing the New Star over a lengthy period to ascertain the annual parallax. He did not attempt this himself, nor of course was he aware that it was impossible with the instruments then available.

2.2. Tycho's nova in relation to the stars of Cassiopeia.

The profound uncertainties raised in 1572 reappeared with the sighting of the spectacular comet of 1577. Comets, unlike the New Star, were well known and Aristotle himself had stated that comets are meteorological in nature. Astronomers eagerly measured the position of the comet to discover whether it exhibited diurnal parallax, and so was close to Earth; but this time only five found none, largely because of the far greater difficulty in determining the precise position of a comet. Tycho's observations were the most complete and it was he who brought together a large body of observation and opinion of others in his *De mundi aetherei recentioribus phaenomenis* of 1588. In agree-

ment with his result were the Landgrave, who published nothing, Mästlin, Gemma and Helisaeus Röslin. In 1577, no astronomer questioned the belief that celestial objects lay in physically real spheres and if the comet were truly a celestial object, then it must either have its own sphere or lie in the sphere of another object. Gemma suggested that the comet moved in the sphere of Mercury, whilst Röslin, the Tübingen-educated physician, offered an entirely separate sphere of 'meteors', with an epicycle to account for the variability of the comet's brightness. The axis of this sphere was inclined to the ecliptic, to provide for the great inclination of the comet's path. For Röslin, how-

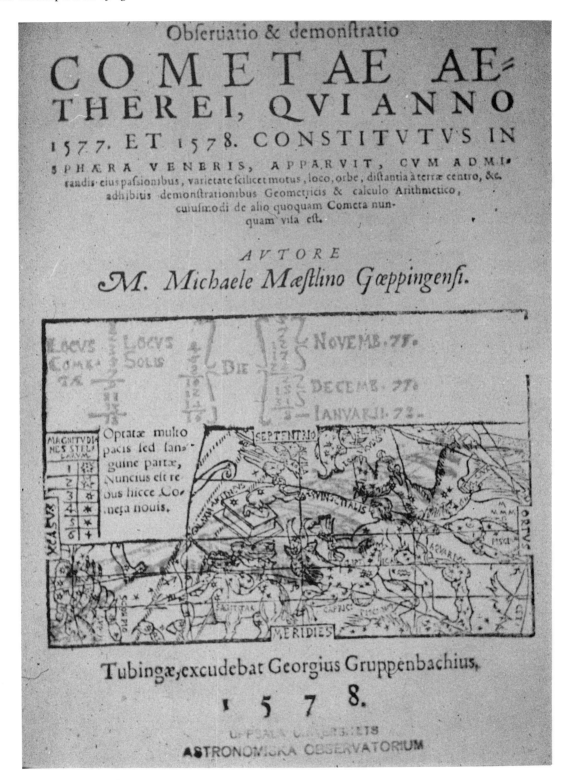

2.3. The title page of Michael Mästlin's treatise on the comet of 1577, in which he proposes an orbit for the comet in the sphere of Venus.

ever, not all comets were necessarily to be found in this sphere; some were no doubt 'meteorological' in origin.

Mästlin, too, sought a convenient sphere but went further than Röslin by attempting to use the cometary path to show the probability of the Copernican theory. With less accurate observational data than Tycho, Mästlin sought amongst Copernicus's planetary models to find one that would fit the comet's motion, believing that the sphere of Venus, with its librational motions in latitude, would account for the highly variable motion and location of the comet. Since he evidently failed to realize that the motions might be described in either Ptolemaic or Copernican terms interchangeably, this 'demonstration' constituted for him a proof of the heliocentric theory. Kepler would later be impressed by his teacher's argument.

The anti-Aristotelian tone of several of the comet treatises, although not directly contributing to the advance of Copernicanism, did have two indirect effects. First, by raising doubts about traditional natural philosophy, such writers as Tycho and Mästlin called authority into question, while reinforcing an attitude of experimentation in thought. It can be no coincidence that the growing anti-Aristotelian currents of the 1570s centred on Protestant Germany, where theological questioning of authority and experimentation with novel ideas – a trend not yet half a century old – was still very much in evidence. Second, such thoughts prepared the way for the dramatic break with the concept of solid planetary spheres. The comet of 1577 eventually persuaded Tycho to abandon them, although other progressive thinkers like Mästlin could not go so far. Kepler, who did complete the transition, owed a great debt to the questioning of that decade. By no means all astronomers would take such a view, for the older natural philosophy remained current for some until the eighteenth century. But the scepticism of 1572 and 1577 endured: Mästlin found no parallax for the comet of 1580 and, unable to find an appropriate orbit, redoubled his attack on the old celestial physics. Tycho and the Landgrave both made careful observations of the comet of 1585. An important minority was now sure that comets were celestial objects that obeyed geometric laws.

The second and third major shifts in attitude, the

acceptance of Copernican technical astronomy and the adoption of the heliocentric system as a true description of the cosmos, were interrelated but not simultaneous. During the first three decades after the publication of Copernicus's *De revolutionibus* it would be difficult to name one astronomer who was a confirmed Copernican. Georg Joachim Rheticus, the theory's first champion, appears to have lost interest after his midwifery of *De revolutionibus* early in the 1540s. There is no doubt, however, that Copernicus's text was widely read with care by astronomers and non-astronomers alike, as attested by the number of extant copies with marginalia (see Figure 2.4). There existed, too, a trade in marginal glosses, for many copies of the book have identical notes or at least comments within a recognizable tradition.

The central tradition was what R.S. Westman has termed the 'Wittenberg Interpretation' of Copernicus, the acceptance of certain of his technical approaches to planetary theory without embracing the theory itself as reality. The University of Wittenberg, heart of Lutheran orthodoxy, derived its strength in mathematics and astronomy from Tübingen, where Johann Stöffler had taught, through the intellectual leadership of its alumnus Philipp Schwarzert Melanchthon. Of the first generation of Wittenberg teachers to assess Copernican astronomy, namely Peucer, Erasmus Reinhold and Rheticus, the first two clearly chose the non-realist route. Reinhold's influence, through his *Prutenic Tables* of 1551, made itself felt well into the seventeenth century. In the public realm, the ephemeris, basis for annual calendars, almanacs and astrological predictions, had to be grounded upon tables, mediaeval variations of those in Ptolemy's *Almagest*. Reinhold's tables, drawn from *De revolutionibus*, but with slightly improved parameters, were generally thought to be more accurate than earlier versions and became the basis for many popular ephemerides. Employment of Reinhold's tables did not commit the user to heliocentric astronomy – even their author was not a Copernican – and as Reinhold did not expound upon the method for deriving his parameters, the user could remain happily ignorant of the theoretical basis for them. The source of the parameters was not made public until Mästlin reconstructed them for Kepler and included them in his "On the dimensions of the heavenly circles and spheres", which was published

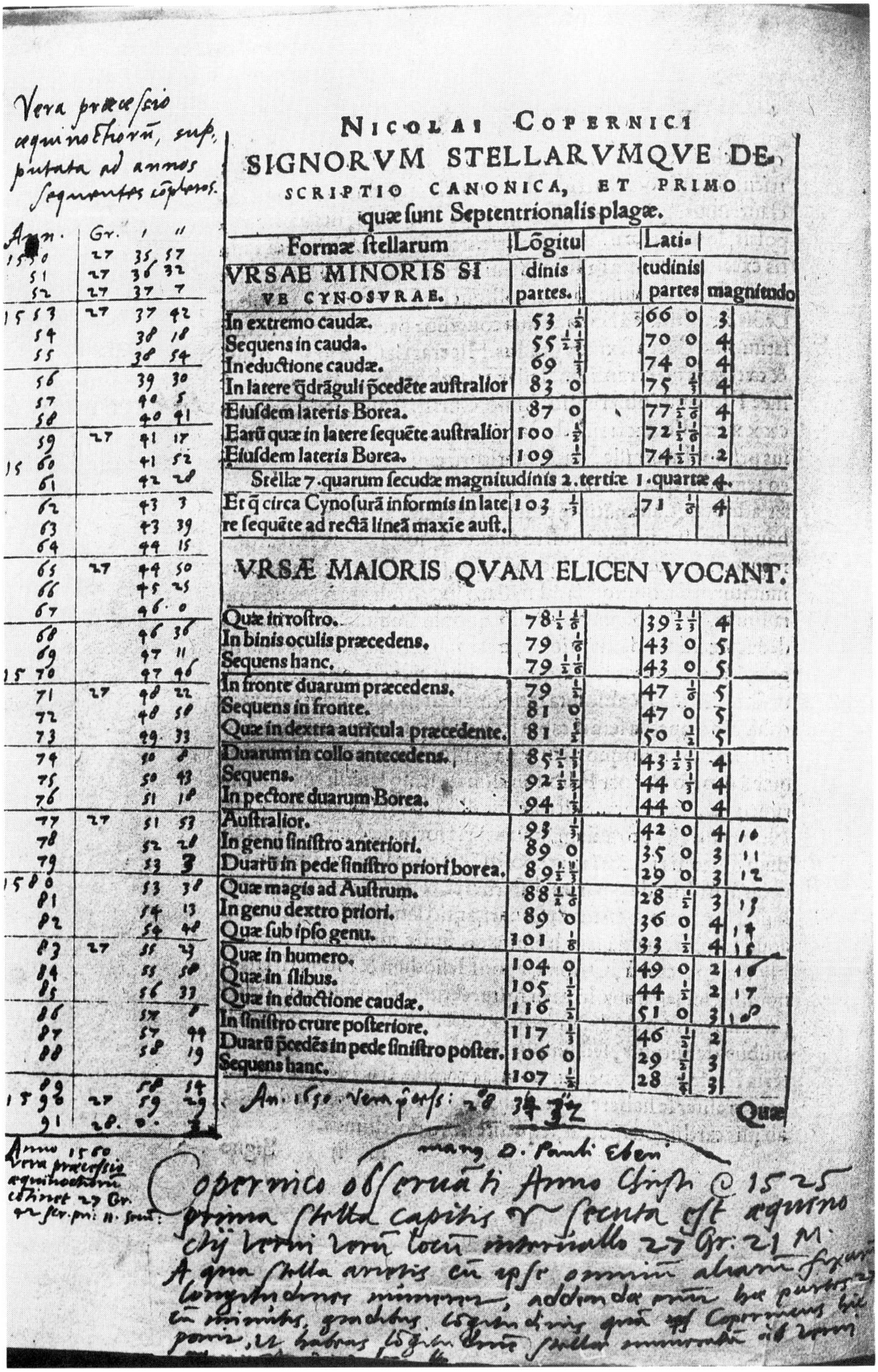

Left margin handwritten:

Vera præcessio
æquinoctiorū, sup,
putata ad annos
sequentes ceteros.

Ann.	Gr.	′	″
1550	27	35	57
51	27	36	32
52	27	37	7
1553	27	37	42
54		38	18
55		38	54
56		39	30
57		40	5
58		40	41
59	27	41	17
15 60		41	52
61		42	28
62		43	3
63		43	39
64		44	15
65	27	44	50
66		45	25
67		46	0
68		46	36
69		47	11
15 70		47	46
71	27	48	22
72		48	58
73		49	33
74		50	8
75		50	43
76		51	18
77	27	51	53
78		52	28
79		53	3
1580		53	38
81		54	13
82		54	48
83	27	55	23
84		55	58
85		56	33
86		57	8
87		57	44
88		58	19
1589	27	58	54
90		59	29
91	28	0	4

Lower left handwritten:

Anno 1550
Vera præcessio
æquinochorum
obtinet 27 Gr.
&c. ſcr. pri: n: ſunt:

Main printed table:

NICOLAI COPERNICI
SIGNORVM STELLARVMQVE DE
SCRIPTIO CANONICA, ET PRIMO
quæ ſunt Septentrionalis plagæ.

Formæ ſtellarum	Lōgitudinis partes.		Latitudinis partes	magnitudo
VRSÆ MINORIS SIVE CYNOSVRÆ.				
In extremo caudæ.	53	½	66 0	3
Sequens in cauda.	55	½ ⅓	70 0	4
In eductione caudæ.	69		74 0	4
In latere q̄draguli p̄cedēte auſtralior	83	0	75 ½ ⅓	4
Eiuſdem lateris Borea.	87	0	77 ½ ⅓	4
Earū quæ in latere ſequēte auſtralior	100	½	72 ½ ⅙	2
Eiuſdem lateris Borea.	109		74 ½ ⅓	2
Stellæ 7. quarum ſecudæ magnitudinis 2. tertiæ 1. quartæ 4.				
Et q̄ circa Cynoſurā informis in latere ſequēte ad rectā lineā maxie auſt.	103	⅓	71 ½ ⅓	4

VRSÆ MAIORIS QVAM ELICEN VOCANT.

Quæ in roſtro.	78	½ ⅙	39 ½ ⅓	4
In binis oculis præcedens.	79	⅙	43 0	5
Sequens hanc.	79	½ ⅙	43 0	5
In fronte duarum præcedens.	79	⅙	47 ⅙	5
Sequens in fronte.	81	0	47 0	5
Quæ in dextra auricula præcedente.	81	½	50 ½	5
Duarum in collo antecedens.	85	½ ⅙	43 ½ ⅓	4
Sequens.	92	½ ⅙	44 ⅓	4
In pectore duarum Borea.	94		44 0	4
Auſtralior.	93	⅓	42 0	4 10
In genu ſiniſtro anteriori.	89	0	35 0	3 11
Duarū in pede ſiniſtro priori borea.	89	½ ⅙	29 0	3 12
Quæ magis ad Auſtrum.	88	½ ⅙	28 ½	3 13
In genu dextro priori.	89	0	36 0	4 14
Quæ ſub ipſo genu.	101	⅙	33 ½	4 15
Quæ in humero.	104	0	49 0	2 16
Quæ in ilibus.	105	½	44 ½	2 17
Quæ in eductione caudæ.	116	½	51 0	3 18
In ſiniſtro crure poſteriore.	117	½	46 ½	2
Duarū p̄cedēs in pede ſiniſtro poſter.	106	0	29 ½	3
Sequens hanc.	107	½	28 ½	3

Handwritten below table: An. 1550 Vera p̄cess: 28 34 32

Lower right: Quæ

Bottom handwritten note (marginalia by Paul Eber):

marg. D. Pauli Eberi

Copernico obſervat Anno Chriſti ☉ 1525
prima ſtella capitis ♈ ſecuta eſt æquino
ctij verni verū locū meridiano 27 Gr. 21 M.
A qua ſtella arietis cū ipſe omnium aliarum fixarum
longitudines minueret, addenda erit hæc præcessio
in minutis, gradibus longitudinum quæ ꝗ Copernicus ibi
ponit, & habras 15 gradu vero ſtellar numeratū ab vero …

as an appendix to Kepler's *Mysterium cosmographicum* of 1596.

The average university student of the second half of the century encountered only descriptive astronomy based upon the elementary *Sphere* of the thirteenth-century writer Sacrobosco. Extant lecture notes and textbooks, whether by Copernicans like Mästlin or non-Copernicans like the Jesuit Christoph Clavius, adhered to the traditional line. Technical astronomy was for a minority of advanced students, who had to work through the *Almagest* and *De revolutionibus*. The importance of the Wittenberg tradition shows in the number of notable technical astronomers educated by Wittenberg professors or their students. Amongst Tycho's contemporaries were Johannes Praetorius, who studied and taught at Wittenberg before moving to Altdorf; Sebastian Theodoric of Winsheim, author of a popular textbook; Paul Wittich of Breslau; Samual Eisenmenger (Siderocrates), who taught Röslin and the Tübingen instructors who first lectured to Mästlin; and Johannes Homelius, later to teach Bartholemaeus Scultetus of Görlitz. Homelius and Scultetus were at Leipzig when Tycho spent time there. Virtually all these men took the same view of Copernican theory; none was truly Copernican.

It should be stressed that the technical astronomy of *De revolutionibus* was not qualitatively different from Ptolemy's astronomy. The cosmic structure set out in Book I of the work, which provided a fixed order for the planets and a method for finding their relative distances, went unappreciated within the Wittenberg tradition. Although recognized as a salient feature of Copernican thought, this cosmic structure could only be accepted if one took a realist stance, which none of them was willing to do. The widely perceived notion that the *Prutenic Tables* were more accurate than their predecessors was already in serious doubt by the 1570s. Tycho and the Landgrave both devised elaborate programmes of observation to improve the basis for prediction of planetary positions. Ephemeris writers such as Mästlin were equally aware that a new observational basis was required before the celestial motions could be 'restored'.

The observation of planets at oppositions and elongations might, if accurate enough, make a sensible difference in the derivation of parameters for planetary models. Tycho led the way with redundant observational data, garnered over extended periods, a practice not yet appreciated by his contemporaries. Most observers, employing simple instruments such as cross-staffs and quadrants and, in some cases, weight-driven clocks, continued to observe occasionally, especially events such as eclipses. Given that these instruments were mostly made of wood or of metal bound to wood, and were often home-made, their accuracy was questionable in any event. The division of arcs by transversals, popularized by Homelius and transmitted via Wittich to Kassel, was not universally adopted. The finest clocks of the period, constructed by Joost Bürgi for the Landgrave (Figure 2.5), were distrusted by Tycho and were unavailable to others. There was no source of systematic observations for any astronomer and each theoretician had to rely upon his own work or had to trade observational data with others. There was a certain commerce in such data, especially to and from Kassel and Tycho's observatory on Hven, and specific observations appeared in treatises on comets and other events, and even in textbooks (such as Mästlin's *Epitome astronomiae*) as examples. Few astronomers had the financial and technical resources of Tycho or Wilhelm IV to undertake more than desultory observation. Even the Landgrave, with his assistant Christoph Rothmann, in attempting to redetermine the positions of more than a thousand stars, accomplished little more than a tenth of the task. Thus, the ambitious approach to systematic observation had little impact upon astronomers of the later sixteenth century, not only because many could not see the need for long-term observation, but also out of practical considerations.

For technical astronomers, the striking feature of Copernicus's theory was the removal of the Ptolemaic equant. Astronomers before Kepler agreed that the geometric basis for planetary theory was uniform circular motion. Although Ptolemy had asserted his adherence to this principle, the equant that he employed in his planetary theories (see Volume 1) was a violation of it: the centre of the planetary epicycle, while moving with *angular* uniformity about the equant point, moved with non-uniform motion on the deferent circle because the equant point was not at the centre of this circle. Copernicus eliminated this violation of principle by

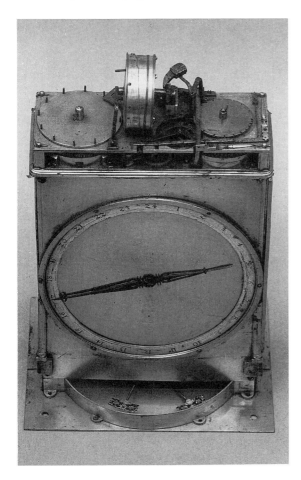

2.5. An astronomical clock by Joost Bürgi, used for timing observations at the observatory of the Landgrave Wilhelm IV at Kassel. Bürgi's clocks represent one of the earliest attempts to use mechanical timepieces in conjunction with astronomical instruments, and their example influenced the work of Tycho Brahe.

using a double epicycle in place of Ptolemy's equant plus single epicycle; in so doing he was employing a construction that had been introduced by Ibn al-Shāṭir in the fourteenth century. The resulting angular motion of the planet agreed very nearly with that produced by the Ptolemaic equant, but was in accord with the principle that all motions be compounded out of uniform circular motions. Moreover, the added epicycle could be represented by a physical sphere, whereas the equant did not admit of being translated into a feasible mechanical model.

Copernican planetary models referred motions to the Mean Sun rather than the true Sun; plan-etary orbital planes passed through the centre of the Earth's eccentric circle, not the Sun, with the result that complex motions had to be introduced to account for planetary latitudes. Thus, the Copernican models are no simpler than the Ptolemaic. They do, however, embody novel devices such as eccentrics with compound, small epicycles, and the Ṭūsī couple. Most technical astronomers within the Wittenberg tradition or sympathetic to it, for instance Praetorius, Reinhold, Wittich, and Tycho, made extensive use of these devices; by the 1570s they were well accepted. And before long several astronomers came to realize that their choices were not limited to the Ptolemaic and Copernican theories: there was an alternative, the geoheliocentric theory, the so-called Tychonic system.

The first to adopt the new approach may have been Wittich, who sketched such a model in his copy of *De revolutionibus*. O. Gingerich and R.S. Westman have argued that Wittich, beginning from Reinhold's annotations, explored means by which one can utilize Copernican techniques to transform models from one form to another, such as from an eccentric circle on an eccentric circle to a double epicycle on a homocentric circle. By February 1578, he hit upon the idea of holding the Earth stationary and soon developed a system with the inferior planets circumsolar and the outer planets circumterrestrial (Figure 2.6). At the same time, Tycho had begun to work towards a similar scheme, examining the ancient model of Martianus Capella, in which Mercury and Venus were circumsolar. By 1584, Tycho's version had progressed to one like Wittich's and four years later had reached its mature form. The Wittichian theory was incomplete in the sense that the motions of the outer planets were not referred to the Sun, but – unless perhaps he was held back by physical considerations – Wittich would seem to have been on the threshold of the concept that the three systems were geometrically equivalent.

In the year that Tycho's final version appeared, Nicolai Reymers Bär (Ursus) published a similar theory with the significant difference that it allowed for the Earth's rotation, leaving the sphere of fixed stars at rest. In 1597, Röslin, in his *De opere dei creationis*, produced another version of Tycho's theory (see Chapter 3), claiming he discovered it independently at the time of Tycho's and Ursus's publications. Tycho believed that both had plagia-

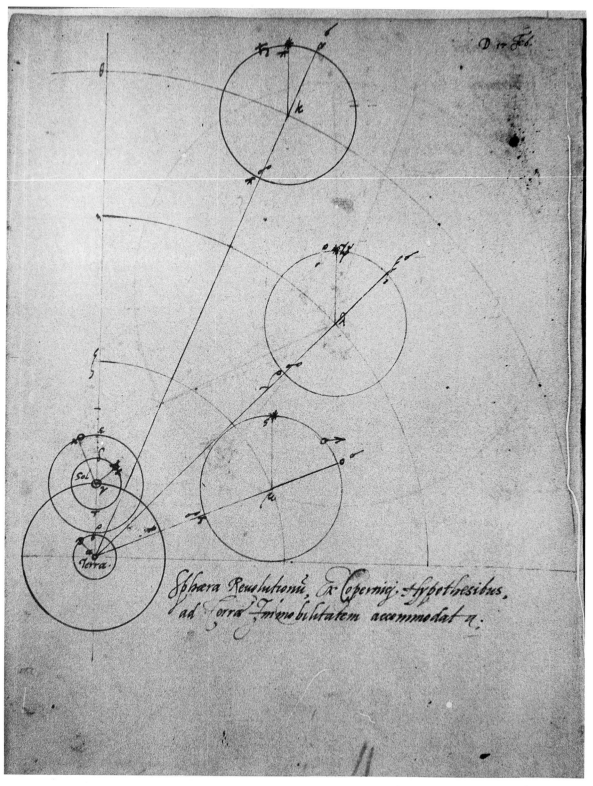

2.6. The geoheliocentric system of Paul Wittich.

rized him. Gingerich and Westman suggest an alternative. Wittich, after his 1578 studies, was in close contact with Hagecius, Praetorius, Scultetus and Craig, members of Tycho's circle of correspondents. In 1580, he spent four months assisting Tycho at Uraniborg and later visited Kassel for discussions with the Landgrave, Rothmann and Bürgi. Ursus arrived at Kassel not long afterwards. The compromise theory was, by then, 'in the air', largely because of discussions prompted by Wittich. Once the idea that Copernican techniques could 'save the appearances' without involving a commitment to heliocentric cosmology, a variety of attempts could be made. The new theory could entertain variations, such as diurnal rotation of the Earth, introduced by Ursus and also employed by Tycho's assistant Longomontanus and by David Origanus. None of the early versions was as radical as Tycho's, for others had not recognized that if the orbit of Mars intersected that of the Sun, then the solid spheres were redundant.

By the last decade of the century, serious astronomers fell into three categories: those loyal to the Ptolemaic order of the universe but willing to introduce Copernican mathematical techniques; those who had adopted some variation of the geoheliocentric framework; and those who were confirmed Copernicans. The last category was the smallest, numbering only Mästlin and Rothmann among established figures. Indeed, the latter appears to have recanted after discussions with Tycho at Hven in 1590 – an inexplicable attitude considering his earlier vehement support for Copernicus – and never returned to Kassel. Both Mästlin and Rothmann took a realist approach. The former had come to Copernicus through doubts about natural philosophy and through his analogical 'demonstration' of the orbit of the comet of 1577. Both agreed on the primacy of the order and distances of the planets as set out in *De revolutionibus*. In his debate with Tycho in the late 1580s, Rothmann had rejected the Tychonic system because although Tycho could show the 'how' of retrograde motion, he could not show the 'why' as could Copernicus. Rothmann, like Mästlin, could not give up solid spheres, so it seemed impossible that the Sun, in Tycho's scheme, could have an attractive power to draw the planets around after it; the Landgrave concurred.

This last problem was common to both Coperni-

can and Tychonic theories, for neither could provide persuasive physical arguments. Ptolemy had been able to unite both mathematical and physical models in his (then lost) *Planetary Hypotheses*, but Copernicus could only offer physical speculation. By the 1590s, the combined assaults on the Aristotelian cosmos resulting from the New Star of 1572, the comet of 1577, and the publication of Tycho's system only exacerbated the problem. The Tychonic theory was rejected by many; the alternatives were to retain solid spheres, to accept Copernican proto-gravity or its logical extension in Tycho's version, as would Longomontanus, or to adumbrate an entirely different (dynamic) system, as William Gilbert would attempt in *De magnete* in 1600.

In traditional astronomy, in which the order and dimensions of the planetary spheres were largely arbitrary, the sphere of the stars was placed beyond, but not too far from, the orb of Saturn. The stars were much more remote for Copernicus because of the failure to detect annual parallax, and his late sixteenth-century successors fastened upon the problem. Tycho believed that placing the stars at a very great distance from the planets was illogical: why should God have created such a large amount of useless space? Rothmann saw no problem with this at all. Mästlin, in his textbook, went so far as to calculate the velocity of a star on the Ptolemaic sphere, to show the absurdity of such a closely packed assemblage. If the stars did lie at great distances from the planets, and if they did not revolve upon a sphere diurnally, then it was a small mental step to assume that they might lie at varying distances, and even extend outward to infinity. Digges (Figure 2.7) suggested both as early as 1576, and was followed by Gilbert. Outside the discourse of astronomers, Giordano Bruno integrated the idea into his own cosmology. But these views, like those concerning planetary dynamics, were based upon faith in the unity of Copernicus's cosmological perspective, not upon hard evidence. Tycho's contemporaries were alive to the possibility of obtaining actual measurements of annual parallax, though they did not realize how far beyond their capacity such measurements actually were.

At the time of Tycho's burial in 1601, the first phase of the Copernican revolution was over, with the paradoxical result that although almost no

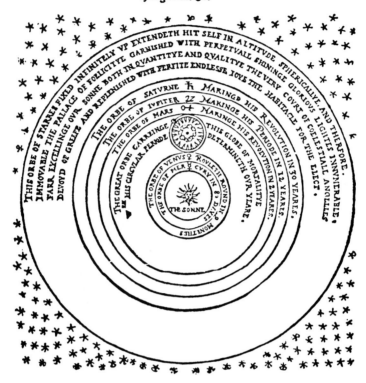

2.7. The Copernican universe according to Thomas Digges, with the stars extended indefinitely outwards in all directions.

recognized astronomer was actually a Copernican (even Kepler had yet to distinguish himself), Copernican technical astronomy had become indispensable. The true successor to the Ptolemaic theory was the Tychonic, a view at once radical in its rejection of Aristotelian physics and conservative in its retention of the geocentric perspective. With the truly radical departures in physics that began with the next generation, astronomers would finally become Copernicans, but would find Copernican technical astronomy totally dispensable.

Further reading

C.D. Hellman, *The Comet of 1577: Its Place in the History of Astronomy* (2nd edn, New York, 1971)

R.A. Jarrell, Mästlin's place in astronomy, *Physis*, vol. 17 (1975), 5–20.

Bruce Moran, Christoph Rothmann, the Copernican theory, and institutional and technical influences on the criticism of Aristotelian cosmology, *The Sixteenth Century Journal*, vol. 13, pt 3 (1982), 85–103

3

The Tychonic and semi-Tychonic world systems
CHRISTINE SCHOFIELD

The world system of Tycho Brahe (1546–1601; see Chapter 1) has not always been accorded its rightful place in the history of astronomy. Whatever its demerits, it excited a great deal of debate in the seventeenth century, and delayed considerably the acceptance of its Copernican rival. Bitter arguments raged as to who had first discovered it. It provoked discussion at all levels, some of its most notable critics being Johannes Kepler, René Descartes and Robert Hooke. It was almost entirely ignored by Galileo Galilei, perhaps unfairly so, for at the time when Galileo's *Dialogue* was published (1632) it was not the Ptolemaic system, but rather the Tychonic, that was widely discussed as the alternative to the Copernican.

Tycho's belief in a physically real geoheliocentric system developed gradually. As early as 1574, when preparing a lecture course to be delivered at the University of Copenhagen, he hints at an 'inverted' version of the Copernican hypothesis, to be used as a mathematical device only, in which the motions of the five planets Mercury, Venus, Mars, Jupiter and Saturn were referred to the Sun, but the Sun moved about the Earth. Such a device may have been considered earlier by Erasmus Reinhold (1511–53), and some aspects of it were depicted (Figure 2.6) in 1578 by Paul Wittich (*c.* 1555–87), who may well have exerted an important influence on Tycho. At this stage the idea had for Tycho no connotations of physical reality, the geoheliocentric compromise being for a time suspect to him because its proportions involved the intersection of the supposedly solid planet-bearing orbs. A study of the comet of 1577 convinced him that these orbs did not in fact exist, and this removed the last obstacle to his proposal of the system as a true representation of the cosmos.

Tycho's non-observational arguments in favour of his system reveal him to a large extent as a man of his time. Torn for a while between traditional views of the cosmos and his own observational work – in which he knew himself to be the final authority – he eventually achieved a compromise. This he regarded as a direct divine revelation, and he prized it above his more lasting achievements. His letters to his friends and fellow-astronomers reveal his fanatical eagerness that they, too, should accept his revelation.

Early indications of the interest in and importance attached to the geoheliocentric system are seen in the various disputes which arose as to the identity of its original discoverer. In fact, as we have seen in Chapter 2, the last years of Tycho's life were marred by an undignified squabble with a German mathematician, who claimed that he had discovered the system in 1585, and that Tycho had stolen it from him. Nicolai Reymers Bär, or Ursus (1551–1600), was a self-taught peripatetic tutor, who visited Tycho's astronomical island of Hven in 1584, and subsequently published two works, in 1588 and 1597, in which he described the system as his own, discovered while reading Copernicus. Tycho's opinion, shared by many of his contemporaries, was that Ursus was a wily, disreputable character, who had purloined the discovery after a surreptitious perusal of Tycho's papers at Hven. The crucial evidence against him was considered to be that, whereas in Tycho's system the orbit of the planet Mars intersected that of the Sun, in Ursus's system Mars's orbit completely enclosed the Sun's (Figure 3.1). This was alleged to be because Ursus had gained his information from a diagram of Tycho's that was wrongly drawn in this respect. Whether or not Ursus was guilty of plagiarism – the available evidence is inconclusive – the very fact that he asserted his claim so vituperatively indicates the importance attached to the new system. In one important feature, however, Ursus did differ

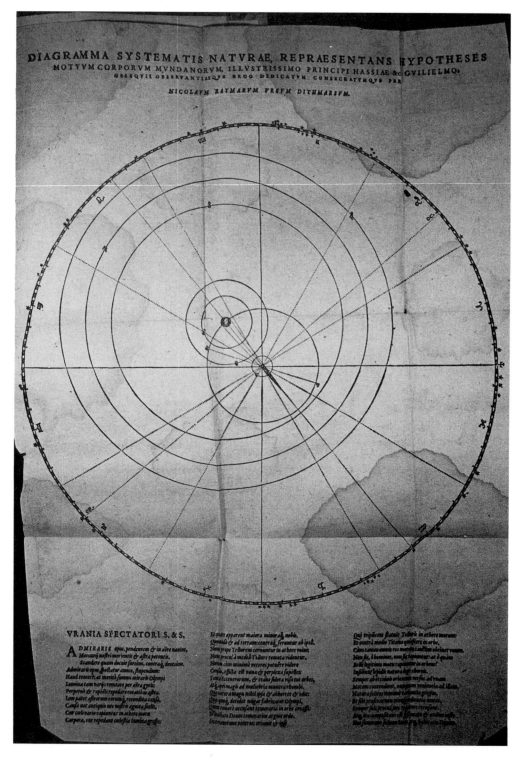

3.1. The geoheliocentric world-system from Ursus's *Fundamentum astronomicum* (1588); the Moon and Sun circle the Earth, and the paths of Mercury, Venus, Mars, Jupiter, and Saturn are shown as circles concentric with the Sun. Note that Mars's orbit is not shown intersecting the Sun's orbit, as it does in Tycho's system.

from Tycho: he unequivocally attributed to the Earth a daily rotation upon its own axis, thus becoming a very early proponent of what came to be known later as the semi-Tychonic system.

Repercussions of this incident were seen in the case of Helisaeus Röslin (1548–1616), an astronomer–physician in Strasbourg, who read Ursus's work of 1588 and in 1597 published details of a world system (Figure 3.2) which he claimed to be his own, but which strongly resembled Tycho's. It differed in that Röslin still believed the individual planets to be embedded in the solid spheres of medieval cosmology, and also in that he made the orbits of Mars, Jupiter and Saturn considerably larger than did Tycho, arguing that God had ordained their radii to be, respectively, once, twice and three times the diameter of the solar orbit about the Earth. It seems clear that Röslin drew his inspiration to a large extent from Ursus. Although fiercely attacked by Tycho, Röslin did not pursue his claim; his case illustrates rather how the new idea, filtering through to men of different backgrounds and beliefs, encouraged further speculation as to the construction of the universe.

Another visitor to Hven in 1587 and 1588 was Duncan Liddel (1561–1613), a Scots physician who studied with Wittich in Breslau (Wrocław) and who held several university posts in Germany. He also incurred the wrath of Tycho for alleged plagiarism. Tycho apparently confided to Liddel pre-publication details of his new system, and was later enraged to learn that Liddel was lecturing in Germany on the Tychonic system and claiming it as his own, displaying a congratulatory letter from Tycho to himself as proof. Evidence from the correspondence of mutual acquaintances suggests that a confusion had arisen over terminology. Liddel had supplemented Tycho's general outline with detailed mathematical hypotheses of his own for some of the individual planets, for which Tycho had praised him. But Liddel's continued emphasis in his lectures that the details were his own gave rise to the rumour that Liddel was claiming discovery of the geoheliocentric system. The rumour seems to have been without foundation, although Liddel was certainly extremely grudging in giving credit to Tycho.

A very late claim to the discovery of the system was made by another German, but this time someone unconnected with Tycho and his circle: Simon Mayr (1573–1624), or Marius, a student of music, mathematics and medicine in Germany and Italy. In 1614 Mayr published a work on the moons of Jupiter which was doubly controversial. In it Mayr not only challenged Galileo's claim to the discovery of the satellites, but also asserted that he had made an independent discovery of the geoheliocentric system in 1595–96 when (like Ursus) reading Copernicus, and quite unaware of Tycho's work. Mayr believed that the existence of the Jupiter system lent plausibility to the idea that the Sun moved about the Earth, carrying with it a five-satellite system of planets.

The enormity of Mayr's claims tends to diminish their credibility, but in any case they were made too late (in the case of Tycho's system) to be considered seriously. The impression given is that the idea was very much 'in the air' at that time, but that the credit for its formulation must belong to Tycho, for when the system acquired a name it was christened Tychonic.

A very early printed reference to the geoheliocentric system appeared in the *Nova de universis philosophia* of the Dalmatian philosopher Francesco Patrizi (1529–97), published in Venice in 1593. Patrizi, who must have had access to a copy of the privately printed 1588 edition of Tycho's book, summarized accurately the development of Tycho's ideas, except that he wrongly attributed to Tycho a continuing belief in solid planetary orbs. Tycho was quick to pounce upon this error, yet expressed gratification that the system was unequivocally attributed to him.

Other references are hard to find until after 1602, when the first public edition of the book containing Tycho's new world system appeared. In that same year the Englishman Nathaniel Torporley wrote in his *Diclides coelometricae* that so far he had been able to teach only the old doctrine ascribed to Copernicus, but that he had high hopes of the recent work of Tycho, which he looked forward to seeing. In the following year his fellow-countryman William Gilbert, in his *De mundo nostro sublunari philosophia nova*, mentioned the system of Tycho, along with those of Ptolemy and Copernicus, and also that of Ursus. In 1611 the poet John Donne, in his anti-Jesuit satire *Conclave Ignati*, has Ignatius address Copernicus thus: "Neither do you agree so well amongst yourselves, as that you can be said to have made a Sect, since,

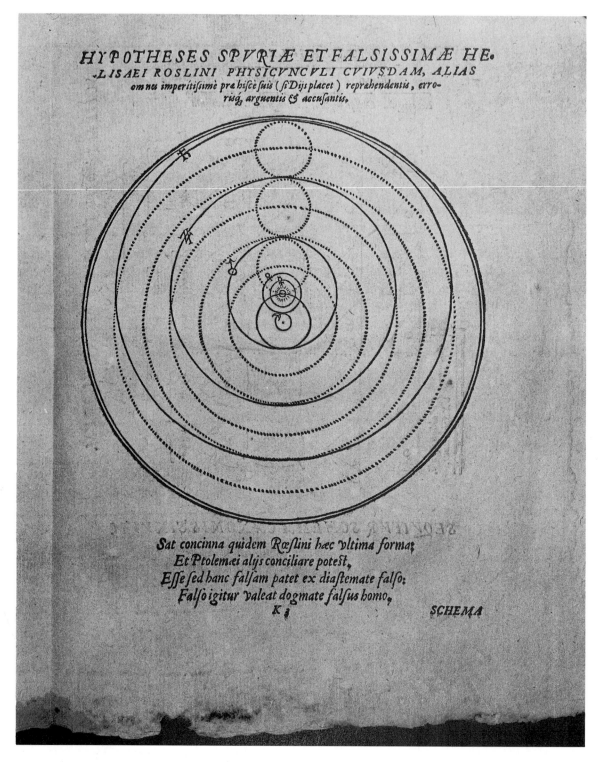

3.2. The geoheliocentric world-system of Helisaeus Röslin, from Ursus's *De hypothesibus astronomicis* (1597). Röslin, like Ursus, avoids an intersection of the orbits of Mars and the Sun: he makes the Martian orbit twice the size of the Sun's circle, instead of about one-and-a-half times as large, as it should be.

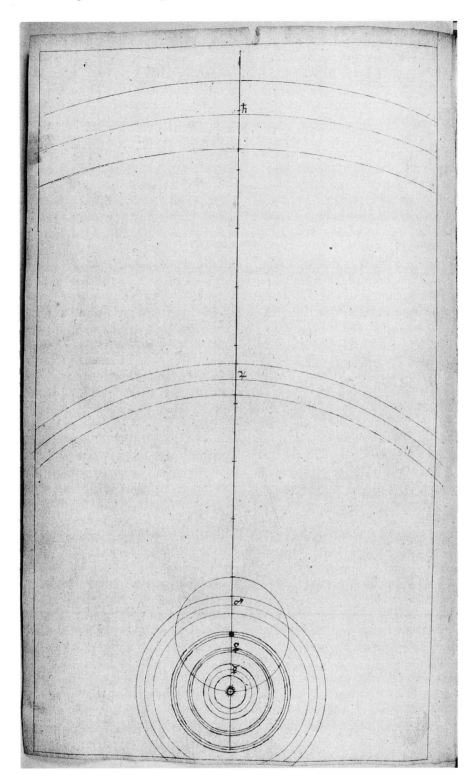

3.3. Horrocks's drawing of the Copernican and Tychonic systems, in his copy of Lansberge's *Tabulae motuum coelestium perpetuae*.

as you have perverted and changed the order and Scheme of others: so Tycho Brachy hath done by yours, and others by his''. Among the Jesuit cosmologists themselves, Johann Baptist Cysat in 1618, and his contemporaries at the College of Pont-à-Mousson in 1622, describe the system without naming its author. In 1621, however, it was described as Tycho's by the English geographer Nathanael Carpenter (1589–1628?), and again in 1622 by Tycho's ex-pupil Longomontanus. From about this time the Tychonic system was fully absorbed into the cosmological literature, and took its place alongside the Ptolemaic and the Copernican as the third system of the world.

It was readily recognized that for practical astronomical purposes the systems of Copernicus and Tycho were equivalent, the Tychonic in fact being for the Earth-bound observer the more easily understood. This similarity was often stressed in discussions on the world system by representing them both on the same diagram. A particularly interesting example of this is furnished by the talented young English astronomer, Jeremiah Horrocks (1618–41), for it is drawn in his own hand (Figure 3.3) on a page inserted into one of the astronomical works he owned – the *Tabulae motuum coelestium perpetuae* of Philip van Lansberge, published in 1632 and acquired by Horrocks in 1635. Horrocks appended a full-page diagram of the Copernican system to the book, yet with the addition of an orbit for the Sun about the Earth, thus giving a shadowy impression of the Tychonic alternative. More frequently, the diagram was of the Tychonic system, with the Copernican orbit of the Earth about the Sun dotted lightly in afterwards, an indication perhaps of where the writer's true sympathies lay. Examples are seen in the works of Tommaso Campanella (1638), Pierre Gassendi (1647) (Figure 3.4), Wolferd Senguerd (1680), Jean Boulanger (1688) and Johannes Luyts (1689), suggesting that the practice was common over a considerable period of time.

Discussion was not limited, however, to three systems of the world, for the Tychonic system paved the way towards further speculation as to a viable alternative to the Copernican. Some participants in the debate felt that Tycho had not yet furnished convincing proof that the three superior planets, Mars, Jupiter and Saturn, moved about the Sun, although they accepted his findings, reinforced by Galileo's telescope, concerning Mercury and Venus. There was, therefore, a revival of interest in the Capellan system in which Mercury and Venus orbit the Sun while the Sun and the other planets orbit the Earth, its supporters including the English philosopher Francis Bacon, the Jesuits Joseph Blancan and Charles Malapert and the Italian astrologer Andreas Argoli. Another Jesuit, the widely respected Giambattista Riccioli, suggested in 1651 his own alternative, in which Mercury, Venus and Mars move about the Sun, and the Sun, Jupiter and Saturn about the Earth. In justification of this idea Riccioli pointed out that Jupiter and Saturn had been shown to possess satellites, and were therefore primary planets like the Sun, centred on the Earth; also, being heavier and slower-moving than the other planets they bore a closer resemblance to the fixed stars, and were therefore more likely to move with these stars about the Earth. The frontispiece of Riccioli's book (see Figure 3.5) shows the system of Copernicus weighed in the balance against that of Riccioli, and found wanting, while Ptolemy's lies rejected on the ground, and Tycho's is nowhere in sight. Riccioli's system was earnestly discussed by his fellow-Jesuits, such as Gabrielo Beati writing in 1662, Christopher Sturm in 1670, and Joseph Zaragoza in 1675.

Although uncertainty prevailed as to the centre of the superior planets' orbits, the opinion was becoming more widespread that the daily motion observable in the cosmos could more plausibly be attributed to the Earth than to the vastly numerous and far-distant 'fixed stars'. Tycho himself hesitated on this point, since it was less a problem in astronomy than in natural philosophy. Many of Tycho's followers, however, explicitly attributed a daily axial rotation to the Earth and the system they expounded was known as semi-Tychonic or, less frequently, semi-Copernican. It is generally attributed in the literature either to David Origanus, who wrote of it in 1609, or to Longomontanus who, although not publishing until 1622, said he had discussed the matter with Tycho himself, and claimed priority over Origanus. In fact, this semi-Tychonic variant was put forward as early as 1588 by the ill-starred Ursus.

As to the influence and following gained by this system, Nicholaus Müler writing in 1617 seems to regard it as of equal status with the Ptolemaic and

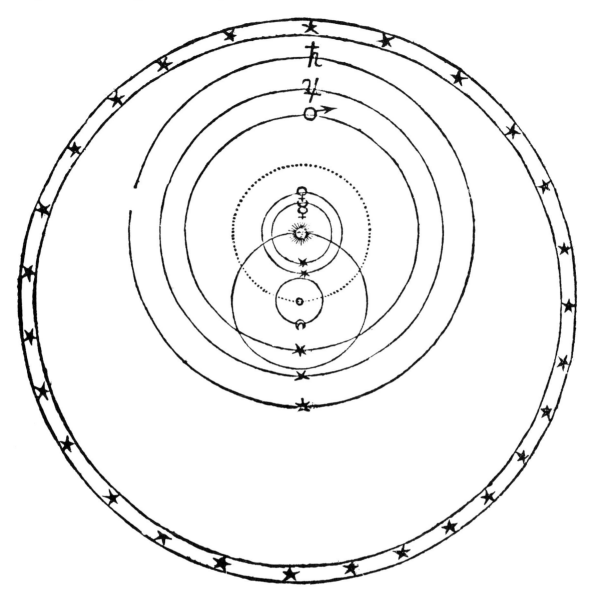

3.4. Gassendi's drawing of the Tychonic system, but with the alternative (Copernican) motion of the Earth added as a dotted curve.

Copernican and gives it greater eminence than Tycho's own. Yet Carpenter in 1625 is unaware of its existence, and explores the idea as if it were his own. News of this particular variant seems to have filtered through but slowly to the English writers. However, because of Gilbert's work on the magnetic nature of the Earth (see Chapter 4) they had ample opportunity to discover it again for themselves, and its adherents included Mark Ridley, Robert Burton, Francis Godwin and William

Pemble. The semi-Tychonic system is referred to frequently until around 1680, and even (scathingly) in 1717 by the Dutch savant Bernard Nieuwientijt, who gives the impression that even at that late date the Tychonic system still commanded an apparent following. Nieuwientijt rejected the semi-Tychonic on the grounds that it possessed neither the simplicity of the Copernican nor the agreement with Scripture of the Tychonic.

One of the most pertinent criticisms of Tycho's

3.5. The frontispiece of Riccioli's *Almagestum novum* (1651), showing Urania weighing Copernicus's and Riccioli's systems against one another. In Riccioli's system, Mercury, Venus and Mars are satellites of the Sun, while Jupiter and Saturn along with the Moon and Sun move in concentric circles about the Earth. Riccioli's system is clearly the weightier one; Ptolemy's system lies discarded on the ground.

system was its inability to explain convincingly the causes of planetary motion. As Kepler pointed out in 1604, Tycho's concept of motion was essentially that of the early Greeks – that the planets circle the Sun naturally, possessing some divinely implanted affinity with it, and reverence for it. Kepler made conscientious public attempts to defend the Tychonic system, because he had promised as much to Tycho, but privately he confessed that he could not believe the motions of the planets to be as intricate as Tycho's system demanded. He was not averse, however, to giving the system a measure of support, because he honestly believed that its acceptance would hasten the transition towards greater acceptance of the Copernican. He possessed none of the cynicism of Galileo, who saw rather that the Tychonic alternative could strengthen the position of the reactionaries.

Galileo shared Kepler's chief objection to the Tychonic system – that it was not based upon any comprehensive physical principle. He refused, however, to be drawn into discussion with Tycho on the subject, in spite of Tycho's strenuous efforts to engage his attention, contenting himself merely, in his work on comets, with scathing asides as to the fanciful and half-baked nature of Tycho's ideas in general. So unsympathetic was Galileo that he appears deliberately to have misrepresented the nature of the controversy over the world system when in 1632 he produced his *Dialogue Concerning the Two Chief World Systems, the Ptolemaic and the Copernican*. By 1632 very few would have bothered to oppose the Copernican system with the Ptolemaic, when there existed the geoheliocentric alternative. Marin Mersenne, the Minim friar, writing in 1636, was so aware of this that he describes Galileo's book as dealing with "the three systems of the world", while Robert Davenport in an unpublished poem written at the time of the publication of the *Dialogue* declared that the present controversy was "twixt Ticho Brahe and Copernicus".

Clearly, the Tychonic system was much more difficult to refute than the Ptolemaic, and, religious considerations apart, one considerable advantage it still possessed over the Copernican was that no-one had yet been able to establish the existence of an annual parallax amongst the fixed stars which would be a necessary result of the annual motion of the Earth. The absolute size of stars was no longer a problem, because the use of the telescope had made

necessary a re-estimation of the true diameter of the fixed stars, which were now known to be far smaller than they appear to the naked eye. Yet as telescopes improved, the difficulty posed by the absence of measurable annual parallax increased, and the only defence of the Copernicans was to maintain that the fixed stars were very far away indeed. Dissatisfaction with this reply grew, and in 1674 the ingenious experimentalist of the Royal Society of London, Robert Hooke, undertook a set of experiments to settle the question. Judging that "whatever mens' eyes were in the younger age of the World, our eyes in this old age of it needed Spectacles", he resolved to obtain as large and as reliable a telescope as possible – an "Archimedean Engine that was to move the Earth". Working with a 36-ft vertical telescope whose objective glass was mounted in an aperture in the roof of his lodgings at Gresham, Hooke took four readings of the zenith distance at meridian transit of the bright star Gamma Draconis, which passed almost overhead, and from these concluded a sensible parallax and "a confirmation of the Copernican system against the Ptolomaick and Tychonick". Hooke's scanty data received a mixed reception amongst his contemporaries, but encouraged others to follow his example, both in England and on the continent of Europe (see Chapter 9, and Volume 3). The Danish astronomer Ole Römer, who made a similar attempt at Copenhagen from 1692, remained throughout his life a convinced follower of Tycho.

Apart from the astronomical problem of no parallax, the strongest pro-Tychonic arguments were religious. The Jesuit astronomers found themselves awkwardly placed in the Copernican controversy, particularly in Italy. Christoph Scheiner, a Jesuit authority on sunspots, working in Rome, was considered to have been coerced into writing in support of Tycho, while Riccioli's anti-Copernican arguments from the motion of falling bodies were considered so futile, for a man of Riccioli's calibre, as to strongly suggest prevarication. Whatever their innermost convictions, the Jesuits produced after the 1633 decree a flood of pro-Tycho literature which continued until the closing decades of the seventeenth century. In 1691 Ignace Gaston Pardies declared that the Tychonic was still the commonly accepted system, while Francesco Blanchinus reiterated this as late as 1728. Although the Jesuits were in a particularly sensitive position, other loyal Roman Catholics were considerably exercised on the question of the motion of the Earth. Belgium, strongly Catholic, had a particularly vociferous band of Tychonians, including Nicolaus Müler writing in 1617, Erik Puteanus in 1619, Arnold Geulincx in 1657, and Johannes Luyts in 1689. The most notable was Libert Froidmond, the only follower of Tycho to earn a measure of approval from Galileo. In 1615 Froidmond had been prepared to discuss the Copernican system, and even to explore the possibility of a reinterpretation of Scripture in the light of it. The following year the Roman Catholic Church ruled that Scripture was still to be interpreted as laid down at the Council of Trent, and that the Copernican system was irreconcilable with this. Having made judicious enquiries to satisfy himself of the existence of this ruling, Froidmond embarked on a work, published in 1631, in which he attempted to justify the Church's action, and gave fulsome support to the system of Tycho for its agreement with Scripture. His Italian contemporary, Bonaventura Cavalieri, after reading Froidmond's book, wrote to Galileo that Froidmond expounded the Copernican arguments so skilfully, and refuted them with so little force, that he seemed still to believe in them. In the last quarter of the seventeenth century Belgians such as Matthias Tombeur and Arnold Deschamps began to reject the Tychonic system, although such a rejection was still fraught with danger, for in 1691 a professor at Louvain University, Martin-Etienne van Velden, was suspended for Copernican teaching. The fact that Guillaume-Walric van Leempoel refers to the Tychonic system as late as 1796 is some indication of how strong a hold it had gained in Belgium.

In France the more stringent decree of 1633 was not officially ratified, yet opinions were guardedly expressed. Mersenne, whose pro-Copernican work of 1634 was not favourably received, had the work reissued in two versions, an expurgated one for general circulation, and one containing Copernican arguments, for distribution to trusted friends. Descartes also found himself on the horns of a dilemma. Strongly persuaded towards the Copernican system, in 1640 he wrote of the internal conflict he was experiencing between faith and reason, and made judicious enquiries as to how far he could express his true opinion with impunity. By 1644 he had effected a prudent revision of his ideas, which

he felt would bear the scrutiny of the Church, his rather surprising conclusion being that according to the true definition of motion, Tycho attributed motion to the Earth, and Copernicus did not (see Chapter 11). Within the context of his vortex theory, and defining the motion of a body as a change in its position with respect to the fluid surrounding it, Descartes argued that in the Copernican system the Earth was merely carried along by the motion of the surrounding fluid (that is, it did not move itself), whereas in Tycho's system it changed position daily with respect to its surrounding fluid, and therefore was actually in motion. The equivocacy of Descartes's position was mocked by many of his contemporaries, who felt that Descartes was laughing up his sleeve at those foolish enough to take him seriously. Others, however, praised this very equivocacy as the strong point of his argument, since it furnished an interpretation of the Copernican system which could be reconciled with a belief in the literal interpretation of Scripture. Certainly, the Tychonic system seemed to have little support in France, a lone voice being that of the resourceful Jean-Baptiste Morin, who defended Tycho on grounds astronomical, astrological, physical and religious, even incorporating the elliptical orbits of Kepler into the system of Tycho.

Lutheran Germany displayed little interest in the planetary system of their fellow-Lutheran, Tycho. Almost its only adherents there were the three men who claimed some personal credit for it – Ursus, Röslin and Mayr. But Christoph Rothmann and Kepler were early Copernicans, as was Michael Mästlin. Lecture courses planned at the University of Hesse-Schaumburg in 1654 included the Copernican system but not the Tychonic. Nor had Dutch writers, such as Lansberge and Christiaan Huygens, much time for it, although Huygens acknowledged its influence elsewhere, remarking bitterly that men who saw the Copernican system confirmed by observation, by argument and almost by the voice of Nature herself still hesitated because they knew that "a different one entered the head of Tycho Brahe". He commented caustically that it should be possible to enjoy the fruits of the labours of such men as Tycho, without repeating their errors. In England also there was early support for Copernicus, by Thomas Digges (*c.* 1546–95), John Dee (1527–1608) and John Feild (1525?–87),

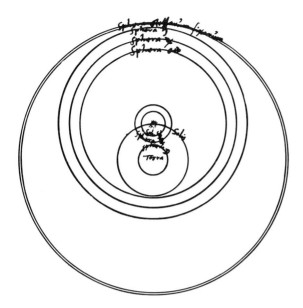

3.6. A (pre-Tychonic) geoheliocentric system due to Rothmann. This system, considered at least a year before Tycho's publication of his system, and derived not from Tycho but from Copernicus, provides yet another example of the prevalence of geoheliocentric theories in sixteenth-century astronomy.

among others, and this was continued into the seventeenth century by Nicholas Hill, Thomas Harriot, Sir William Lower and Horrocks, while the Tychonic system never gained a strong hold. Its most extensive discussion is found in the works of the conservative John Swan, who in 1635 echoed all Tycho's own arguments. Henry More described the system in 1647, but emphasized what an "untoward broken system" it was in comparison with the Copernican, while Seth Ward and John Wilkins, both extolling in 1654 the freedom of thought enjoyed in the English universities, cite as an example their strong support of the Copernican system.

It can therefore be seen that this third system of the world played a major part in seventeenth century discussion on the cosmos. It is widely mentioned in printed books up to 1680, and there are scattered favourable references to it after that date; from 1700 onwards, however, it is normally mentioned only in order to refute it. The system and its variations influenced astrology, and permeated contemporary non-scientific literature, as is shown by the anonymous *Democritus Turned Statesman; or,*

Twenty Queries between Jest and Earnest, Proposed to all True Hearted Englishmen (1659). The first of these enquiries is "whether it be not convenient that the doctrine of Copernicus, who held that the world turned round, should be established by act of parliament, which our late changes, alterations and revolutions in part have verified; and that Tycho Brache, with the gang of the contrary opinion, be declared heterodoxal".

The most serious weakness of the Tychonic system had always been the absurdity of the confused and contrary planetary motions it involved, and it was on these grounds that it was finally rejected. Kepler (see Chapter 5) emphasized the contrasting unity of the Copernican system, by restating it in solar-dynamic terms, while Gilbert's magnetic theories (see Chapter 4) made more plausible the idea that the Earth possessed a daily rotation. The telescopic discoveries of the axial rotation of Mars and Jupiter, and of the satellites of Jupiter and Saturn, strengthened the awareness of the Earth's relationships to these planets. Even the specious criticisms of Descartes told against Tycho. Long before the controversy was conclusively settled by the discovery of true annual parallax, the Tychonic system was rendered obsolete on physical grounds.

It seems, therefore, that its popularity was due largely to the fact that it fulfilled the need for a safe synthesis of ancient and modern, meeting admirably the need occasioned by the Roman Catholic Church's growing opposition to the Copernican system. This is illustrated by the fact that Jesuit writers throughout the seventeenth century presented a united front in favour of Tycho. After 1670 they became less guarded in the expression of their views, many disguising but thinly their preference for Copernicus. Support for Tycho was also markedly stronger in Catholic countries, notably Italy and Belgium, while in Germany, Holland and England his system vanished from the literature much earlier.

A subsidiary reason for the extended influence of the system may perhaps be seen in the personal authority of Tycho. An anonymous English writer of 1642 said that the anti-Copernicans were "blinded by the teaching of Aristotle, preoccupied by the authority of Tycho, or finally terrified by the decree of the ruling authorities", while Gassendi in 1647 expressed the view that those who followed Tycho did so either because of his great name as an observer, or else because they were determined at all costs to oppose Copernicus.

Further reading

J.L.E. Dreyer, *Tycho Brahe: A Picture of Scientific Life and Work in the Sixteenth Century* (Edinburgh, 1890; reprinted New York, 1963)

Owen Gingerich and Robert S. Westman, The Wittich connection: conflict and priority in late sixteenth century cosmology, *Transactions of the American Philosophical Society* (to appear)

Christine Jones Schofield, *Tychonic and Semi-Tychonic World Systems* (New York, 1981)

Edward Rosen, *Three Imperial Mathematicians* (New York, 1985)

Dorothy Stimson, *The Gradual Acceptance of the Copernican Theory of the Universe* (New York, 1917)

4

Magnetical philosophy and astronomy, 1600–1650

STEPHEN PUMFREY

1. Introduction

In the late sixteenth century, several natural philosophers and astronomers became convinced of the physical truth of Nicholas Copernicus's system, and of the wisdom of his injunction that technical astronomers should reform their art upon true physical principles. Copernicus's own sketch of these principles, however, was inadequate, and the currency of Tycho Brahe's arguments against the existence of solid spheres had compounded the problems. We must recall the questions that realist Copernicans had to answer (without Newtonian mechanics!) to gain any credibility.

First, they had to construct a new terrestrial physics, for none existed besides Aristotle's. Scholastic philosophers endowed elemental earth with one natural motion – rectilinear descent to the centre of the universe. This at once denied the possibility of a revolving Earth, and provided persuasive objections to it.

Secondly, they had to make the new terrestrial physics the basis of celestial physics, because Copernicus's Earth behaved as a planet, which destroyed the Aristotelian division between motions above and below the Moon. A new celestial mechanics had much to explain. How were the planets moved, and why were the orbits eccentric? What were the forces that allowed only small changes in the orientation of the Earth's axis with the ecliptic, in the angles of inclination of orbital planes with the ecliptic, and in the positions of perihelion and aphelion (the line of apsides)? The very same principles had to explain the motion of stones on the Earth's surface. In these difficult years for proponents of Copernican (and semi-Tychonic) systems, before the inertial mechanics of Galileo and Descartes became established in mid-century, many sought answers in the 'magnetical philosophy' of William Gilbert (1544–1603).

This may surprise the modern reader since Gilbert, a leading court physician, is usually remembered, through his book *De magnete* (*On the Loadstone*, 1600), for beginning the new science of magnetism, and for doing so with an experimental method of unprecedented thoroughness. Recent historians, however, have recognized the overriding cosmological interests that lie behind Gilbert's work.

The founder of magnetical philosophy, like other philosophers of Renaissance Naturalism, rejected much of the scholastic philosophy he learned at university, because it presented the Earth as an inert agglomeration of corrupt matter at the centre of the world, separated from a perfect heaven. Gilbert believed the Earth was endowed with a soul, and was therefore as noble as the planets. Like Copernicus, he described the scholastic cosmos as a monster, and agreed with him that only a heliocentric universe expressed the divine harmony of creation. In magnetical philosophy, he claimed to have discovered the much needed terrestrial physics that demonstrated the Earth's harmonious motion.

It is, therefore, no coincidence that the main debate surrounding magnetical philosophy in the first half of the seventeenth century concerned the very ideological question of its implications for astronomy. Many Copernicans, notably Johannes Kepler (1571–1630), Simon Stevin (1548–1620) and Galileo Galilei (1564–1642), used and developed Gilbert's claims. Jesuit Aristotelians skilfully inverted the arguments into a defence of orthodox astronomy. Of the leading figures, only the Frenchman Marin Mersenne (1588–1648) denied that magnetic arguments were decisive, though it was Mersenne's position that was to win in mid-century. To follow this strange episode in the history of astronomy, we must begin with *De magnete* itself.

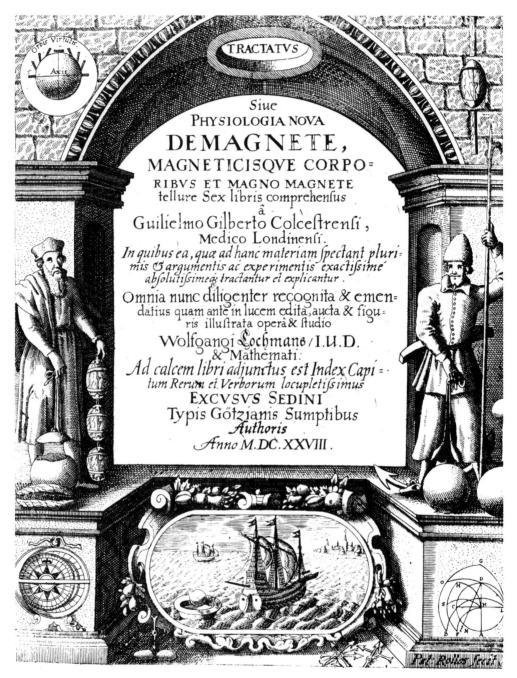

4.1. The frontispiece of the second, pirated edition of William Gilbert's *De magnete* (1628). There were several unauthorized reissues of *De magnete* around 1630, suggesting a renewed interest in magnetical philosophy. Notice that the familiar opening words of the title have been replaced with the grander "Treatise of New Physics of the Loadstone". The representations of experimental apparatus and results are all taken from the first edition. Astern the galleon is Gilbert's basic experiment of the floating *terrella*.

Although the artist's figures make it clear that the book is of interest to both philosophers and mariners, it is the relationship, of navigational use, between magnetic inclination and latitude that is most strongly advertised. Top left shows the inclination of needles on a terrella within its sphere or virtue; bottom left, Gilbert's instrument for measuring inclination; bottom right, the geometry of the relationship.

2. *De magnete* and the foundations of magnetical philosophy

De magnete consists of five 'experimental' books which established the axioms, practices and phenomena of magnetical philosophy, and a sixth in which Gilbert developed its cosmological consequences, to which he had frequently alluded in the preceding books. Two magnetic properties were important. The first was the tendency of magnetized objects to rotate, under the influence of an immaterial (that is, non-contact) force, until they achieved a harmonious (that is, stable) position. Gilbert called this "impulsion to magnetic union" *coition*, and used it to demonstrate the Earth's natural ability to revolve, in contradiction to Aristotle. The second was a magnet's capacity to align itself in a fixed direction, which he called *verticity*. This was to explain the stable orientation of the Earth's axis as it revolved.

Much of Gilbert's empirical knowledge and method came from his acquaintance with the London maritime community (particularly with his collaborator Edward Wright). The cosmological role of magnetism may have been suggested to Gilbert (as it was to Kepler) when he read Jean Taisnier's recent plagiarism of Peter Peregrinus's *Epistola . . . de magnete* (*Letter on the Loadstone*, 1269), where the loadstone was described as a "similitude of heaven", because it had two poles. But Gilbert had to construct the links from common observation to complex cosmology, in order to make them persuasive.

The circularity of magnetic motions was emphasized by Gilbert's apparatus. The exemplary experiments of magnetical philosophy employed small mounted compass needles, or *versoria*, and turned spheres of loadstone, or *terrellae*. Thus Gilbert never saw two loadstones repel. Instead they "wheel round, or the smaller obeys the larger and takes a sort of circular motion". Verticity required no proof, but Gilbert's theory-laden term stressed the turning effect of a magnetic needle.

These properties became cosmological once Gilbert had shown that they were possessed by the planet Earth. Cleverly using a *versorium* and a *terrella* as a model of a compass needle on the Earth's surface, he duplicated all the magnetic phenomena known to navigators and mathematical practitioners. Particularly compelling were his

replication of magnetic dip, described by Robert Norman in 1581, and his account of magnetic variation. Variation was explained as a rotation away from the true North Pole, caused by the Earth's superficial irregularities.

Gilbert's conclusions that elemental earth was magnetic, and that the Earth was itself a giant loadstone, were widely accepted, but he went further: "That the Earth is fitted for circular movement," he wrote, "is proved by its parts, which, when separated from the whole, do not simply travel in a right line, as the Peripatetics [i.e. Aristotelians] taught, but rotate also." This (very Aristotelian) argument by analogy was difficult to deny, and the principle of extrapolating from the behaviour of *terrellae* to the Earth (and beyond) was readily accepted by magnetical philosophers.

Gilbert's account of the vexing problem of variation also had a very necessary astronomical consequence (as well as navigational applications to the problem of finding longitude at sea, which were eagerly pursued). It was necessary because it made the Earth's magnetic and geographical poles the same. Hence they were real, magnetic "limits agreeable to the circular motion, to wit, poles that are not merely mathematical expressions" (that is, not fictional projections of the celestial poles onto a static globe). Consequently, "[the Earth's] verticity holds it in this motion lest it stray into every region of the sky".

Finally, in the least accepted part of his argument, Gilbert experimented to show that, of all alleged attractions (especially electricity), only magnetism acted through matter, and without material contact. For Gilbert, this proved that magnetic motions were caused, not by body, but by soul. The Earth's magnetism was therefore "animate, or imitates a soul".

Gilbert was concerned to provide a physical account of the noble, revolving Earth. Although he was a Copernican, he had little to say about other planetary motions, particularly the annual revolution, except to show that they were the result of celestial souls acting in a "consentient compact", led by the Sun. Furthermore, Gilbert, whom Wright said knew little astronomy, did not think that an exact science of the heavens was possible, and saw no need for realist hypotheses.

Gilbert's animist ontology created two problems for astronomers seeking magnetic solutions. First, to translate the general principles of magnetical philosophy into predictive physical astronomy was tempting but difficult. Second, and more seriously, the extremely unstable political and religious conditions in the early seventeenth century produced a strong move among natural philosophers against occultism and sympathetic magic. In this climate, Gilbert's vitalism proved unacceptable to all but a few such as Robert Fludd (1574–1637) and Johannes Baptista van Helmont (1579–1644). Thus the basic question of what caused terrestrial magnetism was the subject of constant negotiation, and the meaning of magnetic experiments (particularly in the contentious area of astronomy) was left wide open.

Finally, we should note that, although several natural philosophers, such as Mersenne, Pierre Gassendi (1592–1655) and John Wilkins (1614–72), wondered whether gravity was a magnetic force, Gilbert did not. For him, gravity was an electrical effect of the air. In contrast, he stressed the rotational and directional functions of magnetism, which were not analogous to gravity. This emphasis on difference predominated until the mid-century, when much of magnetic cosmology was jettisoned. In Chapter 12 we shall see how the similarity of the two attractions then became useful.

3. Magnetical philosophy and Copernicanism

The early seventeenth-century literature contains many assertions that Gilbert had demonstrated the Earth's motion magnetically. One could list the German physician Brengger, the Italian astronomer G.A. Roffeni, the French savants Nicholas Claude Fabri de Peiresc and Theodore Deschamps, and numerous Englishmen. Perhaps the earliest occurrence is in Nicholas Hill's *Philosophia Epicurea*, published in Paris in 1601.

Of greater importance were the attempts by skilled astronomers to apply the explanatory power of magnetical philosophy more specifically to the heavens. The earliest scheme was that of the Dutch mathematical practitioner Stevin. Although he was one of the first Copernicans, Stevin doubtless read *De magnete* because of its navigational content. Gilbert's theory of variation reinforced the arguments of his own *Havenfinding* treatise of 1599.

Stevin's Copernican textbook *De hemelloop* (*The*

Heavenly Motions, 1608) was written in the early 1600s. Later editions appeared in Latin and French. Stevin maintained the reality of the heavenly spheres, and did not need the magnetic dynamics of rotation. For him, the statical features of fixed planetary axes, planes and apsides showed that Gilbert's verticity existed "not only in the Earth . . . but also (the one serving to confirm the other more fully) in its orbit". He proposed a general principle of 'magnetic rest' for all the spheres, for all exhibited the same apparent stability as a Copernican Earth. To avoid any explanation using heavenly souls, Stevin derived his cosmic magnetism from the sphere of the fixed stars, to the poles of which all the planetary spheres aligned.

While Stevin generalized magnetic philosophy to certain areas of celestial mechanics, Kepler, working independently, completed the new physics. In 1602, he became convinced that planetary orbits were ovoid; at about the same time he acquired a copy of *De magnete*. The book mapped well onto his three astronomical concerns. It seemed to outline the real principles Kepler thought astronomy lacked; it aimed to explain celestial harmonies physically; and, above all, it suggested to Kepler a mechanism for generating eccentricities. He immediately wished that he "had wings with which to travel to England to confer with [Gilbert]. I certainly think I can demonstrate all the motions of the planets with these same principles". By 1605 he had done it.

Kepler's claim that his dynamics was not novel but "built upon . . . William Gilbert's philosophy of magnetism" accurately described his dependence upon Gilbert's proof of the Earth's magnetic power, which caused diurnal rotation around a stable axis. However, two constraints forced him into extensive elaborations which he justified with references to magnetical philosophy, but which his opponents attacked for their lack of any experimental basis.

First, he was constrained metaphysically, by his programme of constructing a cosmology without the traditional planetary souls (for souls were a central feature of both *De magnete*, and his own *Mysterium cosmographicum* of 1596). Celestial mechanics was now to employ only physical, magnetic forces which could be analysed geometrically. A second constraint came from the orbits themselves, for which he demanded a complete physical description. Whereas Stevin had been content with qualitative descriptions, Kepler also wanted quantitative ones.

Kepler's dynamics is discussed in detail in Chapter 5. Like Stevin, he began by applying terrestrial magnetic philosophy to all the planets. The Sun, the divine "prime mover of the universe", was unique. It had a soul, and just one kind of magnetic polarity, diffused over its surface. The other planets had animal faculties of rotation (which were made active by the Sun, as organs of the body were by the soul). The rest of the dynamics is mechanical. The Sun rotates rapidly (a prediction soon confirmed by Galileo's observations of sunspots) and radially extended 'rods' of its magnetic force rotate with it. These move the resisting planets in circular orbits. Each planet contains Gilbertian magnetic poles of opposite polarities, which are made up of magnetic threads. (Kepler misattributed this model to Gilbert himself.) Magnetic philosophy showed that these poles maintained their direction in the heavens, and therefore strongly resisted being attracted (or repelled) into alignment with the Sun. Accordingly, in one half of a planet's revolution, when its unlike pole faces the Sun, the planet suffers a net attraction towards the Sun. In the other half it is repelled, and thus an eccentric orbit is generated. The relative strength of each planet's poles can be calculated from the eccentricity of the orbit, and the direction of its axis from the line of apsides. (They are perpendicular.) The apsides progress because the Sun succeeds in re-aligning the axis very slightly.

The congruence with *De magnete* now became strained.

The axis of the daily rotation of the Earth . . . forever points in longitude towards the beginnings of the Crab and of the Goat . . . . But the threads by which the Earth is repelled from the Sun or attracted pass from sign to sign . . . . Therefore the axis of rotation of the Earth and the thread which changes the interval are different . . . We are therefore compelled to admit that the globe is inside an outer crust: in such a fashion that the crust rotates during the daily movement, while the globe having the threads does not rotate; and the ordinary magnetic virtue belongs to the outward crust.

Kepler had yet to explain the inclination of each orbital plane with the ecliptic. The mechanism was similar to that of the eccentricity. Two poles were needed, this time perpendicular to the line of nodes. Kepler likened them to a magnetic rudder moving through the solar stream: they forced the planet above the ecliptic in one half of the orbit, and below it in the other. The nodes and limits retrograded for the same reason that the apsides progressed, namely that the magnetic threads forming the poles underwent deflection. But these threads were deflected in the opposite direction to those in the stationary inner globe, and therefore could not easily be part of that globe. Kepler hoped the outer threads of the axis of rotation would serve, as the case of the Earth made probable; this being so, he hoped to avoid introducing further complexity in the planets' structures. But even these complexities could be regarded as improvements to magnetical philosophy, for astronomy "clears our way to the inward substance of the globes".

When Kepler's celestial magnetics first appeared in 1609 in his *Astronomia nova* it was little read. In 1620 he tried to gain a wider audience with a simplified and modified account in Bk 4 of the *Epitome astronomiae Copernicanae* (*Epitome of Copernican Astronomy*), and it is this account that we have outlined above. Even so, it excited attention only when the astonishing superiority of his Rudolphine Tables became clear after their publication in 1627, for Kepler explicitly claimed that this superiority resulted from the true (and magnetic) physical principles that lay behind them. J.L. Russell finds that, between 1630 and 1650, the *Epitome* was "almost certainly the most widely read treatise on theoretical astronomy", although few readers accepted the entirety of Kepler's physics. But magnetical philosophy was immediately put to contrasting use by his opponents, the defenders of conservative cosmology.

The unease of religious authorities, particularly Roman Catholic, at the popularity of magnetical philosophy was evident in the trial of Galileo. Galileo's application of magnetical philosophy to Copernicanism in 1632, in his *Dialogue Concerning the Two Chief World Systems*, was not innovative, but it was historically significant. He had certainly been impressed with *De magnete* in the 1600s, and had experimented with capped loadstones. But by 1632 he had developed a theory of circular inertia to defend heliocentricity, which made magnetical philosophy strictly irrelevant.

Nevertheless, during the 'Third Day' of the *Dia-*

logue, Gilbert's work was used by Galileo to establish two points. The proof that true elemental earth is loadstone, and therefore has a circular motion, was employed, as Gilbert intended, to convince the Aristotelian character Simplicio that Aristotelian physics was mistaken. More positively, Galileo's spokesman Salviati asserted the magnetic stabilization of the Earth's axis. Salviati noted that Galileo's inertial mechanics was sufficient explanation, but "let us add next," he continued, "to this simple and natural event the magnetic force by which the terrestrial globe may be kept so much the more immutable".

When Galileo's prosecution for treating Copernicanism as reality was engineered in 1633, part of the evidence brought by the Jesuits was that he "praises exceedingly and prefers to the rest William Gilbert, a perverse heretic and a quarrelsome and quibbling advocate of this opinion".

The convergence of Galileo's trial and Kepler's (posthumous) success made the professors of the Society of Jesus realize that magnetical philosophy belonged to the opposition. There had been some earlier attempts by theologians to sever this link. In 1630 Libert Froidmond, a theology professor in Louvain, had argued that the loadstone's virtue was neither long-range enough, nor sufficiently universal in its attraction, to be a celestial force. Jean-Baptiste Morin, the Parisian physician, astronomer and theologian, thought Froidmond's refutation inadequate, and in 1631 produced a lengthier, but equally ineffective argument. Neither author was trained in modern natural philosophy, and it was left to the Jesuits, who were, to harness the prestige of magnetical philosophy to the interests of orthodoxy.

4. Magnetical philosophy and orthodox cosmology

Jesuit natural philosophers worked within the Society's rich, successful and professionalized education system. They were called to defend Catholicism from heresy, to incorporate the best new science into Aristotelianism, and to meet bourgeois demands for a practical education. Magnetical philosophy engaged all three of these 'neo-Aristotelian' interests. It promised advances in navigation, its prominent champions were anti-scholastic, and it had become popularly associated with Copernicans who (necessarily) claimed that

theology should yield to physics. The Jesuits' efforts between 1629 and 1650 gained them undisputed dominance in magnetical philosophy.

The first publication was the *Philosophia Magnetica* of Niccolo Cabeo (1586–1650), which appeared in 1629. Working in the 1620s, Cabeo was mainly concerned to show that magnetical philosophy complemented rather than contradicted Aristotle. Advocates of the Earth's magnetic motion were dismissed as "so clearly in error that I need not dwell on them". In a brilliant analysis, he exposed gaps in Aristotle's classifications of place, motion and change, which indicated the existence of another prime quality besides the hot and cold, wet and dry, and heavy and light. He 'deduced' that this quality acted by altering the nature of a body, and then by moving it in imperfect circular motions (that is, arcs), directed towards the poles. The quality is, of course, magnetism. In Cabeo's experimental descriptions loadstones induce magnetism in ferrous bodies, which move until they *rest* in their correct alignment.

Cabeo and his followers were fully aware of the cosmological consequences of his integration. Earth had been granted a quality besides gravity, but it was an Aristotelian quality. Therefore magnetism acted like gravity, to make objects rest in their natural place, as magnets were observed to do. It also had the same final cause as gravity. Since magnetic philosophers ascribed the properties of *terrellae* to the whole Earth, magnetism clearly kept the Earth at rest in the centre of the world, for the benefit of its creatures.

Adopting a typical neo-Aristotelian strategy, Cabeo granted that some of the functions of gravity had been correctly dismissed by anti-Aristotelians, but he immediately reinstated them as magnetic phenomena. He admitted that gravity did not prevent the Earth 'wobbling' about its centre, but "this magnetic force exists in the entire globe ... so that it is better held in its place than by gravity alone".

Cabeo's theory was amplified, if not extended, by the renowned Jesuit polymath Athanasius Kircher (*c.* 1601–80). He had published his *Ars magnesia* in 1631, at a time when, like Cabeo, he was untroubled by Copernicanism, and in the book he praised Gilbert more than once. In 1633, however, he was promoted to the Society's Rome headquarters, and adopted the hard party line in his voluminous *Magnes* of 1641. He included extensive and vituperative refutations of the 'heretics' Gilbert, Stevin and Kepler, "because of my zeal for the honour of God, the Holy Mother and the Church, especially since I knew that no-one has publicly refuted these magnetic motions of the heavenly bodies".

Kircher maintained that Copernican magnetic dynamics was contradicted by magnetical philosophy itself. If the Earth's axis aligns magnetically, then the sphere of stars must be magnetic. Copernicans make their magnets revolve, and thus are caught in a paradox. Again, if the Earth's magnetism extends into the heavens, its strength would make iron objects immovable. Kircher, predictably, saw Kepler as the biggest threat. Taking pains, as did all Jesuit astronomers, to praise Kepler's hypotheses and tables, he attacked the unsubstantiated physics of Kepler's solar magnet, and of the manifold planetary threads. He built a model of Kepler's system, using real magnets, to show that the physics did not work. (These magnetic models became very popular with visitors to his museum.)

Kircher's crusade was continued in 1649 by his colleague Niccolo Zucchi (1586–1670) in his *Nova de machinis philosophia . . .: Promotio philosophiae magneticae*. The new argument was drawn from his claimed discovery of a Prime Magnet in the celestial poles, which ended the regress whereby every magnet (such as the Earth) had to align with another.

Proponents of the old astronomy could now claim just as strong support from experimental magnetism as did Copernicans and several neo-Aristotelians did so. Of course, there is no evidence of Copernicans being persuaded by Jesuit magnetical philosophers, but the works of the latter were such complete and useful treatments that they had to be read and recommended (by Walter Charleton and Robert Boyle, for example). Certainly their intervention ensured that the astronomical debate moved into other areas.

5. The collapse of magnetic cosmology

Magnetic arguments in cosmology, and their central principle of the *terra–terrella* analogy, were pursued insofar as they furthered the interests of one side or another. When they ceased to do so, the consensus in favour of the central principle disappeared. This occurred when the Jesuit Jacques Grandami published in 1645 details of a curious experiment he had made in 1641, which he entitled *Nova demonstratio immobilitatis terrae, petita ex virtute magnetica* (Figure 4.2).

His claim, rarely contradicted, was that a suitably poised *terrella* aligned itself not only north–south, but also east–west. He felt able, therefore, to concede the Galilean point that "in no way does [gravity] impede circular motion around a centre in the same place". Instead, "there is no doubt that the magnetic virtue which God gave to [the Earth] not only keeps its poles still and stable, but also its other parts and points. So if one imagined that an axial rotation of the Earth could be produced by some other superior force, then the magnetic virtue would stop the motion."

Grandami's anomalous experiment aroused wide interest in the 1640s, and was still being discussed (e.g. by G.W. Leibniz) in the 1670s. But his conclusion found little favour outside his own college. In 1651 the Jesuits acted to curb their more radical philosophers, and even sympathetic neo-Aristotelians like Zucchi, Gaspar Schott and Vincent Léotaud considered Grandami to have departed from the non-negotiable axiom that a quality (magnetism) only acts towards its end (the poles); therefore magnetical philosophy did not apply to axial rotation. Copernicans, of course, were also committed to disagreeing. Leading magnetical philosophers countered Grandami by restricting the application of the central principle, as Mersenne had consistently argued since 1623.

Mersenne's objection, which stemmed from his epistemology, was taken up by many later seventeenth-century experimentalists. He maintained that fundamental questions in natural philosophy (such as the Earth's motion) could not be resolved by experimental evidence. In the case of magnetic philosophy, he pointed out that every loadstone had to "adapt itself" to the Earth's magnetism. It behaved differently, therefore, from the Earth itself, and behaved the same, whether or not the Earth moved. Mersenne concluded that Grandami's reasoning, "namely that [the loadstone's] parts are analogues with the parts of the Earth from which they draw their power, is not adequate".

Grandami's work provoked several similar objections. In 1663, the Cartesian Henry Power agreed that his argument "is no argument at all, unless that he could prove to us that the *Terrella* could play this trick [if] it were removed out of the

4.2. The frontispiece of Jacques Grandami's *Demonstratio immobilitatis terrae* (1645). Although the Jesuit Grandami's lavishly illustrated book is a complete treatment of magnetical philosophy, the title and frontispiece reflect the Jesuits' preoccupation with magnetic cosmology. At the top two angels symbolize God's providence in imbuing the Earth with a magnetic quality to prevent it moving. The quotation from Ecclesiastes I, 4 emphasizes the unique conformity of Jesuit magnetics with scripture. The central image is of a cherub conducting Grandami's basic experiment to prove magnetic immobility. Notice how similar this is to Gilbert's basic experiment (Figure 4.1). Navigational interests are again represented. The cherub on the right carries Grandami's allegedly non-declining compass needle, with which he claimed to have solved the problem of longitude at sea.

sphaere of the Earth's Magnetisme". In his *Magnetologia* of 1668, the final Jesuit compendium of magnetical philosophy, Léotaud similarly observed that "there is therefore a great difference between the motionless axis, perpendicular to the horizon, of a loadstone placed on a boat [that is, a float], and the axis of the Earth. The former is held in its place by force, the latter is free and natural." In the latter half of the seventeenth century, magnetic arguments for or against the Earth's motion were usually avoided or else declared to be indecisive.

After half a century, the consensus closed against magnetic astronomy. Copernican interests were furthered more profitably through mechanistic philosophy, and magnetic philosophers began to dismantle the elements of cosmological astronomy that Gilbert had built into their discipline.

A good example of this separation of magnetical philosophy from cosmology is the way in which the Earth's magnetic poles lost their astronomical function after Henry Gellibrand (1597–1636) discovered secular changes in magnetic variation, the major empirical development of the century. Gilbert's theory, almost unquestioned in England, did not allow compass needles to change their direction. In 1635 Gellibrand, a Copernican, discussed whether secular variation might be caused, not by a magnetic change, but by a fifth motion of the Earth's axis, which Galileo had reported at the very end of his *Dialogue*. The French engineer and astronomer Pierre Petit agreed that it was, for the Earth's magnetic constancy was "beyond dispute".

But once magnetism and the Earth's motion had been separated, very un-Gilbertian views prevailed. The secular variation led to renewed claims, even in England, that the magnetic poles did not lie on the Earth's axis. In 1657 Joseph Moxon republished the *Certain Errors in Navigation* of Gilbert's collaborator Edward Wright, and criticised Wright for having treated the idea of independent magnetic poles as a serious error. Kircher's view, that secular variation indicated the presence of randomly shifting magnetic fibres, which were confined to the Earth's crust, was also widely held.

These changes placed further obstacles in the way of cosmological analogies, because they undermined the most basic premise of Gilbert's magnetical philosophy – that the Earth's axis of rotation was magnetic. It is ironic that, as we shall see in Chapter 12, the only analogy to persist, that between magnetic and gravitational attraction, was one that William Gilbert had specifically excluded.

Further reading

Martha R. Baldwin, Magnetism and the anti-Copernican polemic, *Journal for the History of Astronomy*, vol. 16 (1985), 155–74

Jean Daujat, *Origines et formation de la théorie des phénomenes electriques et magnétiques*, vol. 2: *Dix-septième siècle* (Paris, 1945)

Gad Freudenthal, Theory of matter and cosmology in William Gilbert's *De magnete, Isis*, vol. 74 (1983), 22–37

William Gilbert, *De magnete*, transl. by P. Fleury Mottelay (Ann Arbor, 1893; reprinted New York, 1958)

John Heilbron, *Elements of Early Modern Physics* (Berkeley, 1982)

Johann Kepler, *Epitome of Copernican Astronomy*, Books IV and V, transl. by C.G. Wallis, in Great Books of the Western World, ed. by R.M. Hutchins, vol. 16: *Ptolemy, Copernicus, Kepler* (Chicago, 1952)

Duane H.D. Roller, *The "De Magnete" of William Gilbert* (Amsterdam, 1959)

J.L. Russell, Kepler's laws of planetary motion: 1609–1666, *British Journal for the History of Science*, vol. 2 (1964), 1–24

5

Johannes Kepler
OWEN GINGERICH

"Divine Providence granted us such a diligent observer in Tycho Brahe that his observations convicted this Ptolemaic calculation of an error of 8′. . . . Because these 8′ could not be ignored, they alone have led to a total reformation of astronomy." Thus Johannes Kepler (1571–1630) wrote of the man who served as his mentor during a short but crucial interval in 1600–01. Kepler had promptly perceived the significance of Tycho's treasury of observations (see Chapter 1), but he also realized that the Danish master observer lacked an architect to erect a reformed astronomical structure. So successfully did Kepler carry out this reform that most of what we now consider to be the Copernican system is the result of Kepler's reconstruction; when Isaac Newton's *Principia* was recorded in the register of the Royal Society, it was introduced as "a mathematical demonstration of the Copernican hypothesis as proposed by Kepler".

The secret of the universe

There was little in Kepler's youth to suggest that he would become one of the foremost astronomers of all time. He was born in Weil der Stadt (near Stuttgart), Germany, on 27 December 1571. A weak and sickly child, but intelligent, he benefited from the widespread educational opportunities recently introduced into Württemberg. Consequently, he easily won a scholarship to the nearby Tübingen University so that he could study to become a Lutheran clergyman. In recommending him for a scholarship renewal, the University Senate noted that Kepler had "such a superior and magnificent mind that something special may be expected of him."

Yet, as Kepler himself later explained (in Chap. 7 of his *Astronomia nova*), although he had done well enough in the prescribed mathematical studies, nothing alerted him to his special talent for astronomy. Hence he was surprised and distressed when, early in 1594, midway through his third and final year as a theology student at Tübingen, he was dispatched to Graz, far away in southern Austria, to become an astronomy teacher and provincial mathematician. It was there that the twenty-three-year-old Kepler hit upon what he believed to be the secret key to the arrangement of the universe. As he soon related in his *Mysterium cosmographicum* (1596):

When I was studying under the distinguished Master Michael Mästlin at Tübingen six years ago, seeing the many disadvantages of the commonly accepted theory of the universe, I became so delighted with Copernicus, whom Mästlin often mentioned in his lectures, that I not only defended his opinions in the debates of the physics candidates, but even wrote a thorough disputation about the first motion, maintaining that it happens because of the Earth's rotation. . . . Thus I have gradually collected, partly through hearing Mästlin and partly by my own efforts, the advantages that Copernicus has mathematically over Ptolemy. . . . At last in the year 1595 [at an intermission in the lectures in Graz] I pondered this subject with the whole force of my mind. . . . And there were three things above all for which I stubbornly sought the causes as to why it was this way and not another: the number, the dimensions, and the motions of the orbs.

As a Copernican, Kepler accepted six planets (rather than the Ptolemaic seven, which included the Sun and Moon). Furthermore, he realized that whereas the Ptolemaic system provided a rationale for the spacing of the orbs – by a plenum nesting of the celestial spheres – no such explanation had been forthcoming in the Copernican system, with its compact but by no means contiguous and hence largely empty arrangement of the orbits. Kepler's restless and inquisitive mind continually sought for

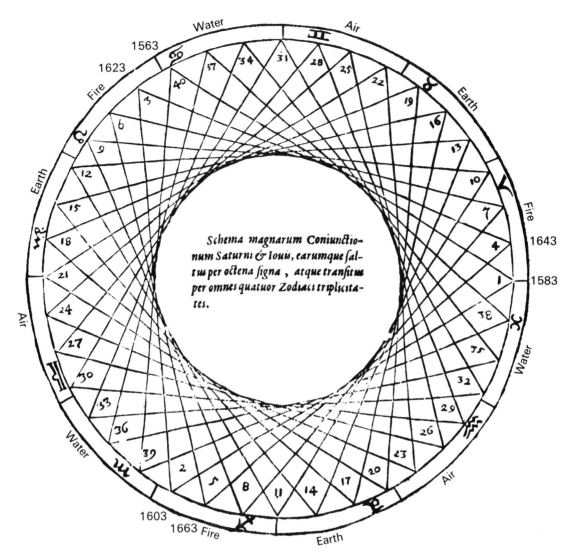

Water Air

1563
1623
Fire

Earth

Earth

Air

Water

Fire Earth

*Schema magnarum Coniunctio-
num Saturni & Iouis, earumque sal-
tus per octena signa, atque transitus
per omnes quatuor Zodiaci triplicita-
tes.*

Fire

1643
1583

Water

1603
1663

5.1. The pattern of successive conjunctions of Jupiter and Saturn, from the *Mysterium cosmographicum* (but with elements and dates added).

the underlying reasons for things, and so it was not surprising that he tried to find a natural explanation for the planetary spacings. He recounted the "trifling occasion" when inspiration struck: during a class lecture in July 1595, in which he was describing how the great conjunctions of Jupiter and Saturn fall sequentially along the ecliptic (Figure 5.1), he drew a series of quasi-triangles whose lines began to form an inner circle half as large as the outer ecliptic circle. The proportion between the circles struck Kepler's eye as almost identical with the proportion between the orbits of Saturn

and Jupiter. Immediately, he began a search for a similar geometrical relation to account for the spacing of Mars and the other planets, but in vain.

"And then it struck me," he wrote. "Why have plane figures among three-dimensional orbits? Behold, reader, the invention and the whole substance of this little book!" Like all mathematicians of his day, he knew that there were five regular solids, that is, three-dimensional shapes, each with faces all of the same kind of regular polygon. By inscribing and circumscribing these five polyhedra between the six planetary spheres (all nested in the

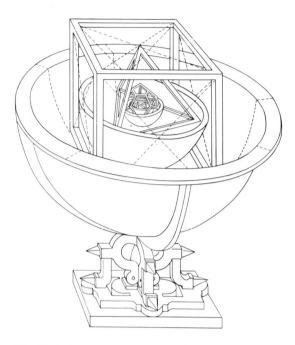

5.2. Kepler's nested polygons and planetary spheres (redrawn for greater clarity in the central portion), from the *Mysterium cosmographicum*.

Table 5.1 *Relative sizes of the adjacent planetary orbits (assuming the innermost part of the outer orbit to be 1000)*

Planet	Intervening regular solid	Computed by Kepler	From Copernicus
Saturn			
Jupiter	Cube	577	635
Mars	Tetrahedron	333	333
Earth	Dodecahedron	795	757
Venus	Icosahedron	795	794
Mercury	Octahedron	707	723

proper order), he found that the positions of the spheres closely approximated the spacings of the planets (Figure 5.2). Since there are five and only five of these regular or Platonic polyhedra, Kepler thought that he had explained the reason why there were precisely six planets in the solar system.

That Kepler's scheme works with fair accuracy when space is allowed for the eccentricities of the planetary paths is all the more astonishing when we notice that the cube and octahedron have the same spacing ratios as do the dodecahedron and icosahedron. This means that Kepler had essentially only three spacing ratios to work with. He was, however, obliged to compromise the elegance of his system by inscribing Mercury in a sphere within a square formed by the edges of the octahedron rather than in the octahedron itself. With this concession, everything fits within 5% – except Jupiter/Saturn, at which "no one will wonder, considering such a great distance".

Not only did the spacing of the planets seem arbitrary in the heliocentric system as proposed by Copernicus, but also the ancient scheme of driving the planetary system through a motive force at the outer edge was untenable when the stars were truly fixed. Kepler, ever eager to find physical causes for the structure and behaviour of the natural world (whether it was the form of snowflakes or the revolutions of the planets), became convinced that the Sun must supply the driving force to keep the planets in motion. Hence the Sun's centrality was essential for his *physical* conception.

Copernicus had already pointed out that the more distant the planet, the slower it circumnavigated the Sun, declaiming in the climax of the cosmological chapter in his *De revolutionibus* that "in this arrangement we find an admirable commensurability and sure bond of harmony between the motion and the size of the orbs that can be found in no other way". After telling his former teacher Mästlin of his nest of spheres and regular polyhedra that seemed to explain the *spacing* of the planets, Kepler mentioned a "moving soul" (*anima movens*) in the Sun whose efficacy diminished with distance from the Sun. He then turned to search for the basis of the regularities in the *periods* (that is, the qualitative harmonious connection between the period and distance already noticed by Copernicus), and for this he sought to discover how the Sun's driving force varied with distance. By examining the planetary periods (P_1, P_2, . . .) and distances (r_1, r_2, . . .) he found that the strength of the driving force is not directly proportional to its closeness to the Sun, but that it weakened with distance more rapidly than $1/r$. Expressed in ratios, the relation he found was

$$P_2/P_1 > r_2/r_1.$$

Kepler then directed his inquiry in terms of *increments* rather than ratios, and he reasoned in effect that the excess driving strength is proportional to the incremental distance between successive orbits:

$$\frac{P_2}{P_1} \cdot \frac{r_1}{r_2} - 1 = \frac{\Delta r_{12}}{r_2}$$

or

$$\frac{\Delta P_{12}}{P_1} = 2 \frac{\Delta r_{12}}{r_1}.$$

The relationship that Kepler found is nearly equivalent to $P_1/P_2 = (r_1/r_2)^2$, rather than the correct $(r_1/r_2)^{3/2}$, but this gave a sufficiently satisfactory result, as seen in Table 5.2.

There is a remarkable parallel between the use of increments in this problem and Kepler's attempt a few years later to find a law of refraction. In his *Ad Vitellionem paralipomena*, or *Supplements to* [the medieval Polish optician] *Witelo* (the first and separately named section of his *Astronomiae pars optica*, 1604), Kepler also sought a relation expressing an incremental quantity, in that case the difference in the incident and refracted light rays. Although he could fit Witelo's inadequate data tolerably well, this approach effectively prevented him from finding the correct functional relationship between incidence and refraction.

In writing to Mästlin and others, Kepler clarified his views about the principles underlying the construction of the world. These were to be sought not with pure numbers but in geometrical relations. For the properties of pure numbers were accidental, whereas those of geometry, such as the regular polyhedra or musical harmonies, were grounded in nature. From Nicolas of Cusa, Kepler took up the symbolism of the curved and the straight, which he used to express the relation of God to creation. With a slight modification of Cusa's ideas, the Sun represented God the Father, the celestial sphere God the Son, and the intervening space the Holy Spirit. In contrast to these curves, creation itself is expressed in magnitudes by the straight lines of geometry. In Kepler's view such a useful distinction must have been contrived in the beginning by God. So that quantities could become real, God created bodies before all other things.

Kepler sent the manuscript of his *Mysterium cosmographicum* to Tübingen, urging Mästlin to see it through the press. When the publisher balked at

Table 5.2 *Mean ratios of the planetary orbits*

Planetary increment	Predicted by by Kepler	Observed by Copernicus
Jupiter/Saturn	0.574	0.572
Mars/Jupiter	0.274	0.290
Earth/Mars	0.694	0.658
Venus/Earth	0.762	0.719
Mercury/Venus	0.563	0.500

printing the work without prior university approval, Kepler petitioned the University Senate for permission to publish. Publication was recommended with the suggestion that Kepler explain the Copernican theory more clearly. Mästlin achieved this in part by appending George Joachim Rheticus's *Narratio prima* (1540) and in part by adding his own supplement and diagram.

Kepler himself argued for the Copernican system by saying this system alone provided reasons for phenomena that otherwise provoke astonishment. For example, the Ptolemaic hypothesis failed to explain why the number, sizes, and times of retrogressions agreed so exactly with the position of the Sun. Yet there must be a reason, said Kepler, and in the notes added to the second (1621) edition of the *Mysterium*, he explained more fully that the beautiful order revealed by Copernicus was the kind that existed between cause and effect, so that the reasons given by Copernicus must be true ones and not just fictitious hypotheses. In his opening chapter Kepler took the opportunity to analyse how true conclusions may sometimes be drawn from false premises, but a false hypothesis will not explain *every* true appearance of the heavens. The idea that a true hypothesis will explain *all* the appearances was in keeping with his concept of physical nature. For nature, according to Kepler, loves simplicity and unity, and often a single cause will produce many effects. Later on, this philosophical approach provided an important motivation for Kepler to continue in his work on Mars when he had a model (his so-called vicarious hypothesis) that generated excellent planetary longitudes, but which did not yield the correct latitudes.

Kepler regarded the Copernican hypothesis as one that not only *described* physical reality, but also gave *reasons* for the appearances. However, whereas Copernicus had recognized the admirable arrangement of the universe *a posteriori* from observationally based geometrical propositions, Kepler claimed that this could have been proved *a priori* from the archetypal ideas of creation, by implication those found in his book.

The full title of his book was *Prodromus dissertationum cosmographicarum, continens mysterium cosmographicum,* the *Precursor to Cosmographical Treatises, containing the Cosmographic Secret.* Quixotic as his polyhedra may appear today, we must remember the revolutionary context in which they were proposed. The *Mysterium* was the first new and enthusiastic Copernican treatise since *De revolutionibus* itself. Without a Sun-centred universe, the entire rationale of the book would have collapsed. Although the principal idea of the *Mysterium* was erroneous, Kepler proposed a group of non-Aristotelian physical causes for celestial phenomena, thereby reintroducing the demand for physical explanations in astronomy. Seldom in history has so wrong a book been so seminal in directing the future course of science. Looking back as a man of fifty, Kepler noted in the second edition of the *Mysterium* that "the whole scheme of my life, studies, and works arose from this one little book".

Even the inquiry into the causes of the number and motions was itself a novel break with medieval tradition, which considered the 'naturalness' of the universe sufficient reason. For Kepler, the theologian–cosmologist, nothing was more reasonable than to search for the architectonic principles of creation. He attributed his success in finding the secret key to the universe to Divine Providence, for, as he announced in his preface: "I have constantly prayed to God that I might succeed if what Copernicus said was true." To Mästlin he wrote: "For a long time I wanted to be a theologian; for a long time I was restless. Now however, behold how through my effort God is being celebrated in astronomy!"

The new astronomy

Kepler sent a copy of his *Mysterium cosmographicum* to a number of astronomers, including Tycho Brahe. Unknown to Kepler, the renowned Danish observer had puzzled over the question of the distance to the Sun, and had considered in a quite different and strictly numerical fashion the possibility that the Platonic polyhedra might hold the

key. Recognizing the ingenuity of Kepler's work and feeling a kinship with his own efforts, Tycho invited the young cosmologist to Denmark, but the distance deterred the newly married Kepler from considering the trip. Kepler describes the sequence of events that followed in Chap. 7 of his *Astronomia nova:*

Tycho Brahe, himself an important part in my destiny, did not cease from then on to urge me to visit him. But since the distance of the two places would have prevented me, I ascribe it to Divine Providence that he came to Bohemia. I thus arrived there just before the beginning of the year 1600 with the hope of obtaining the correct eccentricities of the planetary orbits. Now at that time Longomontanus had taken up the theory of Mars, which was placed in his hands so that he might study the Martian opposition with the Sun in 9° of Leo [that is, Mars near perihelion]. Had he been occupied with another planet, I would have started with that same one. That is why I again consider it an effect of Divine Providence that I arrived in Prague at the time when he was studying Mars; because for us to arrive at the secret knowledge of astronomy, it is absolutely necessary to use the motion of Mars; otherwise it would remain eternally hidden.

Given Tycho's level of observational accuracy, only Mars has a high enough orbital eccentricity and sufficiently close approaches to the Earth to reveal its non-circular orbit.

Although Kepler started to work on the problem of Mars within a few weeks of his arrival at Tycho's Benatky Castle near Prague, he worked simultaneously on an analysis of astronomical hypotheses that provided a philosophical underpinning for his astronomical researches. The occasion arose out of a controversy in which Tycho had become embroiled with the self-taught astronomer Nicolai Reymers Ursus, whom Tycho had accused of plagiarism (see Chapter 3). Ursus had thereupon produced a vicious and uncouth counterattack. Kepler's carefully reasoned but polemical reply has become known as the "Defence of Tycho against Ursus", but Kepler himself called it simply "Tract on Hypotheses". Kepler took the opportunity to analyse the history and nature of astronomical hypotheses, to defend a realistic philosophical position in astronomy (against the instrumentalist view of both Ursus and Andreas Osiander, the author of the anonymous introduction to Copernicus's *De revolutionibus*), and to sharpen his own arguments on the truth of the Copernican premises: "If in their geometrical conclusions two hypotheses coincide, nevertheless in physics each will have its own peculiar additional consequence." In this way the stage was set for the introduction of 'physical' requirements into his researches on Mars.

At the very outset Kepler realized that Copernicus had used the centre of the Earth's orbit, and not the Sun itself, as the fundamental reference centre of his system. Copernicus's own arrangement was therefore helio*static* but not truly helio*centric*. Kepler could hardly have missed this feature of Copernicus's work, because the relevant passage was specifically marked by the previous owner of the copy of *De revolutionibus* that he acquired during his student days. This non-heliocentric arrangement was a natural consequence of the geometrical transformation from Ptolemy's geocentric system, but it offended Kepler's physical intuition about the driving role of the Sun. Hence this became the first point of attack in his "warfare on Mars" during the earliest months at Benatky.

As Kepler might have explained it, Ptolemy had accounted for the variation in speed of a planet (other than the retrograde motion) half by an 'optical part' and half by a 'physical part'. The first was the apparent variation produced by the eccentric placement of the orbital circle, the second the actual speed variation in the orbit, which Ptolemy modelled by an off-centre seat of uniform angular motion, the so-called equant (see Figure 5.3), Copernicus had followed this plan except that he replaced the equant by a small uniformly rotating epicycle, or 'epicyclet'. This had the effect of avoiding the actual speed variation (which violated traditional natural philosophy). In contrast to Copernicus, Kepler had no objections to using the equant as a mathematical device; in fact, he preferred it because it answered to his physical belief that the planets, propelled by the Sun's driving force, should move faster when closer to the Sun, whereas the Copernican epicyclet tended to hide this fact in its combination of uniform circular motions.

Neither Ptolemy nor Copernicus had used an equant in the particular case of the Earth–Sun orbit. The practical consequence was that the Earth moved with uniform speed around the centre

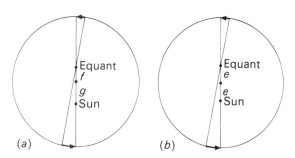

5.3. The orbit of Mars with an equant, the seat of uniform angular motion. In the vicarious hypothesis (*a*), accurate longitudes are obtained by setting $g/f = \frac{5}{3}$. The quasi-Ptolemaic scheme (*b*) with its equal-and-opposite equant satisfies the Sun–Mars distances but errs in longitude by 8′ in the octants.

of its circle. But if the Sun provided the driving force, then, Kepler reasoned, the Earth ought to move faster in its circle when it was at perihelion and slower at aphelion. Furthermore, Kepler understood that any meaningful physical explanation would need to use the same geometry for all of the planets including the Earth.

Kepler set it as one of his initial tasks to ascertain whether the physically desirable variation in the speed of the Earth could be accounted for by halving the Copernican eccentricity for the Earth's orbit and adding an equant, which would have made its geometry equivalent to that of the other planets. Kepler found that the traditional eccentric-only and the experimental half-eccentricity half-equant models would predict virtually the same heliocentric longitudes, but there would be an appreciable difference with respect to distances, such that in the traditional model the variation from the mean solar distance was ± 3.6%, but with the bisected eccentricity and equant only ± 1.8%.

One way to investigate the varying distance to the Sun would be to measure its apparent diameter throughout the year. Eventually, Kepler tried just this method, with inconclusive results. The effort led him to a thorough study of optics, with the resulting work attached to what started out as appendices to Witelo. *Astronomiae pars optica* (1604) became a foundation work for modern geometrical optics. In it he explained, for the first time, how an inverted image is formed on the retina of the eye, he clearly defined the light ray, and he investigated the effects of apertures of various sizes and shapes on the formation of an image.

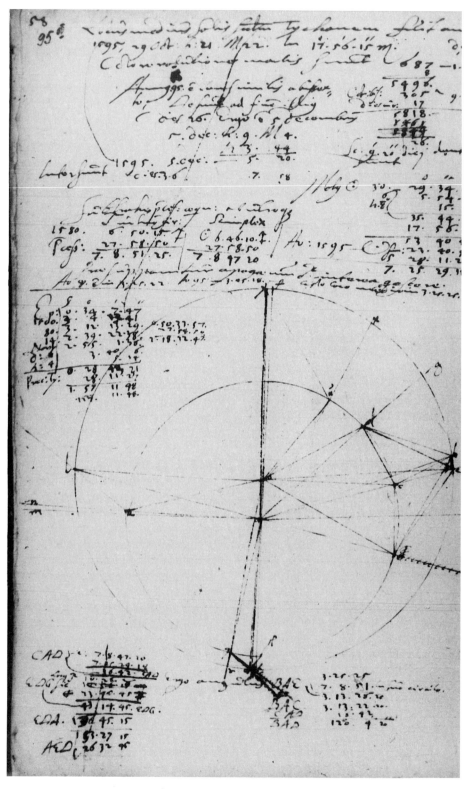

5.4. Kepler's earliest Earth–Mars triangulation attempt.

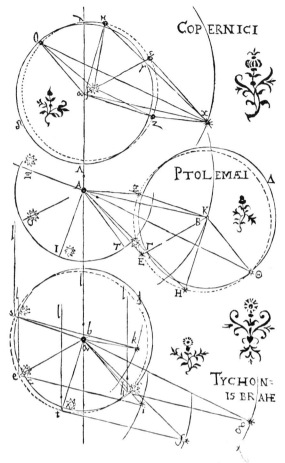

5.5. The Mars triangulation method forces a displacement (broken circles) in the traditionally accepted position of the Earth's (or Sun's) orbit (solid circles), regardless of the cosmological system, as shown in Chapter 24 of Kepler's *Astronomia nova*. Kepler followed Tycho's deathbed wish to give his system an equal opportunity, but only to this point in his treatise.

nated θ, η, ϵ, and ζ. Fortunately Tycho's massive data had provided several pairs of observations separated by 687 days, so that Kepler could undertake the requisite triangulations. Of course, the procedure depended critically on knowing the exact sidereal period of Mars, not just the approximate value of 687 days, so that much of the work involved this essential detail. The method quickly showed that his traditional 'Ptolemaic' model (solid circle) had too large an eccentricity. Kepler therefore bisected Tycho's solar eccentricity of 0.035 85 to make the geometry comparable to that of the other planets (broken circle). The resulting halved value, 0.018, was one of the few major parameters that Kepler did not himself determine from observations, and it was ultimately to be one of the weakest links in his *Rudolphine Tables* (1627).

Motivated by his search for a physically acceptable astronomy, Kepler established a fundamental premise during his initial months with Tycho. First, the traditional eccentricity of the Earth–Sun relationship had to be bisected and the arrangement modifed to include an equant or its equivalent; and, furthermore, the orbital planes of Mars and the other planets had to pass through the Sun itself, not through the centre of the Earth's orbit, as Copernicus had arranged them.

At the same time that Kepler adopted these physical considerations, he began to develop a most refined working orbit for Mars, what he was eventually to call his substitute or vicarious hypothesis. In this model, which would become a major tool for the calculation of Martian positions, he perfected the quasi-traditional circular orbit with its equant. But, unlike earlier analysts, Kepler decided to allow the equant to fall at any arbitrary point along the apsides of the circular orbit. In so doing, he was in effect following Brahe and Longomontanus who, although they used a small Copernican epicycle instead of the equant, had already achieved good results for the Martian longitudes by the equivalent of a 3:5 spacing (found semi-empirically), which agrees exactly with the first-order solution.

By the ingenious method of choosing Martian oppositions, when the Earth and Mars lie in a straight line from the Sun, Kepler was able to eliminate the moving Earth from his geometry. Accordingly, he wished to find four parameters: the two eccentricities (f = Sun–centre distance and g = centre–equant distance), the direction of

Soon after he came to work with Tycho in February 1600, Kepler had tried an ingenious triangulation procedure to locate the Earth's orbit. A long sequence of manuscripts survives for the first two years of his work on Mars, and Figure 5.4 shows the first time he tried this method. He knew that Mars returned to the same point in space every 687 days or so, whereas by that time the Earth would not yet have completed a second revolution. Kepler laid out this geometry in his *Astronomia nova*, shown in our Figure 5.5. In the Copernican component of his diagram, Mars is fixed at x, and the successive 687-day positions of the Earth are desig-

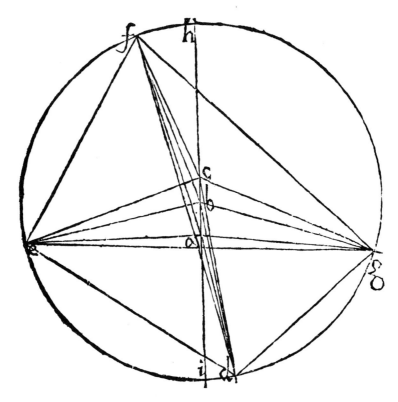

5.6. Geometry of the vicarious orbit solution, from the *Astronomia nova. a, b*, and *c* on the line of apsides *hi* represent the Sun, the centre of the orbit, and the equant point respectively; *d, e, f*, and *g* represent the four opposition positions of Mars.

aphelion, and the time of aphelion; and these required four observed heliocentric longitudes. (In reality, because the sidereal period of Mars was imperfectly known and because of the problems in establishing the precise times of opposition – which depended on a still-imperfect solar theory – Kepler actually had more than four parameters to juggle.) In this scheme, if we neglect terms in e^2, the angular heliocentric motion (true anomaly) as a function of time T is:

$$v = T - (f+g)\sin T + \tfrac{1}{2}g\ (F+g)\sin 2T.$$

For comparison, the motion in an ellipse with the law of areas is

$$v = T - 2e\sin T + \tfrac{5}{4}e^2\sin 2T.$$

Hence if $g = \tfrac{5}{4}e$ and $f = \tfrac{3}{4}e$, the hypothesis satisfies the equations to second order and the remaining error can be shown to be approximately $\tfrac{3}{4}e^3$, which, in the case of Mars with its eccentricity of nearly 0.1, amounts to about $2\tfrac{1}{2}'$.

The analytical equivalent of the problem leads to a 16th-order equation, eight of whose roots are

real, as D.T. Whiteside has shown. Two of the solutions, alien to Kepler's purpose, lay dangerously close – within a degree or so – to the answer he sought. Kepler attacked the problem in a variety of ways, but his favoured method, expounded in the *Astronomia nova*, ingeniously divided the problem into two essentially independent parts. Figure 5.6 shows the heliocentric orbit of Mars, with the Sun at *a*, the apsidal line *ach*, and four observed places at *d, e, f*, and *g*. The four observed angles at true opposition are subtended from *a*, the four times of observation give the angles about the equant point *c*, and *b* is the centre of the circle. The angles about both the Sun and the equant are fixed relatively one to another like irregularly positioned spokes of a wheel, but the sets can be rotated independently to adjust the time of aphelion or the aphelion angle. Kepler took as his primary requirement that the sum of the angles *gfe* and *gde* be 180° to ensure that the four observed points lay on a circle; and to have the centre *b* of this circle lie along the apsidal line, he took as his secondary requirement that the angle *bac* be zero. The first condition was achieved by modifying the time of aphelion, the second by modifying the aphelion angle, and the two eccentricities fell out from the calculation.

Kepler found this iterative method tedious: "If you are wearied by this procedure," he pleaded with his readers, "take pity on me, who carried out at least 70 trials." Kepler started out rather naïvely concerning the accuracy required, and initially he used four-place and then five-place tables of sines and tangents; but his interpolations were accurate to three or four (at the most) and sometimes to only two places, quite inadequate for the subtlety of the problem, and in addition he was not a good arithmetician, making many mistakes in his calculations and copying errors in the observations (often not being able to read his own figures!). He sometimes used a method of proportional parts to speed up the convergence, which was almost fatal because of the nearby false solutions. Eventually, for controlling the errors he employed an exact solution for the simpler equal-eccentricities case ($f = g$), so that he could use the observations three at a time to ferret out an offending observation. Although many of the attempted solutions failed for reasons that Kepler could not discover, he gradually recognized which parameters gave the best results for an extended series of Martian oppositions. Although his vicarious model required years to shape, by the spring of 1601 he had been able, almost miraculously, to achieve a solution for Mars's longitudes nearly two orders of magnitude better than any of his predecessors. Nevertheless, even in the final stage of the published version he bungled a long division, getting 0.113 32 and 0.185 64 for *ab* and *ac* instead of 0.112 19 and 0.185 64. Kepler had based his solution on the oppositions of 1587, 1591, 1593, and 1595, but his computation for eight other oppositions (including two that he had observed himself) showed a maximum discrepancy of 2′ (remarkably close to the theoretical limit). The details of this are shown in Table 5.3.

Kepler was by this time sensitive to the fact that different models could produce similar heliocentric longitudes without predicting the same heliocentric distances. The Martian latitudes offered a decisive test as to whether the distances were satisfactory. Alas! from the disagreement between his predicted latitudes and those observed by Tycho, Kepler quickly realized the inadequacy of

Table 5.3 *Martian longitudes from the vicarious hypothesis*

Time of opposition				Predicted			Observed			Difference	
	d	h	m	°	′	″	°	′	″	′	″
1580 Nov 18	1	31		66	28	44	66	28	35	0	9
1582 Dec 28	3	58		106	57	4	106	55	30	1	34
1585 Jan 30	19	14		141	37	46	141	36	10	1	36
1587 Mar 6	7	23		175	43	16	175	43	0	0	16
1589 Apr 14	6	23		214	26	12	214	24	0	2	12
1591 Jun 8	7	43		266	43	51	266	43	0	0	51
1593 Aug 25	17	27		342	16	42	342	16	0	0	42
1595 Oct 31	0	39		47	31	54	47	31	40	0	14
1597 Dec 13	15	44		92	28	3	92	28	0	0	3
1600 Jan 18	14	2		128	38	18	128	38	0	0	18
1602 Feb 20	14	13		162	25	13	162	27	0	−1	47
1604 Mar 28	16	23		198	36	43	198	37	10	−0	27

his scheme. Unlike previous astronomers, who were satisfied with separate models for longitude and latitude, Kepler sought a unified, physically acceptable description. He designated his unequal eccentricities solution the "vicarious hypothesis" because, although it gave an excellent representation of the longitudes, he knew it failed for the latitudes and therefore at best it could be only a substitute for the consistent model he was seeking.

To obtain the correct distances his physical model demanded, he was obliged to reposition his circular orbit with its centre mid-way between the Sun and the equant; this case is represented by setting $e = f = g$ in the earlier equation to give

$$v = T - 2e \sin T + e^2 \sin 2T.$$

The error in true anomaly (and heliocentric longitude) is therefore $\frac{1}{4}e^2 \sin 2T$, or for Mars, 8′ in the octants. Consequently, this move destroyed the excellent results he had previously found for the longitudes. With errors of 8′, which exceeded the limits imposed by Tycho's data, Kepler returned to the warfare on Mars, saying that "it is only right that we should with a grateful mind accept God's gift [Tycho Brahe's diligent observations] to find the true celestial motions".

In his *Astronomia nova*, Kepler chose to describe the vicarious hypothesis at the outset, reserving the major physical considerations for a later section even though he had worked on them all more or less simultaneously. He concluded his discussion of the vicarious orbit by exclaiming "Who would have believed it? This hypothesis that agrees so well with the observations is nevertheless false.''

Kepler then returned to his earlier speculations on a planetary driving force. With the demise of the crystalline spheres that had transmitted the motions in the old geocentric system, Kepler necessarily sought some power that could act at a distance. In fact, he already had ideas about the physical mechanism involved. Through Jean Taisnier's book on the magnet (1562) and, later, William Gilbert's (1600), he convinced himself that the planetary driving force emanating from the Sun must be magnetic. He envisioned a rotating Sun with rotating emanations that continuously coaxed the planets in their orbits and without which each of the planets would come to a halt. Kepler's revised geometry, with the centre midway between the Sun and equant, enabled him to formulate what we can call his *distance law*: the velocity of a planet is inversely proportional to its distance from the Sun. This relation is correct on the apsidal line, but is only approximate elsewhere. From the distance law he deduced that the strength of the solar emanation decreased in inverse proportion to distance, much like the attenuation of material on the circumference of a circle of increasing radius, and specifically unlike light, which decreased as $1/r^2$, like a substance spreading out on the surface of a sphere of increasing radius. Though Kepler's magnetic emanation filled space, he conceived it as being effective unidirectionally in the sense of the rotation, and hence it behaved like an inverse proportion rather than an inverse square.

The problem of finding the time taken to traverse a segment of the orbit involved summing the effects of the solar emanation that pushed the planet along the tiny elements of the arc. According to Kepler's Aristotelian physics, the ''delays'' needed to traverse tiny equal segments of the orbit were inversely proportional to the impulses, so that a summation of the distances was needed. Since an infinity of distances could not be calculated within his lifetime, Kepler resorted to a tiresome approximation, dividing the circle into 360 equal parts over each of which the distance could be regarded as constant. Searching for a less laborious method and taking inspiration from Archimedes, Kepler had the happy idea of replacing the sums of the lines between the Sun and the planet with the areas

within the orbit. Because of the inverse proportionality between the radius vector and the distance travelled in unit time, he concluded that the radius vector swept out equal areas in equal times. Kepler recognized that replacing a sum of lines by an area was mathematically objectionable, but, like a miracle, it provided a good approximation to the orbital motion predicted by the distance law. This procedure is formulated in Chap. 40 of the *Astronomia nova*, but nowhere in the chapter itself, nor anywhere else in the book, is this relation, the so-called ''law of areas'', clearly stated. (By 1621, however, he finally understood its fundamental nature and in the *Epitome* clearly stated both the area law and a revised distance law.)

At this point Kepler had an accurate but physically inadmissible scheme for calculating longitudes (the vicarious hypothesis) and an intuitively satisfactory physical principle (the distance law) that worked well for the Earth's orbit. But when Kepler applied his distance law and its areal approximation to Mars – a laborious calculation in which he stepped the delays or velocity intervals around the circle – he soon found that the calculated planet moved too fast at the apsides and too slow at the mean positions. Interpreted in areal terms, this meant there was too much area at the quadrants perpendicular to the apsides of the orbit. Although the problem could be cured if the orbit bowed inwards, Kepler was at first hesitant to place this much faith in what he considered only an approximation to his distance law; however, when he established that the distance law gave the same result, he began to suspect that the orbit really was noncircular. The way for this had already been paved by Copernicus's remark that his epicycle riding on an eccentric circle did not generate a circle, and the marginal note ''ellipse'' (possibly by a previous owner) marked the place in Kepler's copy of *De revolutionibus*.

Kepler now turned to a closer examination of the shape of Mars's path. Having established the actual position of the Earth's orbit by triangulation to Mars, he was able to turn the procedure around and investigate a small number of points in the orbit of Mars itself. Although the method did not yield a quantitative result, it clearly confirmed that Mars's orbit was non-circular. Kepler recognized that, despite the quality of Tycho's observations, the data were inadequate to allow him to determine precise distances to the orbit. Because of this scatter he had to use, as he picturesquely described it, a method of "votes and ballots".

The earliest pages of Kepler's Mars notebook show a number of experiments with epicycles. It is fascinating to see that although Kepler was exploring very new ground, he still adapted his tools from traditional astronomy. This was a time-honoured procedure, but one that left Kepler distressed by its absurdity. Just as sailors cannot know from the sea alone how much water they have traversed, he argued, so the 'mind of the planet' will not be able to guide its motion in an imaginary epicycle except by watching the apparent diameter of the Sun. Despite its absurdity, Kepler turned once again to an epicycle as a convenient means for generating a simple noncircular path. The resulting curve was similar to an ellipse, but was slightly egg-shaped with the fat end nearest the Sun. Unfortunately, this ovoid led to a computational labyrinth when Kepler tried to apply his area rule to various segments of the curve. In fact, the difference between the longitudes generated from the distance law and the area law reached 4', precariously close to the 8' discrepancy that had driven him to a renewed assault on the problem. Writing to David Fabricius in July 1603, Kepler remarked: "I lack only a knowledge of the geometric generation of the oval. . . . If the figure were a perfect ellipse, then Archimedes and Apollonius would be enough."

Discouraged by his lack of progress, Kepler turned his attention to a study of atmospheric refraction, which broadened to a complete reworking of traditional optics; his fifteen months of optical studies were eventually published as *Ad Vitellionem paralipomena, quibus Astronomiae pars optica traditur*. In late 1604, however, he returned to the preparation of his *Astronomia nova* or *Commentarius de stella Martis*. At the end of the year he wrote to Fabricius: "I am now completely immersed in the *Commentarius*, so that I can hardly write fast enough." By early 1605 he had drafted fifty-one chapters without yet obtaining the ellipse, although he had unwittingly tried an astonishing number of other non-circular curves. He had, however, framed a physical justification for the ovoid. These physical causes were an extension of the effects of the magnetic emanation that drove the planets around the Sun: the Sun's presumed unipolar magnetism could alternately attract and repel the planet from the Sun. Thus the same mechanism, Kepler hoped, might account for both the varying speed of the planet in its orbit and its varying distance from the Sun. His goal, as he wrote to J.G. Herwart von Hohenburg in February 1605, was "to show that the celestial machine is not so much a divine organism but rather a clockwork . . . inasmuch as all the variety of motions are carried out by means of a single very simple magnetic force of the body, just as in a clock all the motions arise from a very simple weight".

Around this time he did use an ellipse to approximate his 1602 ovoid but since his approximating ellipse had an eccentricity of $\sqrt{2}e$ and thus did not have the Sun at a focus, Kepler would have been unlikely to have taken it as the 'true' solution his inquiries demanded. Furthermore, he would hardly have expected that the dynamical solution would be a curve symmetrical about the minor as well as the major axis, since the Sun was not centred in the orbit. By now, however, he must have suspected that his problems lay not with the inadequacy of the quadrature but in his deficient knowledge of the amount by which the path of Mars departed from a circle. His renewed assault included a revised 'triangulation' to Mars, carried out by coupling the observed geocentric position angles with the heliocentric longitudes predicted by his accurate but physically unacceptable vicarious hypothesis. In fact, the triangulation procedures determined the points on Mars's orbit rather poorly, and his real hold on the problem came through the predicted longitudes and not the distances. Once more he found a discrepancy in the octants, amounting to 8', but in the opposite sense from his earlier results with the eccentric circle. Kepler then realized that the ovoid lay as far inside the true curve as the circle fell beyond it. Thus his ovoid and its elaborately reasoned physical causes "went up in smoke".

Then, in the course of his calculations, he stumbled upon a pair of numbers that alerted him to the existence of another ellipse, a curve answering some of his requirements almost perfectly. These steps are shown in greater detail in Figure 5.8. "It was", he wrote, "as if I had awakened from a sleep."

Kepler was meanwhile searching for a *physical* cause of planetary motion, something quasi-magnetic and connected with the Sun that would ex-

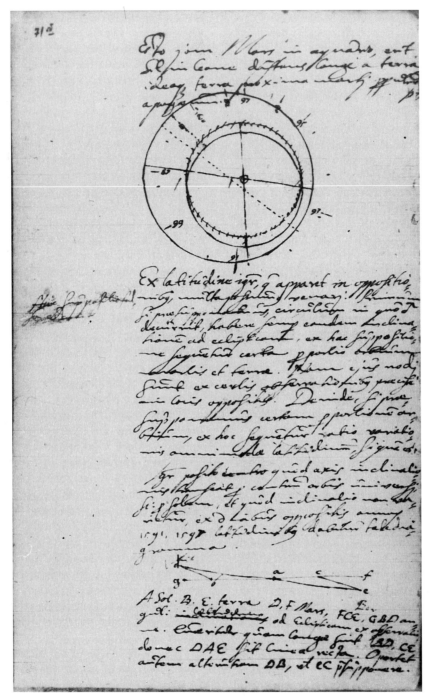

5.7. Page 8 of Kepler's workbook shows that he considered the latitudes at a very early stage. The diagram near the bottom of the page displays a cross-section of the Earth's orbit, *dae*, corresponding to the Martian oppositions of 1591 and 1597. As may be noted from the orbits at the top of the diagram, the oppositions are chosen perpendicular to the line of nodes, presumably for determining the inclination of Mars's heliocentric orbit. Kepler later used a similar diagram to check the predicted distances in his model. The vicarious orbit with its unequal equant made the perihelion distance *e* too short and the aphelion distance *d* too long, so that the predicted latitude at perihelion (in 1593) was too large to the south while at aphelion (in 1585) it was not large enough to the north.

Cap.
XXVII.

5.8. The triangulation that revealed the non-circular orbit of Mars, from the *Astronomia nova*.

plain not only the varying speed of Mars but also its varying distances. He fervently hoped that the oscillations of a hypothetical magnetic axis of Mars would satisfy his requirements. From Gilbert's *De magnete*, Kepler knew about the magnetic axis of the Earth. Such a magnetic axis, he proposed, could act as a rudder in the Sun's magnetic emanation, guiding the planet first near and then far from the Sun by means of alternate attractions and repulsions. If the magnetic axis is fixed in space, then its projection as seen from the Sun will be approximately cos θ. Such a cosine term in fact appears in the polar equation for the ellipse:

$$r = a(1 - e^2)/(1 - e \cos \theta).$$

To the first order in eccentricity, the ellipse satisfies

this physical picture of the magnetic axis governing the advance and retreat of the planet. For Kepler, who of course did not work with the polar equation, the real hurdle was to find the geometrical equivalents between the oscillating magnetic axis and the ellipse. "I was almost driven to madness in considering and calculating this matter", he wrote in Chap. 58 of his *Astronomia nova*.

I could not find why the planet would rather go on an elliptical orbit. Oh, ridiculous me! As if the oscillation on the diameter could not also be the way to the ellipse. So this notion brought me up short, that the ellipse exists because of the oscillation. With reasoning derived from physical principles agreeing with experience there is no figure left for the orbit except a perfect ellipse.

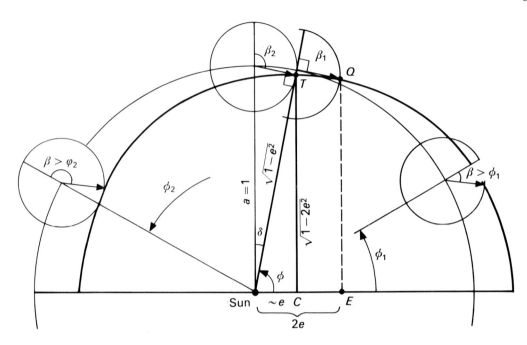

5.9. The epicyclic construction of Kepler's ovoid (the darker curve). The planet moves along the ovoid at the tip of the epicyclic radius vector. The epicycle has radius e. Angle β moves uniformly with time, whereas ϕ moves non-uniformly in order to satisfy the area law, that is, so that the vector from the Sun to the planet sweeps out an area proportional to time (and proportional to angle β). If this construction had an equant, it would fall $2e$ from the Sun at E and Mars would reach Q in a quarter period; hence β_1 in the epicycle must be very close to 90°. As the epicycle centre moves through the angle δ, the epicycle radius vector will also advance by δ since angle ϕ is close to its mean rate in this part of the orbit. Thus, $\beta_2 = \beta_1 + \delta$, and the angle at T is a right angle. Then the line $Sun - T = \sqrt{(1-e^2)}$ and $TC = \sqrt{(1-2e^2)}$. Since the semi-major axis of the ellipse is $(1 - e^2)$, the approximating ellipse to Kepler's ovoid has the eccentricity $\sqrt{2}e$. Kepler, however, was awakened to this relationship in quite another way. He found from an analysis of the errors in longitude that the ovoid gave a maximum departure from a circle too large by a factor of 2; that is, it should have swung in by 660 parts of a semi-diameter (or 0.004 29 for a semi-diameter of 1.000 00) instead of twice that much. He finally realized that an ellipse of eccentricity e gave the required path when he noticed by chance that secant $\delta = 1/\sqrt{(1 - e^2)} \approx 1 + \frac{1}{2}e^2$ was 1.004 29; in other words, that it exceeded unity by precisely the width of the lunula between the circle and the non-circular orbit.

Indeed, Kepler was luckier than he knew. Just as there is an approximating ellipse to the ovoids he originally tried, so there are several approximating ovals to the final ellipse; he cites two: an unnamed lemon-shaped curve, and the *via buccosa* or "puffy-cheeked" curve. Although this latter curve agreed with Tycho's observations, Kepler made a conceptual error in his calculation, and from the apparent disagreement with the data, rejected it. Kepler had confirmed the "observational" agreement of the ellipse with the vicarious hypothesis, but when he tested the *buccosa* against the ellipse, he managed to find a (non-existent) discrepancy of up to $5\frac{1}{2}'$; on that ground he rejected the "puffy-cheeked" curve. (Apparently he never tested the lemon against the ellipse.) In his excitement when he found the el-

lipse, he must have hastily assumed that no other curve would do. Thus, driven by his persistent physical intuition, he had continued until he almost accidentally hit upon the right curve.

Although the work on Mars was completed by the end of 1605, publication did not follow immediately. Tycho Brahe had long since died (in 1601) and Kepler had in turn received the position of Imperial Mathematician from Emperor Rudolph II. Tycho's heirs had never been paid by the emperor for the observation books, however, so they demanded censorship rights over materials based on Tycho's data. Furthermore, they were displeased that Kepler had chosen a Copernican basis rather than Tycho's fixed-Earth, geoheliocentric arrangement, a system that of course made no sense in the

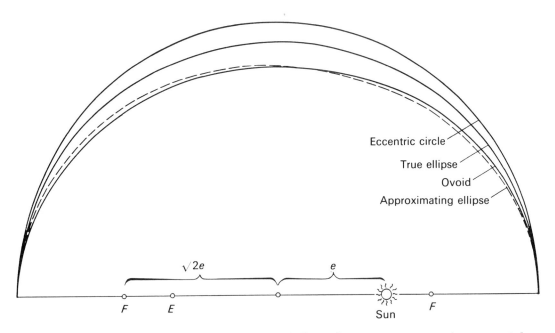

Eccentric circle

True ellipse

Ovoid

Approximating ellipse

$\sqrt{2e}$　　　　e

F　　E　　　　　　　　　F
Sun

5.10. Kepler's ovoid orbit compared with the final ellipse; the eccentricity is greatly exaggerated.

framework of Kepler's physical ideas. Eventually a compromise was reached, primarily with respect to the dedicatory materials at the beginning of the volume, and the book was at last published in 1609. Kepler added to the work a long and interesting introduction defending his physical principles, and he also described how the Copernican system could be reconciled with the Bible. This latter part he had already written in 1596 for inclusion in his *Mysterium cosmographicum*, but the Tübingen University Senate, on being asked to referee his first publication, had objected to such theological material. Now relatively independent, Kepler was no longer under such restrictions.

His greatest work, the *Astronomia nova* broke the two-millennium spell of perfect circles and uniform angular motion. Its subtitle emphasized its repeated theme: "Based on Causes, or Celestial Physics, Brought out by a Commentary on the Motion of the Planet Mars". Although his magnetic forces were soon discarded by his successors, his requirement for a celestial physics based on causes had a lasting influence. It truly was the "New Astronomy". Never had there been a book like it. It was an order of magnitude more complete and complex than anything that had gone before, and for the first time we see an astronomer explicitly wrestling with the ambiguities of imperfect data: both Ptol-

emy in the *Almagest* and Copernicus in *De revolutionibus* had carefully concealed the scaffolding by which they had erected their mathematical models. The book nevertheless hides many of Kepler's own meandering attempts to find a satisfactory orbit of Mars; in places its tables are ingenuously based on calculations that come later in the sequence, and at other times there are astonishing gaps in the arithmetic, leading to a well-justified suspicion that parts of its complexity are determined by rhetoric rather than logic. If so, it succeeded in its aim, for generations of astronomers and historians have passed by these flaws without comment.

It is difficult to gauge what impact Kepler's new astronomy would have had if he had stopped with the *Astronomia nova*. His cleansing and reformulation of the heliocentric system had been worked out in theory for only a single planet, Mars, and he had not provided any practical tables for calculating its motions. As Imperial Mathematician, Kepler had been explicitly charged with the preparation of new planetary tables based on Tycho's observations, an arduous task that he still faced. "Don't sentence me completely to the treadmill of mathematical calculations", Kepler responded to one correspondent. "Leave me time for philosophical speculations, my sole delight."

The nova of 1604, the telescope, and comets

Besides the great works on heliocentric astronomy and cosmology, Kepler wrote prolifically on a wide variety of other astronomical, astrological, and chronological topics. His dozens of books and pamphlets ranged from ephemeral prognostication almanacs to a science fiction account of a dream voyage to the Moon (*Somnium seu astronomia lunari*, 1634) and to a treatise on the nova of 1604 of such thoroughness that the star is today commonly known as 'Kepler's nova'.

Jupiter and Saturn, the slowest moving of the naked-eye planets, come into conjunction every twenty years. Such a conjunction took place at the end of 1603, and it was viewed as a significant omen because it began a series of conjunctions within the triplet of zodiacal signs designated by the astrologers as fiery (Sagittarius, Leo, and Aries). In the months that followed, Mars, too, drew near Saturn and Jupiter, and then, on the morning of 11 October, Kepler was told that a brilliant new star had been sighted in the same vicinity through a gap in the clouds. Kepler hesitated to believe the report, but a week later clear weather allowed him to view this astonishing spectacle.

Kepler promptly published an eight-page tract in German on the new star, comparing it with Tycho's nova of 1572. A larger collection of observations and opinions appeared in his *De stella nova* in 1606. A monument of its time, it is the least significant of Kepler's major writings. Although it broke no new astronomical ground, modern astronomers investigating supernovae have found it the best reference for the nova of 1604. In Chap. 21 Kepler argued that stars are not suns; a discussion of Kepler's views of the sidereal universe, a recurring theme in his work, will be found in Volume 3 of the *General History*.

The fact that the nova had been so closely associated with the first Jupiter–Saturn conjunction to occur in the fiery trigon in several centuries triggered a series of splendid speculations by Kepler, who noted that a related event had taken place in 5BC. Kepler drew an analogy between the nova of 1604 and the star of the Magi, and then entered into an extensive chronological investigation in which he concluded that Christ had been born, not in AD1, but five years earlier. He presented his argument in an appendix to *De stella nova* and later

elaborated it in several works including his definitive account, *De vero anno quo aeternus Dei Filius humanam naturam assumpsit* (1614).

On 29 May 1607 Kepler caught between clouds two short glimpses of what he assumed to be a transit of Mercury. He was so excited by the observation, made by means of a pinhole projection, that he ran all the way to the top of the castle hill to inform the emperor. His report appeared in *Phaenomenon singulare* (1609). After the discovery of sunspots by Galileo Galilei (1564–1642), Mästlin pointed out that he had undoubtedly mistaken a sunspot for the planet, and Kepler finally printed a retraction in the introduction to his *Ephemerides* for 1617, saying: "Lucky me, the first in this century to see a sunspot!"

In April 1610, Kepler received both a copy of Galileo's *Sidereus nuncius*, with its celestial discoveries revealed by the telescope, and the author's request for an opinion. By now Kepler was known as a leading astronomer whose reaction had to be taken seriously. He responded at once, publishing his long letter of approval as *Dissertatio cum Nuncio sidereo*. He not only accepted the new observations with enthusiasm, but he reminded its readers of many of his own researches. A few months later Galileo wrote to thank Kepler, saying: "You were the first one, and practically the only one, to have complete faith in my assertions."

Despite Galileo's warm acknowledgement, it was not until August 1610, that Kepler's wish to try out a telescope was finally satisfied – not through Galileo's generosity but by the loan of an instrument from Elector Ernest of Cologne. It was typical of Kepler's eagerness to publish that with less than two weeks' observations of Jupiter's satellites, he sent to press yet another booklet, *Narratio de Jovis satellitibus* (1611). Swiftly reprinted in Florence, the tract helped to authenticate Galileo's new discoveries.

Meanwhile, Kepler turned his attention to the theory of the telescope. In his *Astronomiae pars optica* (1604) he had discussed lenses only in passing, but his method of analysis using light rays established the essential background by which the formation of images could be explained. With characteristic energy he completed his *Dioptrice* (1611) during August and September 1610. Not only did he discuss the optics of lenses with great thoroughness, but he described the theory of a new kind of telescope with two convex lenses but which gave an inverted image. Because the Keplerian telescope allowed greater magnification more conveniently than the Galilean refractor, Kepler's form became the standard in astronomy.

By 1611 Kepler was at the peak of his career, the prestigious Imperial Mathematician, and the author of books touching nearly every branch of astronomy. But then his world suddenly collapsed. His three children were stricken with smallpox, his favourite son died, and then his wife became ill and died of typhus. The Thirty Years War had begun, and his patron, Emperor Rudolph, was forced to abdicate, although he still demanded Kepler's presence in wartorn Prague. With the death of the emperor at the beginning of 1612, Kepler was at last free to leave, and thus he began a fourteen-year residence in Linz. Under these troubled circumstances it is not surprising that he published no astronomical works from 1612 to 1616.

Yet Kepler did produce the *Stereometria doliorum vinariorum* (1615), which is regarded as one of the significant works in the prehistory of calculus. Using an extension of the type of small interval approximation that had proved so successful in the *Astronomia nova*, Kepler applied the technique to determining the volume of Austrian wine casks as well as a variety of other solids.

The appearance of a bright comet in 1618 turned Kepler's attention to these objects, which he considered in *De cometis libelli tres* (1619). Reflecting on their ephemeral nature, he proposed a strictly rectilinear trajectory, which appeared more complex because of the Earth's motion. Besides the comet of 1618 he discussed in detail the comet of 1607; these latter observations were of special interest to Edmond Halley, who, at the end of the century, showed its periodic nature. The comet of 1618 aroused a considerable controversy among Italian astronomers including Galileo, and Kepler entered the fray in 1625 with his *Hyperaspistes*, a polemical defence of Tycho's comet theories against the Aristotelian views expressed by Scipione Chiaramonti in his *Antitycho*. In the appendix, Kepler took Galileo to task for some of his erroneous views on comets, and he drew Galileo's attention to the fact that the phases of Venus could be as easily explained in the Tychonic system as in the Copernican.

The Harmony of the World

Soon after completing his *Mysterium cosmographicum* Kepler had drafted an outline for a work on the harmony of the universe, but his plan had lain dormant while he grappled with the intricacies of Mars. Then, in the autumn of 1616, after he had completed the first in a series of ephemerides based on his work, he revived his cosmological programme, swiftly putting to paper ideas that had long gestated. Despite domestic distractions such as travelling to Württemberg to arrange the defence for his mother's witchcraft trial, the 225-page treatise, *Harmonice mundi*, was finished by the spring of 1618.

Max Caspar, in his biography *Kepler*, gives an extended and perceptive summary of the *Harmonice*, concluding:

Certainly for Kepler this book was his mind's favourite child. Those were the thoughts to which he clung during the trials of his life and which brought light to the darkness that surrounded him . . . . With the accuracy of the researcher, who arranges and calculates observations, is united the power of shaping of an artist, who knows about the image, and the ardour of the seeker for God, who struggles with the angel. So his Harmonice *appears as a great cosmic vision, woven out of science, poetry, philosophy, theology, mysticism . . . .*

Kepler developed his theory of harmony in four areas: geometry, music, astrology, and astronomy. His archetypal ideas of harmony were hierarchically arranged with the trinitarian Godhead in the highest place, and each level an image of the one above it. But equally high is geometry itself: "Geometry is coeternal with the Mind of God before the creation of things; it is God himself, it has supplied God with the models for the creating of his world, and it has been directly transferred to man with the image of God", Kepler wrote in Bk IV of the *Harmonice*. "Geometry was not received into the mind of man through his eyes."

Book I of the *Harmonice* concentrates on the "regular figures that give rise to harmonic proportions", essentially on the constructability of polygons that serves as the geometrical (rather than the traditional arithmetical) basis for the musical harmonies of Bk III. The "congruence of the harmonic figures", that is, their fitting together into polyhedra, forms the substance of Bk II, which in turn provides the basis for the discussion in Bk IV of the soul's faculty for perceiving harmonies. This theoretical statement of the nature of astrological influences – the effect of cosmic geometrical harmonies on the soul – contains very little practical astrology. In fact, Bk IV is entitled "On the Harmonic Configurations of Stellar Rays at the Earth, and their Effect on the Weather and Other Natural Phenomena". In keeping with the hierarchical arrangement of the harmonies, he remarked that "meteorology and music are, so to speak, different peoples both stemming from the same fatherland".

Today it is the treatment of astronomy, in Bk V of the *Harmonice*, that commands most attention. Here Kepler returned to the problem of planetary spacings that he had addressed in his youthful *Mysterium cosmographicum*. Whereas the approximate planetary spacings predicted by his nested polyhedrons and spheres had satisfied him in 1596, now, greatly matured in his scientific outlook and filled with a powerful respect for the efficacy of observation after his encounter with Tycho's data, he could no longer dismiss the 5% error in the spacing predictions of the *Mysterium*. Kepler remained convinced that the regular solids provided the fundamental blueprint for the planetary system, but he now sought the secondary principles by which God had modified the original archetypal model. The analysis of musical scales in Bk III of the *Harmonice* provided a harmonic argument not only for the adjusted planetary distances but also for their orbital eccentricities. The ratios of the extremes of the velocities of the planets corresponded to the harmonies of the just intonation. Of course, the planets would not necessarily all be at their perihelia or aphelia at the same time. Yet, as they wheeled in their generally dissonant courses around the Sun, occasionally the harmonies would occur. Swept on by the grandeur of his vision, he exclaimed:

It should no longer seem strange that man, the ape of his Creator, has finally discovered how to sing polyphonically, an art unknown to the ancients. With this symphony of voices man can play through the eternity of time in less than an hour and can taste in small measure the delight of God the Supreme Artist by calling forth that very sweet pleasure of the music that imitates God.

In the course of this investigation, Kepler stumbled upon the relationship now called his third or

Table 5.4 *Kepler's harmonic law demonstrated*

	Period (years)	Mean distance	Period squared	Distance cubed
Mercury	0.242	0.388	0.0584	0.0580
Venus	0.616	0.724	0.3795	0.3795
Earth	1.000	1.000	1.000	1.000
Mars	1.881	1.524	3.540	3.538
Jupiter	11.86	5.200	140.61	140.73
Saturn	29.33	9.510	860.08	867.69

harmonic law: The ratio that exists between the periodic times of any two planets is precisely the ratio of the $\frac{3}{2}$ power of the mean distances, or

$$a^3 = kP^2.$$

If you want the exact time [Kepler candidly remarked], it was conceived on March 8th of this year, 1618, but unfelicitously submitted to calculation and rejected as false, and recalled only on May 15, when by a new onset it overcame by storm the darkness of my mind with such full agreement between this idea and my labour of seventeen years on Brahe's observations that at first I believed I was dreaming and had presupposed my result in the initial assumptions.

Strangely enough, Kepler, the lover of tables and calculations, never demonstrates the accuracy of the relationship for the planets in a table (although he did publish the numbers for the Jovian satellites in his *Epitome*). Using his own data, we can calculate the table (Table 5.4) he failed to exhibit.

The harmonic relation pleased him greatly, for it neatly linked the planetary distances with their velocities or periods, thereby fortifying the *a priori* premises of the *Mysterium* and the *Harmonice*. The ecstatic Kepler immediately added these rhapsodic lines to the introduction to Bk V:

Now, since the dawn eight months ago, since the broad daylight three months ago, and since a few days ago, when the full sun illuminated my wonderful speculations, nothing holds me back. I yield freely to the sacred frenzy; I dare frankly to confess that I have stolen the golden vessels of the Egyptians to build a tabernacle for my God far from the bounds of Egypt. If you pardon me, I shall rejoice; if you reproach me, I shall endure. The die is cast, and I am writing the book – to be read either now or by posterity, it matters not. It can wait a century for

5.11. The silent notes of the planets in their eccentric courses, from the *Harmonice mundi* (1619). Kepler has transposed the notes to fit them onto the staffs – actually he considered Saturn to be much lower than Jupiter, and Mercury much higher than Earth.

a reader, as God himself has waited six thousand years for a witness.

Nevertheless, Kepler buried his harmonic relation as the eighth of thirteen points "necessary for the contemplation of celestial harmonies". It was seen by Kepler not as a fundamental law in itself, but simply as a clear and accurate manifestation of the more fundamental principles underlying the cosmos – both physical and archetypal. Thus, although Kepler gave the harmonic law relatively little emphasis, it represents the culmination of a life-long search and illustrates his imaginative approach to the mysteries of the universe. It remained for later scientists, however, to single out the importance of the harmonic law.

The Epitome of Copernican Astronomy

At the same time that Kepler was preparing his planetary ephemerides and his *Harmonice mundi*,

he also embarked upon his longest and perhaps most influential book, an introductory textbook for Copernican astronomy in general and Keplerian astronomy in particular. Cast in the catechetical form of questions and answers typical of sixteenth-century textbooks, the *Epitome astronomiae Copernicanae* gave a systematic treatment of all of heliocentric astronomy, including the three relationships now called Kepler's laws. Its seven books were issued in instalments; the first three were printed late in 1617 and dated 1618, the fourth in 1620, and the final three in 1621. Bks I–III dealt mainly with spherical astronomy, only occasionally going beyond the conventional subject matter. Bk IV took up theoretical astronomy, and Bks V–VII explained practical geometrical problems arising from the new astronomy.

Although Bk IV came last conceptually, it was published in sequence. Subtitled "Celestial Physics, that is, Every Size, Motion, and Proportion in the

Heavens Explained by a Cause Either Natural or Archetypal'', it is the most remarkable section of the *Epitome*. To a large extent it epitomized both the *Harmonice* and Kepler's new lunar theory, completed just before this part was sent to press.

Book IV opened with one of his favourite analogies, one that had already appeared in the *Mysterium* and which stressed the theological basis of his Copernicanism: the three regions of the universe were archetypal symbols of the Trinity – the centre, a symbol of the Father; the outermost sphere, of the Son; and the intervening space, of the Holy Spirit. Next Kepler took up a consideration of final causes, seeking archetypal reasons for the apparent size of the Sun, the length of day, and the relative sizes and distances of the planets. From first principles he attempted to deduce the distance from the Earth to the Sun by assuming that the Earth's volume is to the Sun's as the radius of the Earth is to its distance from the Sun. His result is twenty times greater than that assumed by Ptolemy and Copernicus, but he showed from a perceptive analysis of the observations that such a size was not excluded. Subsequently he argued that the sphere of fixed stars must be 2000 times larger than the orbit of Saturn, thereby advocating a size of the universe vastly greater than that considered by the ancients or even by Copernicus.

Kepler's harmonic law, which he had discovered just as the *Harmonice* was going to press, now received a far more extensive treatment. His explanation of the $P \propto a^{3/2}$ law is based on the relationship

$$\text{period} \propto \frac{\text{path length} \times \text{matter}}{\text{magnetic strength} \times \text{volume}}.$$

Clearly, the longer the path, the longer the period; the greater the magnetic strength, that is, the magnetic emanation reaching the planet from the Sun (which furnished the driving force), the shorter the period. The matter in the planet itself provides a resistance to continued motion: the more matter, the more inertia, and the more time required. Finally, with a larger volume of material, the magnetic emanation or 'motor virtue' can be soaked up more readily and the period proportionately shortened. According to Kepler's distance rule, the velocity, and hence the driving force, varied as $1/a$, and of course the path length was proportional to r. Therefore, in order for the period to vary as the

power $\frac{3}{2}$, the ratio of matter/volume or density had to vary as $1/a^{1/2}$. Consequently he argued that the density of each planet depends upon its distance from the Sun, a requirement quite appropriate to his ideas of harmony. To a limited extent he defended this arrangement from telescopic observations, but generally he fell back on vague archetypal principles.

In the third part of Bk IV Kepler continued his discussion of the physical causes of planetary motions and, in particular, the irregularities of speed and shape of orbit. Kepler had grown up in an age when philosophers still attributed the heavenly motions in part to the individual intelligences or 'souls' of the planets: "I deny that the celestial movements are the work of Mind", he stated in opposition to traditional opinion, and by implication he accepted that their movements, either elliptical or circular, are compelled by material necessity after having been so arranged by the Creator. In his *Astronomia nova* Kepler had been equivocal on this point, and certainly in his youthful *Mysterium* he had endorsed the idea of animate souls as moving intelligences for the planets. Now, in the *Epitome*, he sought an explanatory foundation based strictly on the solar magnetic emanations and their interactions with the magnetic poles of the Moon and planets. In this framework he introduced the magnetic influence of the Sun, likening it to the laws of the lever and balance, and then he used its interaction with oscillating 'magnetic fibres' within the planets to explain physically the motion toward or away from the Sun (the 'librations'), which resulted in the elliptical orbits. Similarly, another form of the magnetic fibres interacted with the solar emanations to drive the planets up and down in latitude. Thus, Kepler envisioned three distinct tasks for the solar emanations in conjunction with the magnetic fibres of the planets: the circular driving motion, the libration in distance, and the libration in latitude.

Book V of the *Epitome* dealt with certain practical geometrical problems arising from elliptical orbits, and it is in this section that Kepler clarified and reformulated his 'distance law'. Because the magnetic emanation from the Sun tended to drive the planet in a simple circular path, Kepler reasoned, it must be the circular component of the velocity, the part perpendicular to the radius vector, that is inversely proportional to the solar distance, rather than the full orbital velocity as he had originally assumed. He candidly apologized for the imprecise statement of the law of areas in his *Astronomia nova*, saying: "I confess that the thing was given rather obscurely there, and most of the trouble comes from the fact that there the distances are not considered as triangles but as numbers and lines."

With the way prepared by the discussion of the magnetic forces and the elliptical orbit, Kepler turned in Bks IV and VI to the most complex case of all, the lunar theory. In addition to the general features of planetary motion (the elliptical path and the variation in speed expressed by the law of areas), the Moon's motion exhibited three further inequalities: the evection (discovered by Ptolemy), the so-called variation (found by Tycho Brahe), and the annual equation (found independently by Tycho and by Kepler). These latter three inequalities depended on the Sun's position with respect to the Earth and Moon, so Kepler realized that any physical theory must involve a double interplay of the Earth and Sun. In other words, secondary bodies such as the Earth must also have some physical properties such as the magnetic emanations or 'motor virtue' that he had previously associated with the Sun.

Kepler postulated the existence of a magnetic emanation rotating with the Earth to drive the Moon in its orbit, just as the emanation from the Sun drove the planets. However, because the Moon travelled near the ecliptic and not along the celestial equator, the solar emanations were clearly dominant. He supposed that the Sun's influence created synodic variations in both the Earth's circular driving force and in the librations of distance and latitude. In addition, he believed that a small solar-induced inequality affected the *rotation* of the Earth, and therefore Kepler treated the annual equation not as a term in the Moon's motion but as an effect on the equation of time.

Kepler was particularly proud of his solution to the lunar evection problem. Ptolemy had modelled the lunar orbit with an unrealistically large variation in the Moon's distance, and subsequently Islamic astronomers and Copernicus had tried to reduce this discrepancy in lunar distance. Kepler based his explanation of the evection on the physical interaction of the Sun's emanation with the magnetic fibres of the Earth, and hence he did not need to introduce any new mechanism that changed the lunar distance. His ingenious solutions for the evection and other complications of lunar motion offered reasonable predictions for the Moon's positions. Nevertheless, his scheme remained essentially *ad hoc* and failed to offer any foundation for further advances.

Precisely how influential Kepler's *Epitome* was is difficult to assess. His reputation was eventually much enhanced by the publication of his *Rudolphine Tables*, and meanwhile the Copernican approach was rapidly gaining in acceptability. Thus, by 1635, there was sufficient demand for the *Epitome* to warrant reprinting it, and for many years it remained one of the few accessible sources for the details of the Copernican system (including, of course, those essential revisions introduced by Kepler).

The *Rudolphine Tables*

The *Rudolphine Tables* (*Tabulae Rudolphinae*) were finally published in 1627. This monumental work, based on Kepler's reforms of the Copernican system, furnished working tables from which planetary positions could be computed. Kepler apologized for the long delay, mentioning among other things, "the novelty of my discoveries and of the unexpected transfer of the whole of astronomy from fictitious circles to natural causes, which were the most profound to investigate, difficult to explain, and difficult to calculate, since mine was the first attempt". Among the novelties that had postponed the completion of the tables was what Kepler termed a "happy calamity" – his learning about Napier's logarithms.

Kepler had been so filled with admiration when he eventually saw John Napier's *Mirifici logarithmorum canonis descriptio* (1614) that he dedicated his 1617 *Ephemerides* to the Scot, taking the opportunity to praise the potential advantages of the new invention. Although his former teacher Mästlin cautioned that "it is not seemly for a professor of mathematics to be childishly pleased about any shortening of the calculations", Kepler plunged into an extensive examination of the logarithms. Because Napier had not yet published any description of their construction, Kepler created a different form by a new geometrical procedure. Kepler's logarithms are related to modern natural logarithms by

$$\log_{\text{Kepler}} x = 10^5 \ln(10^5/x).$$

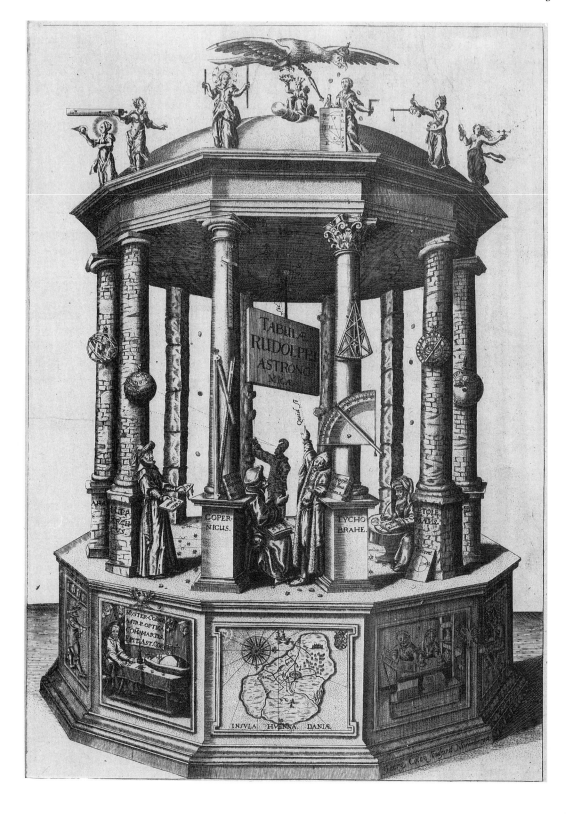

Although Kepler had already used logarithms implicitly in calculating his later *Ephemerides*, the *Rudolphine Tables* was the first book explicitly to require logarithms in a scientific application. Kepler used his own form, quoting from Ovid that it is better for a farmer to serve the produce of his own gardens.

In several ways Kepler's *Epitome* served as a theoretical handbook to his *Rudolphine Tables*. In Bk V was introduced what is today called Kepler's equation,

$$E = M - e \sin E,$$

where e is the orbital eccentricity, M is the mean angular motion about the Sun, and E is an auxiliary angle related to M through the law of areas. Given E, Kepler's equation is readily solved for M; the more useful inverse problem has no closed solution in terms of elementary trigonometric functions, and he could only recommend an approximating procedure. In the tables, Kepler solved the equation for a uniform grid of E values and provided an interpolation scheme for the desired values of M. This helped make his *Rudolphine Tables* far different from any that had gone before, a feature that must have bewildered many astronomers, and induced some of them to publish a variety of simplified versions.

Nevertheless, the success of the tables ultimately provided the chief vehicle for recognition of Kepler's accomplishments. Today we know that his predicted positions were generally about thirty times better than those of the prior or competing tables, but this was not immediately obvious when the tables were published. However, Kepler had used the tables to predict a transit of Mercury, and in 1631 (the year following his death) Pierre Gassendi in Paris dramatically verified the prediction with a successful observation: for the first time the planet Mercury was seen to cross the face of the Sun (Figure 7.2). Kepler's prediction erred by only 10′, compared to 5° for tables based on Ptolemy, Copernicus, and others. This graphic demonstration was a forceful testimony to the efficacy of the heliocentric system, and in particular to Kepler's version of it.

Meanwhile, the printing of the *Rudolphine Tables* had barely begun when the Counter-Reformation swept into Linz; the town was blockaded and the press went up in flames. Ultimately the tables were printed in Ulm, and Kepler began to search in earnest for a new residence. Some years earlier he had dedicated his *Harmonice mundi* to James I, and later had received an invitation to England. But to his Strassburg friend Matthias Bernegger he confided: "Therefore shall I cross the sea. . . . I, a German? A lover of firm land, who dreads the confinement of an island?"

Seven years later, however, he wrote to Bernegger: "As soon as the *Rudolphines* are published, I desire to find a place where I can lecture about them to a large audience, if possible in Germany, otherwise in Italy, France, Holland, or England, provided the salary is adequate for a traveller." Torn between his desire to find religious toleration and his reluctance to lose his official salaries, Kepler journeyed to Prague at the end of 1627 to arrange for further employment. There the imperial commander-in-chief, Albrecht von Wallenstein, a superstitious man who had earlier applied anonymously to Kepler for a horoscope, agreed to support Kepler in a newly acquired fiefdom, the duchy of Sagan.

Kepler spent the two final years of his life in Sagan. The most notable accomplishment was the

5.12. The frontispiece of the *Rudolphine Tables* (1627). It depicts an allegorical Temple of Urania, based on one painted in the foyer of Tycho's Stjerneborg Observatory on Hven. The most ancient Chaldean observations are represented by columns made of rough-hewn logs in the background. Hipparchus and Ptolemy rest beside pillars of brick, whereas Copernicus has a fine Ionian column and Tycho an even more splendid Corinthian column. Addressing Copernicus, Tycho points out his geoheliocentric system on the ceiling, asking "How about that?" ("*Quid si sic*"). Around the dome of the temple can be seen six goddesses, all recalling aspects of Kepler's own work: Physica (physics of light and shadows), Optica, Logarithmica (these tables being the first explicit use of logarithms), Doctrina Triangulorum (with the ellipse), Stathmica (with the lever and balance for the law of areas), and Magnetica. (Invisible but listed in the introductory poem are six more divinities standing above the other six zodiacal columns, all equally germane to Kepler's interests, including Computus, Chronologia, and Archetypica.) In the panel at the bottom sits Kepler, working by candlelight, with a model of the dome before him, as if saying to his readers, "Indeed, Tycho deserves an elaborate column, but if I had not laboured long into the night, the Temple of Urania would not yet be completed."

completion of his *Somnium seu astronomia lunari* (published posthumously in 1634), a pioneering piece of science fiction that invoked a fantasy journey to the Moon. Its perceptive description of celestial motions as seen from the Moon made it a subtle but effective polemic for the Copernican system.

In October 1630 the fifty-eight-year-old Kepler set out for the electoral congress in Regensburg, apparently intending to consult with the emperor about yet another residence. There he became ill, and, on 15 November 1630, died.

Evaluation

With Ptolemy and Copernicus, Kepler shared a profound sense of order and harmony. In Kepler's mind this was linked with his theological appreciation of God the Creator. Repeatedly, Kepler stated that geometry and quantity are co-eternal with God and that mankind shares in them because man is created in the image of God. From these principles flowed his ideas on the cosmic link between man's soul and the geometrical configurations of the planets; they also motivated his indefatigable search for the mathematical harmonies of the universe.

Contrasting with Kepler's mathematical mysticism, yet growing out of it through the remarkable quality of his genius, was his insistence on physical causes. In Kepler's view the physical universe was not only a world of discoverable mathematical harmonies but also a world of phenomena explicable by mechanical principles. The result of his work was the mechanization and the cleansing of the Copernican system, setting it into motion like clockwork and sweeping away the vestiges of Ptolemaic astronomy.

It remained for Isaac Newton to banish the last traces of Aristotelian physics and to place the heliocentric system on a consistent physical foundation. Although Newton indirectly owed much to Kepler's insistence on physical causes, he rejected the type of physical arguments so dear to Kepler's mind, and with it, he tried to withhold credit for Kepler's achievement. As a result, Kepler is nowhere mentioned in Bk I of Newton's *Principia*. A fairer evaluation of Kepler has come from Edmond Halley in his review of the *Principia*: Newton's first eleven propositions, he wrote, were "found to agree with the *Phenomena* of the Celestial Motions, as discovered by the great Sagicity and Diligence of *Kepler*".

Further reading

This chapter is based in part on the author's entry on Kepler in *Dictionary of Scientific Biography*, vol. 7 (New York, 1973), 289–312, and on his "Johannes Kepler and the new astronomy", *Quarterly Journal of the Royal Astronomical Society*, vol. 13 (1972), 346–73

Carola Baumgardt, *Johannes Kepler: Life and Letters* (New York, 1951)

Arthur Beer and Peter Beer (eds.), *Kepler: Four Hundred Years* (*Vistas in Astronomy*, vol. 18) (Oxford, 1975), a large and uneven compendium containing key articles by E.J. Aiton, A. Beer, O. Gingerich, and E. Rosen

Max Caspar, *Kepler*, transl. by C. Doris Hellman (London, 1959)

Gerald Holton, Johannes Kepler's universe: its physics and metaphysics, *American Journal of Physics*, vol. 24 (1956), 340–51

N. Jardine, *The Birth of History and Philosophy of Science: Kepler's A Defence of Tycho against Ursus* (Cambridge, 1984)

Johannes Kepler Gesammelte Werke (Munich, 1938–)

Johannes Kepler, *Epitome of Copernican Astronomy*, Books IV and V, and *The Harmonies of the World*, transl. by C.G. Wallis, in Great Books of the Western World, ed. by R.M. Hutchins, vol. 16: *Ptolemy, Copernicus, Kepler* (Chicago, 1952)

Johannes Kepler, *Mysterium Cosmographicum: Secret of the Universe*, transl. by A.M. Duncan (New York, 1981)

Johannes Kepler, *The New Astronomy*, transl. by William Donahue (Cambridge, 1989)

Alexandre Koyré, *The Astronomical Revolution*, transl. by R.E.W. Maddison (Ithaca, NY, 1973)

Edward Rosen, *Kepler's Conversation with Galileo's Sidereal Messenger* (New York, 1965)

Robert Small, *An Account of the Astronomical Discoveries of Kepler* (London, 1804; reprinted Madison, 1963)

Curtis Wilson, Kepler's derivation of the elliptical path, *Isis*, vol. 59 (1968), 5–25

PART II

The impact of the telescope

6

Galileo, telescopic astronomy, and the Copernican system
ALBERT VAN HELDEN

1. Introduction

By 1600 astronomers and natural philosophers had several world systems to chose from: the traditional geocentric system of Ptolemy (see Volume 1), the heliocentric system of Nicholas Copernicus (see Volume 1), and the compromises between the two, most notably the geoheliocentric system recently put forward by Tycho Brahe (see Chapters 1 and 3 above). Each had points in its favour. Ptolemy's was broadly based on Aristotelian physics, whatever the tensions between them on points of detail, and had the weight of tradition behind it. Copernicus's system as outlined in Bk I of *De revolutionibus* was elegant and simple and accounted systematically for relations that had remained merely coincidental in the Ptolemaic system. Tycho's compromise, though clearly a hybrid, retained much of the elegance and simplicity of the heliocentric system while avoiding the stumbling block of the motion of the Earth. Cosmology, then, was in a period of transition, and in the middle of this transition the telescope was brought to bear on the debate between rival systems. This dramatic chapter in the history of astronomy was dominated by Galileo Galilei (1564–1642), no mathematical astronomer but a brilliant observer and thinker, who became a passionate advocate of the heliocentric world picture.

Upon entering the University of Pisa as a student in 1581, Galileo was instructed in the standard mathematical curriculum which included Ptolemaic astronomy. Unlike Johannes Kepler (1571–1630), who was a Copernican virtually from the beginning of his astronomical studies, Galileo showed no early interest in the Copernican system, even though we may assume that he became familiar with it during his student days. His earliest cosmological compositions, written when he began teaching mathematical subjects, in 1589, show him in an entirely traditional light. Shortly after moving to Padua in 1592, however, Galileo became intrigued with the tides, and now he used the Copernican theory to give an explanation of this phenomenon based on the Earth's annual and diurnal motions (see below). Thus, when in 1597 he wrote to thank Kepler for a copy of *Mysterium cosmographicum*, he told Kepler that he had been of the Copernican conviction for some years because that theory allowed him to explain certain physical effects. He preferred, however, not to publish these views for fear of being ridiculed.

It has been assumed by historians of science that the debates about the location of the New Star of 1604 strengthened Galileo in his Copernican convictions, but lately it has been argued that this supernova may, in fact, have made Galileo much more tentative about the Copernican hypothesis. At any rate, it appears that long before 1609, when the telescope was brought to bear on the heavens, Galileo had been a cautious Copernican in private and among friends. It appears also that his investigations into the science of motion were powered by cosmological speculations: a non-Aristotelian cosmology would need the underpinning of a non-Aristotelian physics. By 1609, when he turned his full attention to telescopic astronomy, he had already arrived at most of the crucial concepts and relations that were to become the foundation for the new physics.

2. Telescopic discoveries

It was in the summer of 1609 that Galileo first heard about a new device for seeing faraway things as though they were nearby, and he quickly made one for himself. By the late autumn of that year he had a 20-powered instrument and was beginning

6.1. Galileo (describing himself as a member of the Lincean Academy and Philosopher and Mathematician to the Grand Duke of Tuscany), as depicted on the frontispiece of his *Letters on Sunspots* (1613).

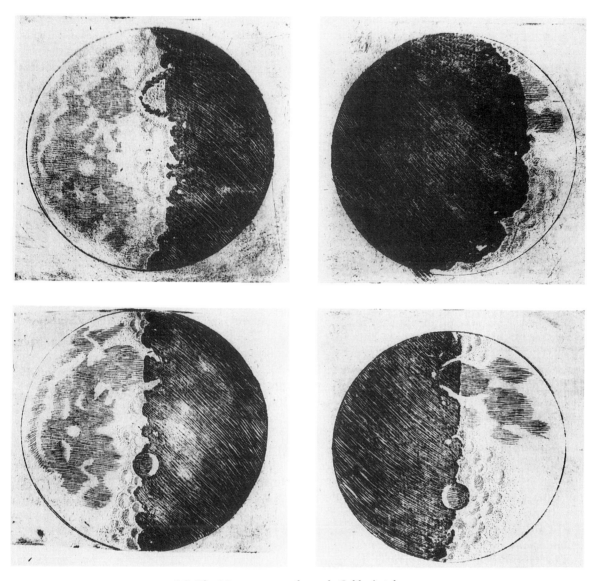

6.2. The Moon as seen through Galileo's telescope.

a survey of the heavens (see Volume 3). Quite naturally, his first target (see Chapter 8) was the Moon. According to Aristotle, the Moon, as a heavenly body, should be perfectly smooth and spherical, but non-Aristotelian speculations from Antiquity were being revived and seriously entertained in some quarters around the turn of the seventeenth century. One of these was that the Moon is much like the Earth. When Galileo examined the Moon at significant magnifications he found that it was anything but perfectly smooth and spherical, as Aristotle would have it. Although its surface had a few smooth patches, it appeared for the most part uneven and mountainous, much like the Earth. To Galileo this was clear evidence that Aristotle was wrong about the nature of at least one heavenly body.

One of the first reports about spyglasses emanating from The Hague in the autumn of 1608 already mentioned that the new device showed many fixed stars that ordinarily are invisible because of their smallness or the weakness of our vision. Galileo examined a number of constellations and found that the traditional catalogues of 'fixed' stars (so named to distinguish them from the 'planets' or wandering stars) accounted for only a very small proportion of those he could now see. Many areas in the Milky Way could be resolved into myriads of very small stars, whose light blends together when seen with the naked eye. This confirmed a speculation first made by Democritus *c.* 400BC.

On 7 January 1610 Galileo wrote a letter to a patron in which he outlined some of the wonderful new phenomena revealed to him by his new instruments. As one example, he drew a sketch of the planet Jupiter on a line with what he thought were three stars on either side, like beads on a string. Thinking this to be a fortuitous alignment, Galileo expected Jupiter to move away from the three little 'stars'. Over the next several evenings (Figure 6.3) he observed the three, and then four, 'stars' to move with respect to each other and to Jupiter, but as a group they followed the planet on its course through the heavens. Galileo quickly realized that these four 'stars' were 'moons' of Jupiter. This was a major discovery, and he realized that he must move quickly to safeguard his claim to it. Two months later *Sidereus nuncius* (or *The Starry Messenger* by which his celestial discoveries were announced) came off the press in Venice. It created a sensation in learned circles, and Galileo became an international celebrity overnight. In an attempt to secure patronage from Cosimo II de' Medici, Grand Duke of his native Tuscany, he had named the four satellites the "Medicean stars". Such a tribute did not go unrewarded, and later that year Galileo moved to Florence to become "Chief Mathematician and Philosopher" of the Medici Court.

In *Sidereus nuncius* Galileo described the new spyglass (baptised 'telescope' a year later) and announced four discoveries: the Earthlike nature of the Moon, the multitude of fixed stars, the difference between the telescopic appearances of planets and fixed stars, and the satellites of Jupiter. None of these discoveries falsified Ptolemy or proved Copernicus correct, but several undermined Aristotle. The Moon appeared not to be perfect and spherical, and the satellites of Jupiter demonstrated that there was more than one centre of motion, no matter what cosmic system one held. Moreover, the fact that the telescope resolved planets into disks, like little moons, while it could not similarly enlarge the stars, tended to support the huge distance to the stars postulated by Copernicus. We may be sure that Galileo's Copernicanism and anti-

Aristotelianism were greatly strengthened by these discoveries. In *Sidereus nuncius* he pointed out that the appearance of the Moon supported those who wished to revive the Pythagorean idea that the Moon is another Earth; he also pointed out that the moons of Jupiter silenced those who would reject the Copernican system because in it the Earth was the only planet to have a moon. He did not, however, go so far as to declare himself a follower of Copernicus. That would have been highly inappropriate and counterproductive in a book designed, at least in part, to glorify the House of Medici, and thus to obtain patronage for Galileo.

Galileo continued to observe the satellites of Jupiter and made several further discoveries in the heavens. In July 1610 he discovered that Saturn does not show a single disk, as do the other planets but, rather, three disks, a central one flanked by two smaller ones. These two lateral bodies turned out not to be satellites like Jupiter's companions, for they did not move with respect to the central disk. Galileo went on to discover their disappearance in 1612, and, following their return in 1613, their expansion into *ansae*, or 'handles' (1616). Saturn's strange appearances would remain a celebrated puzzle in astronomy (see Figure 6.7) until Christiaan Huygens (1629–95) finally published a satisfactory solution in 1659.

A more important discovery came in the autumn of 1610, when Galileo verified that the phases of Venus go through an *entire* cycle, as seen from the Earth, just as does the Moon. This implied that Venus goes around the Sun, so that from our vantage point it is sometimes 'above' (i.e., beyond) and sometimes 'below' the Sun. Thus, the Ptolemaic system, in which Venus was always 'below' the Sun and so could never appear 'full', had to be in error, and only the world systems of Copernicus and Tycho Brahe could save the phenomena. When this discovery was announced by Galileo in semi-public correspondence, late in 1610, and published by Kepler in *Dioptrice* the following year, the Ptolemaic system soon became obsolete among practising astronomers. However, Aristotelian philosophers, who could regard instruments and measurements as irrelevant in describing reality, continued to subscribe to Ptolemy's cosmic scheme for some time.

By the end of 1610 other observers had telescopes (often supplied by Galileo himself) good enough to verify Galileo's discoveries. The satellites

[Handwritten manuscript page — Galileo's journal of observations of Jupiter's satellites, in Italian cursive]

Adi 7. di Genaio 1610 Gioue si uedeua cõ l'cannone cõ
3. stelle fisse cosi   * ⊕ *   delle quali sé il cannone
minore si uedeua. à d. 8. appariua cosi ⊕ ++* era dũq
diretto et nõ retrogrado come segono i calculatori.
Adi 9. fù nugolo. à dì 10. si uedeua cosi ─── * * ⊕ ciò è sõ
giũto à la più ouidetale si che la reultauano quãto è puo credere.

Adi 11. era in questa guisa ** ⊕ et la stella più uicina
à Gioue era l'ametà minore dell'altra, et uicinissima all'altra
doue che le altre sere erano le dette stelle apparite tutte tre
di egual grandessa et trà di loro egualmẽ lontane: dal che
appare intorno à Gioue esser 3. altre stelle errãti inuisibili ad
ogn'uno sino à questo tepo.

Adi 12. si uedde in tale costituzione * *⊕ * era la stella
ouidentale poco minor della orientale, et gioue era ĩ meso lontano
da l'una et da l'altra quãto il suo diametro ĩ circa: et forse era
una terza picciolissª et uicinissª à ♃ uerso oriẽte; anzi pur ui era
veramẽ hauendo io cõ più diligẽza osseruato, et hauedo più imbrunita la
notti.

Adi 13. hauẽdo benissª fermato lo strumẽ si ueddono uicinissª à Gioue
4. stelle in questa costituzione *⊕ * ** ò meglio cosi ⊕ * **
e tutte appariuano della medª grandezza, lo spazio delle 3. ouidẽtali
nõ era maggiore del diametro di ♃. et erano frà di loro notabilmẽ
più uicine che le altre sere; ne erano in linea retta esquisitamẽ come
à quãti ma la media delle 3. ouidẽtali era ĩ poco eleuata, ò uero la
più ouidẽtale alquãto depressa; sono queste stelle tutte molto lucide bẽ che
picciolissª et altre fisse et appariscono della medmª grandẽ nõ sone
cosi splendẽti.

Adi 14. fù nugolo. Adi 15. era cosi ⊕ * * * *. la prossª à
♃. era la minore et le altre di mano ĩ mano maggiori: gl'interstitii
tra ♃ et la 3. seguẽti erano quãto il diametro di ♃. ma la 4.ª era di-
stante dalla 3.ª il doppio ĩ circa: nõ face
uano iteramẽ linea retta, mã come mostra
l'esempio, erano al solito lucidissª bẽ che picco
le, et niente scintillauano come âc fĩ

♃.
♃ long. 71. 38 Lat. 1. 13.
Mei:
2. 30
1. 13
1. 13

6.3. The first page of Galileo's journal of observations of Jupiter's satellites, copied and amplified from two earlier sheets on the night of 15 January 1610 and continued (in Latin) for that and subsequent nights on the verso.

SIDEREVS
NVNCIVS

MAGNA, LONGEQVE ADMIRABILIA

Spectacula pandens, suspiciendaque proponens
vnicuique, præsertim verò

PHILOSOPHIS, atq̃ ASTRONOMIS, quæ à

GALILEO GALILEO

PATRITIO FLORENTINO

Patauini Gymnasij Publico Mathematico

PERSPICILLI

Nuper à se reperti beneficio sunt obseruata in LVNÆ FACIE, FIXIS IN-
NVMERIS, LACTEO CIRCVLO, STELLIS NEBVLOSIS,
Apprime verò in

QVATVOR PLANETIS

Circa IOVIS Stellam disparibus interuallis, atque periodis, celeri-
tate mirabili circumuolutis; quos, nemini in hanc vsque
diem cognitos, nouissimè Author depræ-
hendit primus; atque

MEDICEA SIDERA

NVNCVPANDOS DECREVIT.

VENETIIS, Apud Thomam Baglionum. M DC X.

Superiorum Permissu, & Priuilegio.

6.4. The title-page of Galileo's *Sidereus nuncius* (1610), with a summary of the discoveries announced in the work. Special
prominence is given to the satellites of Jupiter, named for Galileo's patron, Cosimo de' Medici.

RECENS HABITAE.   23

dentalis proxima min. 2. ab hac vero elongabatur oc-.

Ori.   * ◯ *   *   Occ.

cidentalior altera min: 10. erant præcisè in eadem recta, & magnitudinis æqualis.

   Die quarta hora fecunda circa Iouem quatuor fta-
bant Stellæ, orientales duæ, ac duæ occidentales in

Ori.   * *◯   *   *   Occ.

eadem ad vnguem recta linea difpofitæ, vt in proxima figura. Orientalior diftabat à fequenti min. 3. hęc
verò à Ioue aberat min. o. fec. 40. Iuppiter à proxima
occidentali min.4. hæc ab occidentaliori min. 6. magnitudine erant ferè ęquales, proximior Ioui reliquis
paulo minor apparebat. Hora autem feptima orientales Stellæ diftabant tantum min. o. fec. 30. Iuppiter

Ori.   ** ◯   * *   Occ.

ab orientali viciniori aberat min. 2. ab occidentali verò fequente min. 4. hæc verò ab occidentaliori diftabat min.3. erantque æquales omnes, & in eadem recta
fecundum Eclypticam extenfa.

   Die quinta Cœlum fuit nubilofum.

   Die fexta duæ folummodo apparuerunt Stellæ me-

Ori.   * ◯ *   Occ.

                                    dium

6.5. Observations of Jupiter and satellites, from *Sidereus nuncius*.

PLEIADVM CONSTELLATIO.

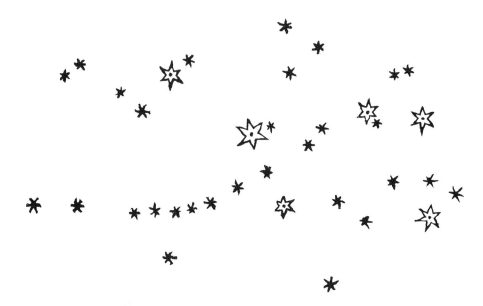

6.6 The Pleiades as depicted in *sidereus nuncius*. Galileo could rightly claim that most of the stars shown had been hidden from human eyes since the creation of the world.

6.7. The problem of Saturn's appearances, 1610–75.

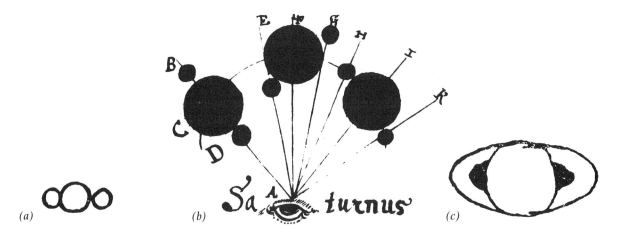

(*a*) In July 1610 Galileo first directed a telescope to Saturn and saw not one but three bodies: "Altissimum planetam tergeminum observavi."
(*b*) As shown here in Christoph Scheiner's unpublished "Tractatus de tubo optico" (1616), early explanations of Saturn's appearances involved satellite models.
(*c*) But, as shown here in Galileo's observation of 1616, some appearances did not fit a satellite model.

(*d*)

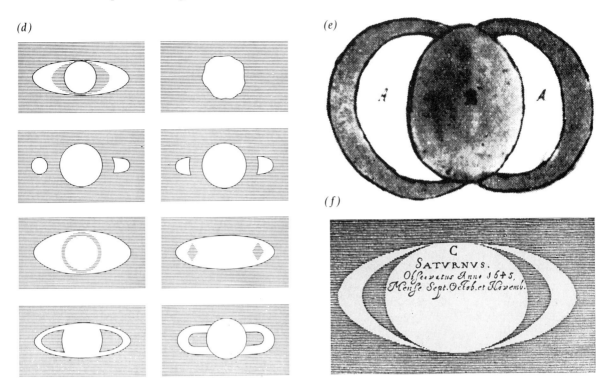

(*e*)

(*f*)

(*d*) Not until after about 1635 did astronomers begin to make frequent observations of Saturn in order to establish the full range of its appearances. But often these observations did little to clarify the situation, as is shown by these observations made by Pierre Gassendi between 1636 and 1652.

(*e*) and (*f*) During this period, however, the 'handled' appearance became the basis for explanatory models. Yet observers often disagreed on the exact shapes of planet and *ansae*, as is shown by the observations of Francesco Fontana (*e, top*) in 1636 and of Johannes Hevelius (*f, above*) in 1645.

(*g*)

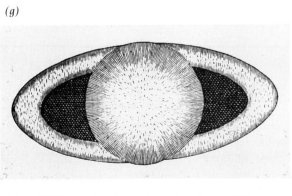

(*h*)

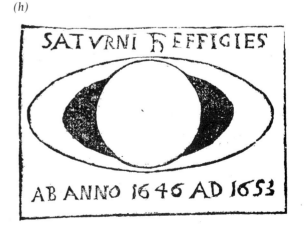

(*g*) and (*h*) Even when (as we know in retrospect) the raw observation was excellent, the depiction was usually misleading, as is shown in the renditions by Eustachio Divini (*g, above*) and Gioanbatista Odierna (*h, right*).

(*i*) and (*j*) When Saturn again appeared edgewise, in the mid-1650s, astronomers were formulating fully-fledged theories to explain the planet's strange appearances. Christopher Wren (*i*, *right*) in 1657 supposed that an infinitely thin elliptical 'corona' was attached to the planet, while the entire formation rotated or librated around its long axis. In 1656 Christiaan Huygens (*j*, *below*) supposed that the planet "is surrounded by a thin flat ring which does not touch him anywhere and is inclined to the ecliptic". The thickness of Huygens's ring was not negligible.

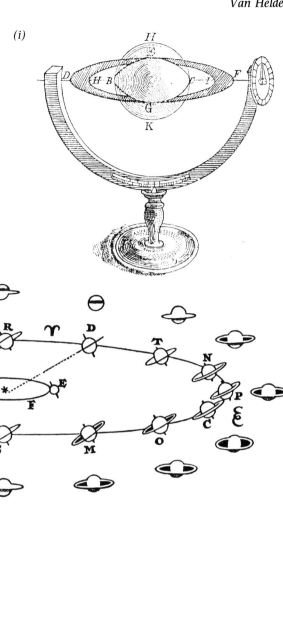

(*j*)

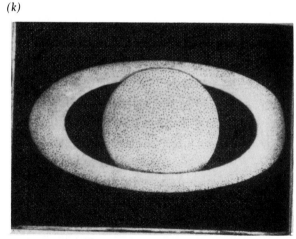

(*k*)

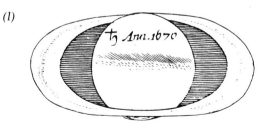

(*l*)

(*k*) Between 1660 and 1675 the ring theory (as well as better telescopes) led to the discoveries of shadow effects that in turn confirmed the theory, as is shown here in the 1664 observation by Giuseppe Campani.

(*l*) Finally, in 1675, Gian Domenico Cassini discovered that the ring had a gap in it.

6.8. Galileo's letter to Giuliano de Medici in Prague, dated 11 December 1610, and giving at the top an anagram intended for Kepler, in which is concealed the discovery of the phases of Venus. The anagram reads: "Haec immatura a me jam frustra leguntur o y". Its deciperment is: "Cynthiae figuras aemulatur mater amorum", that is, the mother of loves (Venus) emulates the figures of Cynthia (the Moon).

were now being observed in London by Thomas Harriot (*c.* 1560–1621) and his friends, and in Aix en Provence by Nicholas Claude Fabri de Peiresc (1580–1637) and his group. Kepler had seen them in September 1610, when he briefly had access to an instrument made by Galileo. Meanwhile, in Rome, the implications of these claimed discoveries were not lost on Church officials, who were anxious for 'in-house' verification. Not until the end of 1610, however, could Christoph Clavius (1537–1612) and his fellow Jesuit mathematicians at the Collegio Romano obtain a telescope good enough

to show Jupiter's satellites. Their observations enabled them, in the spring of 1611 when Galileo was visiting Rome, to certify to Cardinal Robert Bellarmine that the new phenomena claimed by Galileo were indeed to be observed in the heavens with the telescope. They did not, however, agree with every interpretation put forward by Galileo: they hesitated to affirm that the Milky Way is made up of myriads of small stars whose light blends together, and Clavius preferred to think that the Moon's surface was smooth and spherical and that its rough appearance was due to differences in the

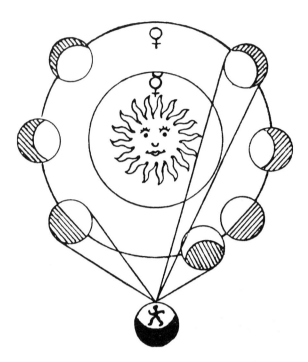

6.9. The observed phases of Venus explained as the consequence of Venus's orbit around the Sun. From the third part (1621) of Kepler's *Epitome astronomiae Copernicanae.*

density of the body of the Moon. Others would repeat this last argument, and some went so far as to argue that the Moon had a rough interior covered by a perfectly smooth and spherical layer of some transparent substance. But these were details: Galileo had won the day in Rome, as he had in the rest of Europe, and his discoveries were celebrated in the eternal city. The mathematicians of the Collegio Romano held a banquet in his honour, and Prince Federigo Cesi initiated him as member of his scientific academy, the Accademia dei Lincei, at a dinner at which the new instrument received the name 'telescope'. Galileo was now the most celebrated man of science in Europe.

3. Sunspots, Copernicanism, and theology

By the close of 1610, Kepler had published a number of openly Copernican books, including several commentaries on Galileo's discoveries. Yet Galileo, working much closer to Rome, remained cautious. Although he considered the new phenomena discovered with the telescope compelling evidence for the Copernican system, and although there can be little doubt that in public debates he often sup-

ported the Copernican system openly, in his publications he dealt with the subject only obliquely. It was in the debate about sunspots that Galileo's Copernicanism finally emerged in print.

The ancient Chinese annals make mention of a number of sightings of dark spots on the Sun, spots sufficiently large to be visible with the naked eye. Because of their Aristotelian belief in the perfection of heavenly bodies, Westerners had usually interpreted such appearances as transits of Mercury or Venus. Indeed, in 1607, a year before the advent of the telescope, Kepler had observed (by means of a primitive camera obscura) what he thought was Mercury on the Sun. He realized some years later, however, that it had been an unusually large sunspot. The first telescopic observation of a sunspot was made by Harriot late in 1610. Galileo showed sunspots to observers in Rome during his stay there in the spring of 1611. At about the same time, the astronomer David Fabricius (1564–1617) and his son Johann (1587–1615), in East Frisia (North-west Germany), discovered them for themselves and began making regular observations. In June Johann published a tract on these observations,

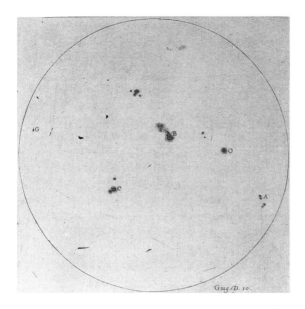 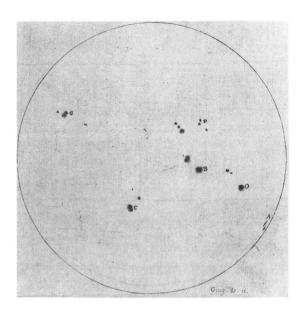

6.10. Sunspots as drawn by Galileo on successive days. He interpreted the changes as most probably caused by the rotation of the Sun. From his *Letters on Sunspots* (1613).

arguing that the spots were on the Sun and that the Sun rotates on its axis. This little book, however, attracted very little attention.

Early in 1612, Galileo received from Marc Welser, a banker in Augsburg, a little printed tract entitled *Tres epistolae de maculis solaribus* (*Three Letters about Sunspots*), written under the pseudonym Apelles. Up to this point no one had disputed any of Galileo's claims to priority in discoveries made with the telescope. Doubly sensitive about his reputation, now that he was in the employ of a powerful patron, Galileo lost little time in countering this challenge to his priority in all matters telescopic. What followed was an exchange of letters in which Galileo laid the foundation for the scientific study of sunspots and argued convincingly that they were phenomena on the Sun or in its atmosphere. Apelles, quickly identified as Christoph Scheiner (1573–1650), professor of mathematics at the Jesuit university of Ingolstadt, argued unconvincingly that they were caused by a swarm of satellites of the Sun located between the Sun and the Earth. Galileo's letters, together with (in some copies) reprints of Scheiner's letters, were published by the Accademia dei Lincei, in 1613, under the title *History and Demonstrations concerning Sunspots and their Properties*, with a long introduction in which Galileo's priority of discovery was

affirmed and Scheiner's claim ridiculed. In his last letter, Galileo cautiously stated that his telescopic discoveries "harmonized admirably with the great Copernican system", and suggested that this system was now in the ascendancy.

Up to this point the central issue surrounding the telescope had been the reality of Galileo's telescopic discoveries, and their meaning for Aristotle's philosophy and Ptolemy's astronomy. Now, however, Galileo's Copernicanism, argued brilliantly and openly, became the central issue. More importantly, his opponents were increasingly able to dictate the terms of the argument. These terms were not only scientific but, more significantly, theological: could the Copernican system be made to harmonize with biblical passages that denied the motion of the Earth and the stability of the Sun? It is important to note that it would have been perfectly acceptable to argue (following Andreas Osiander's unauthorized and anonymous preface to Copernicus's *De revolutionibus* (see Volume 1)) that as a mere mathematical hypothesis (which made no claims about reality) the Copernican scheme accounted for the appearances (especially the phases of Venus) better than the Ptolemaic. However, Galileo (as indeed Kepler had done before him) argued that the Copernican system represented the universe as it was really constructed – that the Sun really stood still in the centre and that the Earth really moved about it. Thinking, perhaps, that he could win Church officials over to his view, Galileo spent the next two decades in a crusade for the acceptance of the Copernican theory.

In the years after Galileo took up his appointment at the Medici Court in Florence, an informal 'league' of opponents, made up of clerics, scholars, and others, became increasingly active. In 1611 its leader Lodovico delle Colombe (after whom the league was often called the 'pigeon league') circulated a manuscript tract entitled *Contro il moto della Terra* (*Against the Motion of the Earth*). Here Colombe used biblical passages to combat the Copernican position, challenging Copernicans (though without mentioning Galileo) to explain these passages along Copernican lines. Two years later Galileo took up this challenge in what was at first a private letter written to his disciple Benedetto Castelli, a professor at the University of Pisa, who had supported the Copernican position at a court dinner. Galileo argued that the Bible was not written to instruct us in science but rather in religion, and that to interpret every word in it literally would mean many absurdities, such as having to ascribe to God feet, hands, and eyes, as well as human emotions. The famous passage (Joshua 10:12–14) in which Joshua commanded the Sun to stand still could not be explained straightforwardly in the Ptolemaic system, for stopping the Sun's own motion would in fact merely hasten its setting. Joshua's goal would have to be achieved by stopping the 24-hour diurnal rotation of the sphere of the fixed stars (a motion in which all heavenly bodies partake in the Ptolemaic system), thus stopping the entire system of celestial spheres. Assuming the truth of the Copernican system, however, and knowing from his study of sunspots that the Sun rotates on its axis, Galileo argued that by ordering the Sun, the central source of motion in the solar system, to stop rotating, Joshua brought the entire solar system to a temporary halt. The passage was thus explained more naturally in the Copernican system than in the Ptolemaic. Copies of this letter circulated widely, reaching as far as London, and Galileo had thus more or less publicly entered upon the theological battlefield. Over the next several years he familiarized himself with theological literature bearing on this issue.

The 'league' was not content to let the matter rest. It continued its agitation against Galileo, constantly attempting to draw Church authorities in Florence and Rome into the quarrel. Late in 1614 a Dominican priest in Florence preached a sermon against Galileo, and early in 1615 a version of Galileo's letter to Castelli was sent to Rome with accusations of heresy against Galileo. In order to protect himself Galileo quickly sent a correct copy of the letter to Rome. Church officials found nothing objectionable in this letter but, increasingly, they had to take formal note of the problem posed by the realist interpretation of the Copernican theory, which up to this point they had always regarded as a mere mathematical hypothesis. One of the cardinals of the Holy Office (the Inquisition) caused this office to start an investigation that would eventually lead to a pronouncement on this issue.

4. The decree of 1616

Galileo was well connected in court and Church circles, and he was therefore well aware of the pressure that was building in Florence and Rome. He was in a difficult position. As a scientist he was now convinced that the Copernican system best represented reality, but he was also a devout son of the Church. How could he convince the Church that one ought to be able to be at the same time a Copernican and a Catholic? Only by straying further into theology could he hope to convince the Church of the dangers of using scriptures and theology to endorse a particular scientific viewpoint. As he was girding his loins for this task, help of a sort arrived from an unexpected quarter.

Early in 1615 a Neapolitan theologian of the Carmelite order, Paolo Antonio Foscarini, published an open defence of the reality of the Copernican system, arguing that this system does not contradict the Scriptures. Galileo was mentioned in this work, but only for his telescopic discoveries, not for his Copernican sentiments. Foscarini sent a copy to Cardinal Bellarmine and asked for his opinion. Bellarmine responded that Foscarini and Galileo should treat the Copernican system as a mathematical hypothesis only, because the realist interpretation conflicted with many biblical passages and their interpretation by the Church Fathers. To go against the unanimous interpretation of the Fathers had been forbidden by the Council of Trent. Only if there were convincing proof that the Sun really does stand still and the Earth moves about it would the Church be forced to reassess the meaning of these scriptural passages. But Bellarmine would not believe that such proofs existed until they were shown to him. And, indeed, the observation most damaging to Ptolemy – the phases of Venus – was as compatible with the Tychonic as with the Copernican system.

When Galileo read Foscarini's book and Bellarmine's reaction to it, he had already begun an expansion and revision of his letter to Castelli. By the summer of 1615 he had finished this task, giving the product the title *Letter to Madame Christina of Lorraine Grand Duchess of Tuscany Concerning the Use of Biblical Quotations in Matters of Science*. This classical statement on the relationship between science and theology circulated widely in manuscript and was eventually printed in Strassburg in 1636. In it, Galileo argued for a recognition of two languages, the precise language of science and the vague everyday language in which God had composed the Bible in order to make it comprehensible to the ordinary people. And Galileo consciously strove to make the issue itself comprehensible to ordinary people: the *Letter to the Grand Duchess Christina* (like all his works after *Sidereus nuncius*) was written in Italian, so that any literate person could read it. This was an invitation to laymen to make up their own minds about the meaning of the Scriptures, a practice common among Protestants, but one specifically forbidden to Catholics by the Council of Trent.

Foscarini's and Galileo's books, along with the continued agitation by Galileo's enemies, brought the Copernican issue to a head. Afraid that the Copernican theory might be condemned, Galileo decided to go to Rome, arriving there in December 1615. While busy promoting his cause, he now also wrote a tract on his theory of the tides (see below), arguing that the tides prove the motion of the Earth. Perhaps here was the physical proof that Bellarmine had demanded. His efforts were, however, in vain. Upon being shown this purported proof, in February 1616, Pope Paul V asked the theological consultants of the Holy Office for a judgment on Copernicanism. Within a few days the consultants (who may not have seen the 'proof') had their answer ready:

Proposition 1: *the Sun is the centre of the world and completely immovable by local motion.*

Censure: *all declared the said proposition to be foolish and absurd in philosophy and formally heretical, because it expressly contradicts the doctrine of the Holy Scripture in many passages, both in their literal meaning and according to the general interpretation of the Fathers and the Doctors of the Church.*

Proposition 2: *the Earth is not the centre of the world, nor immovable, but moves according to the whole of itself, and also with a diurnal motion.*
Censure: *all declared that this proposition receives the same censure in philosophy, and that, from a theological standpoint, it is at least erroneous in faith.*

It is remarkable that the consultants had not the least compunction about pronouncing on the philosophical, that is, scientific, merits of the Copernican theory!

This decision was quickly approved by the Holy Office, and an official decree to this effect was sent to branches of the Holy Office all over the world. The decree also stated that Copernicus's *De revolutionibus* was suspended until corrected. It, Foscarini's book, and an earlier commentary on the Book of Job by Diego de Zuñiga (1584) were placed on the *Index of Forbidden Books*. In 1620, very exceptionally, a dozen specific corrections were ordered to be made in existing copies of *De revolutionibus*, to make it appear more hypothetical and not a description of physical reality; over half the copies in Italy were so corrected (Figure 6.11), but very few in other countries.

Although the 1616 decree was not a public document, and still less was it binding on Catholics as of faith, the Church authorities had taken a significant step towards the very mistake from which Galileo had hoped to save them: an anti-Copernican stand, in a question he considered an essentially scientific issue. The proceedings had not involved Galileo personally, and he was not mentioned in the decree. Cardinal Bellarmine, as Head of the Holy Office, did, however, call him in to acquaint him with the decree and to instruct him on what would henceforth be permissible. Because of their importance in the later process against Galileo, the exact proceedings of this audience have been a subject of much speculation among historians. Upon Galileo's request, Bellarmine provided him with a signed affidavit saying that during the audience Galileo had not abjured his convictions or been forced to do penance, but had merely been hac ordinatione admirandam mundi symmetriam, ac certum harmoniæ nexum motus & magnitudinis orbium: qualis alio modo reperiri non potest. Hic enim licet animaduertere, non segniter contemplanti, cur maior in Ioue progressus & regressus appareat, quàm in Saturno, & minor quàm in Marte: ac rursus maior in Venere quàm in Mercurio. Quodq́ frequentior appareat in Saturno talis reciprocatio, quàm in Ioue: rarior adhuc in Marte, & in Venere, quàm in Mercurio. Præterea quòd Saturnus, Iupiter, & Mars acronycti propinquiores sint terræ, quàm circa eorū occultationem & apparitionem. Maximè uero Mars pernox factus magnitudine Iouem æquare uidetur, colore dumtaxat rutilo discretus: illic autem uix inter secundæ magnitudinis stellas inuenitur, sedula obseruatione sectantibus cognitus. Quæ omnia ex eadem causa procedit, quæ in telluris est motu. Quod autem nihil eorum apparet in fixis, immensam illorum arguit celsitudinem, quæ faciat etiam annui motus orbem siue eius imaginem ab oculis euanescere. Quoniam omne uisibile longitudinem distantiæ habet aliquam, ultra quam non amplius spectatur, ut demonstratur in Opticis. Quòd enim à supremo errantium Saturno ad fixarum sphæram adhuc plurimum intersit, scintillantia illorum lumina demonstrant. Quo indicio maximè discernuntur à planetis, quodq́ inter mota & non mota, maximam oporteat esse differentiam.

De hypothesi triplicis motus terræ, & eiusq́ demonstratione. — De triplici motu telluris demonstratio.

Cvm igitur mobilitati terrenæ tot tantæq́ errantium syderū consentiant testimonia, iam ipsum motum in summa exponemus, quatenus apparentia per ipsum tanquam hypotesim demonstrentur, quem triplicem omnino oportet admittere. Primum quem diximus νυχθημερινὸν à Græcis uocant, diei noctisq́ circuitum proprium, circa axem telluris, ab occasu in ortum uergentem, prout in diuersum mundus ferri putatur, æquinoctialem circulum describendo, quem nonnulli æquidialem dicunt, imitantes significationem Græcorum, apud quos

NICOLAI COPERNICI

net, in quo terram cum orbe lunari tanquam epicyclo contineri diximus. Quinto loco Venus nono mense reducitur. Sextum denique locum Mercurius tenet, octuaginta dierum spacio circu currens. In medio uero omnium residet Sol. Quis enim in hoc

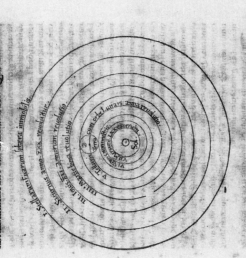

pulcherrimo templo lampadem hanc in alio uel meliori loco poneret, quàm unde totum simul possit illuminare? Siquidem non inepte quidam lucernam mundi, alij mentem, alij rectorem uocant. Trimegistus uisibilem Deum, Sophoclis Electra intuentem omnia. Ita profecto tanquam in solio regali Sol residens circum agentem gubernat Astrorum familiam. Tellus quoque minime fraudatur lunari ministerio, sed ut Aristoteles de animalibus ait, maximam Luna cum terra cognationem habet. Concipit interea à Sole terra, & impregnatur anno partu. Inuenimus igitur sub hac

6.11. Galileo's copy of the 1566 edition of Copernicus's *De revolutionibus*, with two of the ten changes demanded by the Roman authorities in 1620: in Book I, Chapter 10, the final sentence ("So vast, without question, is the divine handiwork of the Almighty") is deleted, and the title of Chapter 11 is amended by the addition of the phrase "De hypothesi", so that it reads "On the hypothesis of the three-fold motion of the Earth and its explication".

notified that, in view of the decree, he could no longer defend or hold the Copernican doctrine – that is to say, he could no longer believe that the Copernican system was true and defend this belief. An unsigned minute of transactions in the file of the Inquisition, however, contains the statement that during the audience Galileo was told by Michelangelo Segizi, Commissary-General of the Holy Office, no longer to hold, teach, or defend the Copernican doctrine in any way whatsoever, orally or in writing. According to this document, then, he could no longer teach or discuss the Copernican theory even as a mathematical hypothesis. When Galileo's file was brought out again seventeen years later, this last document became an important weapon against him.

5. The *Dialogue*

The battle for the Copernican theory had been lost in Italy, and Galileo had been effectively silenced. It now seemed that his book on cosmology, mentioned as long ago as 1610, would remain unwritten. Galileo kept his silence on the Copernican issue for eight years. During this time he became embroiled in a dispute about the nature of comets, occasioned by three comets that appeared in 1618 and 1619. While earlier parts of the exchange were written by his students, and merely supervised by Galileo himself, the final blast *The Assayer* came from his own pen in 1623. *The Assayer* is often considered Galileo's greatest work on the philosophy of science.

In 1624 Pope Paul V died. His successor, Urban VIII, was Maffeo Barberini, who as a cardinal had shown great interest in Galileo's work and had supported him in the Church. Galileo now went to Rome and had a number of audiences with the Pope, during which he was given to understand that he might discuss the Copernican theory in print as long as he treated it only as a mathematical hypothesis and refrained from theological arguments. He first tested the waters in 1624 with a little treatise *Letter to Ingoli*, which was a reply to the manuscript *De situ et quiete Terrae contra Copernici systema*, written in 1616 by Francesco Ingoli, the director of the new office of the Propagation of Faith. Ingoli used both theological and physical arguments, but Galileo limited himself to a critique of the latter. While he argued that the rightness of the Copernican system appeared in-

controvertible, Galileo prefaced his arguments with a disclaimer that he would never maintain that the Copernican doctrine was true. This reply circulated widely in manuscript form in Florence and Rome. The authorities in Rome, including Urban VIII, who read it with interest, did not raise the least objection. Galileo, now sixty years old, took courage and began to put together his long planned book on cosmology. For various reasons, not the least of which had to do with his health, Galileo did not finish his *magnum opus* until the beginning of 1630. It took eighteen months to get the approval from the censors in Rome and Florence, and the book finally appeared in 1632 under the title *Dialogue Concerning the Two Chief World Systems, the Ptolemaic and the Copernican*.

As its title suggests, the *Dialogue* is cast in the form of a conversation. Galileo used three protagonists: Salviati, the spokesman for the new science, Sagredo, the intelligent layman, and Simplicio, the Aristotelian. Their conversation takes place over four days in Sagredo's house in Venice. The First Day was a systematic attack on Aristotelian natural philosophy and cosmology, which was the foundation on which Ptolemy's astronomy was built. Galileo started with a general discussion of Aristotle's notions about the universe, arguing that his ideas such as the perfection of the heavens and the corrupt nature of the earthly region do not stand up to logical analysis. He next moved on to some general considerations of motion, showing by means of experiments (or at least thought experiments) that the Aristotelian distinctions between rest and motion, and between natural and violent motion, cannot be maintained. Then, with the help of recent discoveries and measurements, Galileo launched an experiential attack against the perfection and immutability of the Aristotelian heavens. A long, detailed discussion of the telescopic appearance of the Moon showed the reader that there can be no doubt that the Moon's surface appears to us just as the Earth's surface would appear from a distance. A similar discussion of sunspots demonstrated that they are indeed on the Sun or close to it, and that their generation and dissipation proves that there is change in the heavens, a fact also proved by the new stars of 1572 and 1604 and by the parallax measurements of comets.

The Second Day began with an onslaught on the authority of Aristotle and a ridiculing of his slavish

6.12. The frontispiece of *Dialogue Concerning Two Chief World Systems*. Those depicted are not the characters in Galileo's dialogue, but (from left to right) Aristotle, Ptolemy, and Copernicus.

DIALOGO

DI

GALILEO GALILEI LINCEO

MATEMATICO SOPRAORDINARIO

DELLO STVDIO DI PISA.

E Filofofo, e Matematico primario del

SERENISSIMO

GR.DVCA DI TOSCANA.

Doue ne i congreffi di quattro giornate fi difcorre
fopra i due

MASSIMI SISTEMI DEL MONDO
TOLEMAICO, E COPERNICANO;

*Proponendo indeterminatamente le ragioni Filofofiche, e Naturali
tanto per l'vna, quanto per l'altra parte.*

CON PRI VILEGI.

IN FIORENZA, Per Gio:Batifta Landini MDCXXXII.

CON LICENZA DE' SVPERIORI.

6.13. The title-page of Galileo's *Dialogue Concerning Two Chief World Systems* (1632). The two systems are the Ptolemaic (against which Galileo had powerful arguments) and the Copernican: he makes little mention of the Tychonic and semi-Tychonic systems which by then had superseded the Ptolemaic.

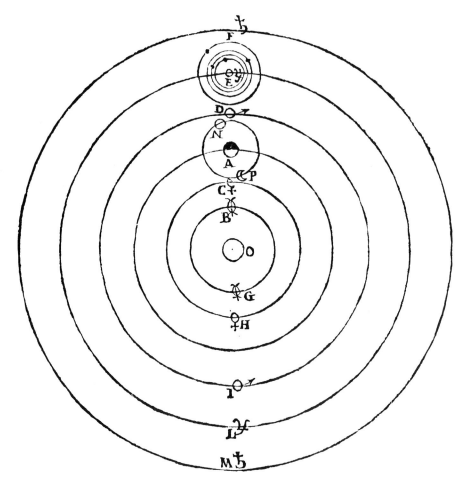

6.14. The Copernican system as depicted in Galileo's *Dialogue Concerning Two Chief World Systems*. On the Copernican view the Earth had been anomalous in having a satellite, but Galileo's telescope had shown that at least one other planet also had satellites.

followers. One can interpret Aristotle in many ways; why, one can even find the invention of the telescope in his texts by careful reading and creative interpretation! Then ensues a long discussion of the relativity of motion, showing that the inhabitants of the Earth would not be aware of its diurnal motion, and that this motion served the principle of economy better than having the entire heavens rotate once every twenty-four hours. Galileo here disposed of all the traditional arguments against the Earth's diurnal rotation: that clouds and birds would be left behind, that a powerful wind would constantly blow from the east, that a cannon ball shot toward the west would travel much further than the same ball shot towards the east, and that this whirling Earth would throw off bodies. He did

this by destroying Aristotle's ideas of motion, especially projectile motion, which had always been the Achilles's heel of his natural philosophy. Among other novelties, Galileo stated here, for the first time in print, the time-squared law for falling bodies and the isochronous nature of the pendulum.

Galileo began the Third Day with a detailed investigation of the parallax measurements of the New Star of 1572, disposing of the claims of traditional astronomers that these measurements showed that the New Star was below the Moon. He next marshalled the arguments for the Copernican system, restating all the advantages Copernicus himself had already mentioned and also now adducing the evidence of the telescope: the phases of Venus, the varying apparent sizes of Venus and

Mars as revealed by the telescope, the satellites of Jupiter (which showed that the Earth was not the only planet to have a moon), and the complicated motions of sunspots which could be explained with reasonable simplicity only if one assumed the motion of the Earth. Galileo also disposed here of the traditional overestimation of the apparent sizes of the planets and fixed stars (see Chapter 7). He argued that the lack of a detectable annual stellar parallax was merely due to the inaccuracy of instruments. This part of the book ended with a discussion of the magnetic philosophy of William Gilbert (see Chapter 4).

The Fourth and final Day of the *Dialogue* contained Galileo's 'proof' of the motion of the Earth. He argued that while all other terrestrial events were indifferent as to the Earth's rest or motion, the tides were not. After brushing aside all the previous attempts at explanation, including ones involving the Moon, he stated his theory: a point on the Earth's surface that is at midnight moves with a linear velocity that is equal to the velocity of annual motion plus the linear velocity due to the diurnal rotation; a point that is at noon, however, moves with a linear velocity equal to the difference between the two. This alternate acceleration and deceleration causes an oscillation in the water covering the Earth's surface. If the Earth had no motions, there would be no tides and, conversely, the existence of the tides proved the complex motion of the Earth. Ingenious as this explanation was, it was not to prove influential (nor was it of course correct).

The *Dialogue* is a brilliant attack on Aristotle and Ptolemy, and a marshalling of the evidence for the Copernican system. The dialogue form allowed Galileo to make the fullest use of his persuasive power. As he had so often done in conversation, he built up the strength of the traditional arguments, in the mouth of Simplicio, until they seemed invincible, and only then did he destroy them. It is fair to say that after the publication of the *Dialogue* Aristotle and Ptolemy were never the same. But the book's persuasiveness also brought trouble to its author.

6. The trial of Galileo

Throughout the *Dialogue* Galileo avoided theological arguments, and there were a few passages in which he *pro forma* distanced himself from the question of the ultimate reality of the Copernican system. Thus in the Second Day, Salviati, who clearly represents Galileo himself, interrupts the proceedings as follows:

Before going further I must tell Sagredo that I act the part of Copernicus in our argument and wear his mask. As to the internal effects upon me of the arguments which I produce in his favour, I want you to be guided not by what I say when we are in the heat of acting out our play, but after I have put off the costume, for perhaps then you shall find me different from what you saw of me on the stage.

Although Urban VIII had encouraged Galileo to write a book on cosmology and had been looking forward to reading it, he had not retracted the decree of 1616. He had made it clear to Galileo that God's power could not be limited by scientific theories: whatever theories we might construct to explain the appearances, God could easily have attained the same effects by an entirely different and unknown construction. Natural philosophy could, therefore, never arrive at certain conclusions, but only probable ones. This meant that Galileo could not write a book in which he *demonstrated* the Copernican system to be true and certain; he had to limit himself to hypothesizing. Galileo had originally planned to entitle his book *On the Ebbing and Flowing of the Sea*, but the Pope had insisted that the title should not feature a purported proof of the reality of the Copernican system. Galileo had also been told by the censors to add a preface in which the hypothetical nature of his argument was stressed, and to add at the end of the book the Pope's own argument on the uncertainty of scientific demonstrations. Galileo had done so, but strictly *pro forma*, with the result that these additions, as well as the few other disclaimers in the book, struck the reader as disingenuous. Thus, at the very end of the dialogue, Galileo put the Pope's position into the mouth of Simplicio, the Aristotelian who throughout the book had been made to look foolish. After complimenting the other two interlocuters on the persuasiveness of their arguments, Simplicio stated that he was nevertheless not entirely convinced, because,

. . . keeping always before my mind's eye a most solid doctrine that I once heard from a most eminent and learned person [that is Urban VIII] and before which

one must fall silent, I know that if asked whether God in His infinite power and wisdom could have conferred upon the watery element its observed reciprocating motion using some other means than moving its containing vessels, both of you would reply that He could have, and that He would have known how to do this in many ways which are unthinkable to our minds.

The publication of the *Dialogue* raised an outcry of protest from conservative academics and clerics. Pope Urban VIII now turned on Galileo, perhaps because he had been acquainted by the Holy Office of the existence of the (unsigned) record of an admonition supposedly given to Galileo in 1616, not to hold, defend, or teach the Copernican doctrine in any way. Presumably the Pope was also furious about the book itself. Early in October 1632 the sale of the *Dialogue* was suspended and Galileo ordered to appear before the Inquisition in Rome, but because of illness and the infirmities of his advanced age (sixty-eight) he did not arrive until February 1633. For two months he stayed at the Tuscan embassy, and then he was moved into a spacious apartment in the Holy Office; he was never in a prison cell.

Galileo was charged with disobeying the order of 1616, but he was able to produce the affidavit, in the late Cardinal Bellarmine's own hand and dating from shortly after the events in 1616, certifying that he had merely been admonished not to hold or defend the Copernican doctrine. His defence was that he had treated the heliocentric theory only as an hypothesis. This defence could not save him, however. There ensued 'plea-bargaining' negotiations and an agreement was reached whereby Galileo would admit that although he did not believe in the reality of the Copernican system, out of vanity or ambition perhaps he had been too enthusiastic in its defence; to someone ignorant of his intentions, it might appear, therefore, that his arguments for the Copernican system were victorious. In return he was promised a lenient sentence.

In June, on orders of the Pope, Galileo was examined as to his intentions when he wrote the *Dialogue*. He denied that he intended to persuade the reader of the truth of the Copernican system. His examiners went through the formality of showing him the instruments of torture, but there was no question of actually proceeding with torture. Within a week he was brought again before the

Inquisition to hear his sentence read. The sentencing document charges Galileo with being "vehemently suspected" not only of the heresy of having held and defended the stability of the Sun and the motion of the Earth, but also the heresy of believing "that any opinion may be held and defended as probable after it has been declared and defined contrary to the Holy Scripture". Of these charges he would be absolved, provided that he abjured in the appropriate manner. He was, therefore, not convicted of heresy as is sometimes stated. However, his *Dialogue* was prohibited, and he was sentenced to life imprisonment by the Holy Office, with the (trivial) added task of having to recite once a week for three years the seven penitential psalms.

According to the formula agreed on, Galileo then knelt and read his abjuration. In order to remove the vehement suspicion of heresy, Galileo stated that he abjured, cursed, and detested all his errors and heresies, and that in the future he would never again give the Holy Office cause for similar suspicion. This ended the process, and a few days later Galileo was released to the custody of the Archbishop of Siena, who was his personal friend. His daughter, Sister Maria Celeste, was given permission to recite the penitential psalms in his stead. Galileo remained in Siena until the end of 1633 and was then moved to a spacious house in Arcetri, a village in the hills overlooking Florence. Here the ageing Florentine spent the remainder of his days under house arrest. Over the next several years he finished his last book, *Discourses on Two New Sciences*, which was published (with an imprimatur) in Leiden in 1638. In this work Galileo finally summed up his researches in the science of motion, laying the foundations for the mechanics that would eventually reach maturity in the *Principia* of Isaac Newton (see Chapter 13).

7. The aftermath of the trial

The scientific climate in Catholic areas, especially Italy, was, of course, profoundly affected by the decree of 1616 and the sentence of 1633. Technically speaking, neither decision had been made dogmatic by the Pope or a Council of the Church, and therefore the Copernican theory was not a matter of faith, and believing in its reality was not heresy. Catholics merely owed external obedience to the decree of 1616 and could therefore not hold or defend the Copernican theory. In practice, in

Italy henceforth cosmological discussions involving the Copernican theory were to be found only in books written by clerics to attack and often ridicule heliocentric astronomy. Scriptural arguments were used freely in these diatribes. In this tradition we must place the *Almagestum novum* (1651) of Giambattista Riccioli (1598–1671) although its sophistication and balance otherwise sets it apart.

Riccioli, a professor at the Jesuit college in Bologna, set himself the task of reviewing all of astronomy from the ancient Greeks to his own times. He completely mastered the astronomical literature and treated all aspects of it in his two massive volumes. In Bk IX, "On the System of the World", Riccioli reviewed the recent cosmological arguments. He began with an astronomical exegesis of the first four days of creation, and then examined some fundamental Aristotelian propositions, concluding, for instance, that the visible heavens are corruptible, that the heavens containing the planets are fluid while the sphere of the fixed stars is solid, and that heavenly bodies are not animated. In his arguments he cited ancients such as Aristotle along with such moderns as Kepler, René Descartes (1596–1650), and Galileo, and he freely mixed scientific and theological arguments.

Section 3 of Bk IX contains a complete review of all the geocentric systems put forward since Antiquity. Not surprisingly, Riccioli rejected the Ptolemaic system in favour of the Tychonic, with one modification (see Chapter 3): Riccioli made the Earth, not the Sun, the centre of the motions of Jupiter and Saturn. The next section is an exhaustive examination of geokinetic astronomy. Here Riccioli gives all the arguments for the Copernican system and marshals all possible evidence, scientific and scriptural, against it. As his final argument he presents the documents of the decisions against Copernicus: the decree of 1616 including the surprisingly short list of the corrections that had to be made to *De revolutionibus*, Galileo's sentence, and his abjuration. Riccioli ended this section (Vol. 2, p. 500) with the inevitable conclusion: "It is entirely to be asserted that the Earth is situated naturally immobile in the centre of the world, and that the Sun moves about it with a diurnal as well as an annual motion."

Thorough and up-to-date though it was, Riccioli's treatment of the cosmological controversy was a sterile exercise. Astronomical issues could no longer be settled by a preponderance of scientific and scriptural authority or by any number of decrees from Rome. Astronomers all over Europe used Riccioli's book not for its arguments against the Copernican hypothesis but for its encyclopedic treatment of technical astronomical issues.

Italy was the only Catholic region in the forefront of science where cosmological speculation was severely hampered by the Galileo affair. In Catholic France the intellectual climate was considerably less restricted. Two of the foremost defenders of the reality of the Copernican system, Pierre Gassendi (1592–1655) and Ismaël Boulliau (1605–94), were both clerics, and they published their Copernican astronomical works without serious interference. Yet Descartes, whose vortex cosmology was based on the Copernican theory, chose to spend much of his productive career in the Protestant Dutch Republic, where freedom of conscience and expression were, at this time, the most advanced.

Whereas originally the Copernican theory had been criticized by Protestant leaders such as Luther and Melanchthon for scriptural reasons, by the middle of the seventeenth century the Copernican issue hardly raised an eyebrow in Protestant regions. The overwhelming majority of Protestant astronomers, or at least the creative Protestant astronomers, were now Copernicans. The condemnation of Galileo and the prohibition of the Copernican doctrine was, to them, further evidence of the perfidy of Rome, while to the Church hierarchy in Rome, the Protestant swing to the Copernican system was merely further evidence of the danger of individual interpretation of the Scriptures. Astronomy had now become divided along religious lines in a way it had never been before. Not until the eighteenth century did this religious polarization of astronomy begin to disappear.

8. Telescopic astronomy after Galileo

A number of the phenomena discovered by Galileo in 1609–10 lay at the limit of visibility with the naked eye. There is a long, if occasional, record of observations of sunspots before the telescope; the Moon exhibits large spots to even a dull eye, and a sharp eye can detect considerable detail in our neighbour; and there is evidence that the satellites of Jupiter (stellar magnitudes 5–6 at their brightest) and even the phases of Venus were seen under special conditions. In one leap, the telescope transcended this limit and allowed these phenomena to be studied systematically and thoroughly. The initial harvest was therefore spectacular, but it was not sustained.

The useful limit of magnification of the so-called Dutch or Galilean telescope is about twenty, and this had already been achieved by Galileo in the autumn of 1610. This instrument produced the first wave of discoveries, but its potential had been exhausted very quickly. For the time being, no higher magnifications were possible for technical reasons (see Volume 3), and therefore only a few minor further discoveries were made in the next several decades. Among these we may count the description by Simon Mayr of the telescopic appearance of the Andromeda Nebula in his *Mundus Iovialis* of 1614, and that by Johann Baptist Cysat of the nebula in the sword of Orion in his book on the comet of 1618–19. This last 'discovery' (itself anticipated by Peiresc in November 1610) was, however, so little known that Huygens thought he had been the first to see the nebula half a century later.

There were, of course, some research programmes that sprang from these early discoveries. Galileo quickly recognized the importance of Jupiter's satellites for the problem of measuring longitude at sea and, whenever possible, during the rest of his career, he made observations and worked at making accurate tables of their motions (see Chapter 9). But he was not an astronomer for whom regular observations of the heavens are a way of life. After about 1612 he made observations of phenomena other than Jupiter's satellites only occasionally. Others, such as Harriot, Mayr, and Peiresc, also made studies of Jupiter's satellites. The study of sunspots was pursued systematically only by Scheiner, who, despite his disastrous start, quickly became the expert on this subject. His massive *Rosa Ursina* of 1630 was, for all its vituperation against Galileo, the standard work on that subject for over a century. The Moon, subject of the first telescopic observations by Harriot and Galileo, languished in neglect until the 1630s, when lunar mapping became an important research area (see Chapter 8). Kepler's admonition to astronomers to observe the transits of Mercury and Venus of 1631 drew only a handful of observers. Three witnessed

Mercury's transit; the Venus transit occurred in Europe after the Sun had already set. The subsequent Venus transit of 1639 was witnessed only by Jeremiah Horrocks and William Crabtree in Lancashire (see Chapter 7).

Significant new discoveries became possible only after the Dutch telescope had been replaced by the astronomical telescope whose large field of view allowed great magnifications (see Volume 3). Francesco Fontana (1580–1656) was the first to claim new phenomena revealed by his more powerful telescopes, in the late 1630s. His observations of the Moon and Saturn circulated in letters, and it is interesting to note that Galileo, now well over seventy, denied that anything new was revealed by these more powerful instruments. Early fantastic descriptions of Fontana's observations can be found in Matthias Hirzgarter's *Detectio dioptrica* of 1643. Fontana finally published his observations in 1646 in a book entitled *Novae coelestium terrestriumque rerum observationes*, or *New Observations of Celestial and Terrestrial Things*. This is the first 'picture book' of telescopic astronomy, with large illustrations, accompanied by a minimum of text. Fontana showed the phases of Mercury, belts on Jupiter, many spectacular (if imperfect) aspects of Saturn and its ansae, and perhaps a surface marking on Mars. After Fontana, detailed illustrations of the telescopic appearance of heavenly bodies became more common.

As the power and quality of telescopes increased, new phenomena became detectable. By the middle of the seventeenth century a number of observers in Europe had telescopes (of varying qualities) with magnifications of 50 or more, with a useful field of view. At this point telescopic astronomy moved decisively out of the infancy that had been so utterly dominated by Galileo. Now the race was on again: more powerful instruments led to new discoveries which brought their authors prestige and patronage.

The first major discovery comparable to those of Galileo came in 1655, when Huygens found a satellite of Saturn (now called Titan), using a 50-powered telescope. Others quickly verified his discovery, and Johannes Hevelius in Poland and Christopher Wren in England testified that they had seen it before, but had taken it to be a fixed star. In 1659 Huygens published *Systema Saturnium* in which he gave refined elements of the orbit of the satellite and revealed his ring theory to explain the enigma of Saturn's appearances.

By far the greatest discoverer, however, was Gian Domenico Cassini (Cassini I, 1625–1712), who began his astronomical career in Bologna and moved to Paris in 1669 to become director of the new Royal Observatory. Using instruments by the Roman telescope maker Giuseppe Campani (1635–1715), Cassini discovered surface markings on Mars and Jupiter and managed to deduce the rotation periods of these planets in the 1660s. After taking up residence in Paris, Cassini discovered two satellites of Saturn at its edgewise appearance of 1671–72 (Rhea and Iapetus), the division named after him in that planet's ring (in 1675), and two more Saturnian satellites (Tethys and Dione) during the planet's edgewise appearance in 1684. Cassini was also the first to observe Jupiter's oblateness and belts or zones on Saturn. His moonmap of 1679–80 (see Chapter 8, Figure 8.16), one of the most detailed and accurate produced in the seventeenth century, gives a good indication of the exceptional level of sophistication to which Cassini had brought telescopic astronomy.

Cassini's discoveries in Paris were made with very long telescopes with compound eyepieces. His discovery of Tethys and Dione was made with telescopes without tubes (see Volume 3). By the end of the seventeenth century these very long anachromatic refractors, with aperture ratios in excess of f/100, had exhausted their potential, and this second wave of discovery came to an end. The next important discovery came almost a century later, when William Herschel was able to exploit the potential of the reflecting telescope.

Further reading

Giorgio de Santillana, *The Crime of Galileo* (Chicago, 1955)

Stillman Drake, *Telescopes, Tides, and Tactics* (Chicago, 1983)

Galileo Galilei, *The Starry Messenger*, in *Discoveries and Opinions of Galileo*, transl. by Stillman Drake (Garden City, NY, 1957). This book also contains Galileo's *Letter to the Grand Duchess Christina* and excerpts from his *Letters on Sunspots* and *The Assayer*

Galileo Galilei, *Dialogue Concerning the Two Chief World Systems*, transl. by Stillman Drake (Berkeley, Calif., and Cambridge, 1953)

Galileo Galilei, *Sidereus Nuncius or the Sidereal Messenger*, transl. by Albert Van Helden (Chicago, 1989).

Jerome J. Langford, *Galileo, Science and the Church* (Ann Arbor, 1966)

Ernan McMullin (ed.), *Galileo: Man of Science* (New York, 1967)

Olaf Pedersen, Galileo and the Council of Trent: the Galileo affair revisited, *Journal for the History of Astronomy*, vol. 14 (1983), 1–29

Albert Van Helden, The invention of the telescope, *Transactions of the American Philosophical Society*, vol. 67, pt 4 (1977)

Ewen A. Whitaker, Galileo's lunar observations and the dating of the composition of *Sidereus nuncius*, *Journal for the History of Astronomy*, vol. 9 (1978), 155–69

7

The telescope and cosmic dimensions
ALBERT VAN HELDEN

1. Pretelescopic cosmic dimensions

Copernicus's generation inherited from the Middle Ages, and through them from Antiquity, a consensus as to the sizes and distances of the celestial bodies. The values – supposedly known with great accuracy – resulted from applications of the geometry developed by the Greeks.

Some of the techniques involved were straightforward: Eratosthenes in the third century BC had shown how to derive the size of the Earth from observations of a star from two locations at a known distance north–south of each other; Aristarchus, his older contemporary, had shown how observations of the angle Moon–Earth–Sun when the Moon was at quadrature could give the relative distances of the Sun and the Moon, and had concluded that the ratio lay between 18 and 20. Other applications were more complex. Taking advantage of the fact that the apparent diameters of the Sun and Moon are roughly equal, Aristarchus constructed an eclipse diagram that could yield the solar and lunar distances. This eclipse method was further refined and applied by Hipparchus (*c.* 125BC) to determine the absolute lunar distance, while Ptolemy (*c.* AD150) used it in his *Almagest* to find the absolute solar distance (Figure 7.1).

Ptolemy first found the absolute distance of the Moon from parallax observations. Then in his eclipse diagram (Figure 7.1) he adopted the following parameters:

1. The Moon's apparent diameter is equal to the Sun's when the Moon is at its apogee distance, $64\frac{1}{6}$ Earth radii (e.r.).
2. The Moon's and Sun's apparent diameters are then 31′ 20″.
3. The width of the Earth's shadow cone at the Moon's distance is 'very little' less than $2\frac{3}{5}$ times as wide as the Moon.

These parameters produced a solar distance of 1210 e.r., a figure that (whether by accident or design) confirmed Aristarchus's ratio of solar to lunar distances. Ptolemy now knew the absolute sizes and distances of the two luminaries.

He took up the problem of planetary distances and sizes in a more speculative and cosmological work, his *Planetary Hypotheses*, written after the *Almagest*. According to Aristotelian natural philosophy, the cosmos is a plenum, and there could thus be no spaces between the spherical shells carrying the planets. From the planetary models developed in the *Almagest*, the ratio of apogee to perigee distance for each heavenly body could be determined, and thus, if any one absolute distance was known, all absolute distances in the cosmos could be calculated. This is what Ptolemy did. Knowing that the Moon's apogee distance was $64\frac{1}{6}$ e.r., and knowing that (in his convention) Mercury follows immediately above the Moon, he knew that Mercury's perigee distance had to be $64\frac{1}{6}$ e.r. From the Mercury model in the *Almagest* he found the ratio of apogee to perigee distances of Mercury to be 88:34. This made Mercury's apogee distance 166 e.r., which was also Venus's perigee distance, and so on, right up to the sphere of Saturn whose apogee distance he found to be 19 856 e.r. He assumed the fixed stars, lying beyond the sphere of Saturn, to be removed from us by 20 000 e.r. For the Sun this nesting-sphere method produced distances that chanced to be close to the mean distance found by the eclipse method in the *Almagest*.

As to the sizes of heavenly bodies, the apparent diameters of the Sun and Moon were easily determined to be about $\frac{1}{2}$°, but the apparent diameters of the planets and fixed stars were not so easily determined. It appears that Hipparchus was the first to give naked-eye estimates of some of these bodies. Ptolemy followed his example, making Venus's

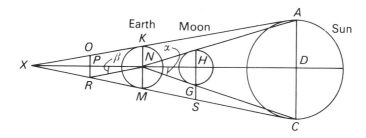

7.1. Ptolemy's use of an eclipse diagram to determine the solar distance.

Table 7.1 *Cosmic dimensions according to Al-Farghānī (c. AD 850)*

Body	Absolute distance in e.r.			Apparent Diameter at mean distance (Sun = 1)	Actual Diameter (Earth = 1)	Volume (Earth = 1)
	Least	Greatest	Mean			
Moon	$33\frac{1}{2}+\frac{1}{20}$	$64\frac{1}{6}$	$48\frac{5}{6}$	$[1\frac{1}{3}]$	$\frac{1}{3\frac{2}{5}}$	$\frac{1}{39}$
Mercury	$64\frac{1}{6}$	167	$115\frac{1}{2}$	$\frac{1}{15}$	$\frac{1}{28}$	$\frac{1}{22000}$
Venus	167	1 120	$643\frac{1}{2}$	$\frac{1}{10}$	$\frac{1}{3\frac{1}{4}}$	$\frac{1}{37}$
Sun	1 120	1 220	1 170	1	$5\frac{1}{2}$	166
Mars	1 220	8 876	5 048	$\frac{1}{20}$	$1\frac{1}{6}$	$1\frac{1}{2}+\frac{1}{8}$
Jupiter	8 876	14 405	11 640	$\frac{1}{12}$	$4\frac{1}{2}+\frac{1}{16}$	95
Saturn	14 405	20 110	17 258	$\frac{1}{18}$	$4\frac{1}{2}$	91
Fixed stars						
1st mag	—	—	20 110	$\frac{1}{20}$	$4\frac{1}{2}+\frac{1}{4}$	107
2nd mag	—	—	20 110	not given	not given	90
3rd mag	—	—	20 110	not given	not given	72
4th mag	—	—	20 110	not given	not given	54
5th mag	—	—	20 110	not given	not given	36
6th mag	—	—	20 110	not given	not given	18

1 e.r. = 3 250 miles.

angular diameter about one-tenth the Sun's (without specifying at what distance), Jupiter's one-twelfth, those of first-magnitude stars one-twentieth, and so on. Knowing the distances and the apparent sizes of all bodies in the cosmos, he could now also compute their absolute sizes.

Ptolemy's Muslim successors tried, in various ways, to rid his scheme of sizes and distances from some minor imperfections. In doing so, they made small changes in Ptolemy's numbers. The variant scheme of sizes and distances ascribed to Al-Farghānī (ninth century AD) became the common currency of the Latin West, beginning in the twelfth century. In whole or in part it can be found at all levels of literate European culture during the

Middle Ages and Renaissance. We find it in Christopher Clavius's commentaries on *The Sphere* of Sacrobosco, and we can be sure that Galileo learned it as an undergraduate. Al-Farghānī's variant of Ptolemy's scheme of sizes and distances is shown in Table 7.1.

Although Copernicus changed cosmology profoundly, it chanced that cosmic dimensions were not greatly affected by this change (with the exception of the distance of the fixed stars). In Copernicus's universe there were empty spaces between the successive planetary spheres, but since he followed Ptolemy closely in his solar distance – he used the eclipse diagram to determine it – the sphere of Saturn came out to be about 40% smaller than Ptolemy had made it. Tycho Brahe followed Copernicus for his planetary distances and Ptolemy for the apparent sizes (on which Copernicus had nothing to say), so that his system of cosmic dimensions was also not very different from the Ptolemaic system. The deeply ingrained Ptolemaic expectations about sizes and distances in the cosmos had thus changed very little by 1600. The telescope, however, caused a radical change.

2. Galileo on sizes and distances

In *Sidereus nuncius* (1610), in which he announced his first and epoch-making telescopic discoveries, Galileo Galilei (1546–1642) commented on what the 'fixed' stars and planets had in common and how they differed when seen through the telescope. Planets and stars, he wrote, are not magnified to the same extent as the Sun and Moon. While the naked eye is fooled by the adventitious rays surrounding these bodies into believing them to be large, the telescope strips away the haloes and reveals their actual and much smaller bodies, which it then magnifies in the usual manner. Galileo was saying here that everyone since Hipparchus had been mistaken in their estimates of the apparent diameters of the planets and stars. He went on to point out that while the telescope resolved the planets into little disks, it did not do this for the stars.

Galileo's verification of the Moon-like phases of Venus, later that year, added another important piece of information. Not only did Venus exhibit a full cycle of phases, but its apparent size varied according to its geocentric distance. Over the next several years he verified that Mars's apparent diameter varied in a like manner. Although Ptolemy's models for Venus and Mars predicted the same variations in the apparent diameters, in the traditional scheme (see Table 7.1) the apparent sizes had always been given at mean distance, and, with one or two exceptions, the variations of these sizes with distance had always been ignored. Galileo never tired of pointing out that the traditional apparent diameters were much too large, and that their size variations should be taken into account. Thus, in his sunspot letters of 1613 he took Christoph Scheiner (1573–1650) to task for estimating that Venus's apparent diameter would be at least 3′ (the traditional value) at apogee. He found it to be less than a sixth of a minute at that

distance. Just how Galileo arrived at this astonishingly precise figure for Venus's angular diameter at apogee, in the absence of a micrometer, is a mystery. Ten years later he produced an equally good figure for Jupiter's angular diameter.

In his *Dialogue* (1632) Galileo discussed at length the subject of apparent diameters. Again he hammered home the point that the telescope showed that the traditional estimates of planetary and stellar apparent diameters were grossly in error, and he refused to excuse his predecessors who did not have the telescope: they could have known this by observing Venus at twilight, when it is stripped of its crown of adventitious rays. He argued that the size variations revealed by the telescope were evidence for the Copernican theory, despite the fact that Kepler had pointed out the error of that claim. The full force of his argument about apparent sizes was, however, blunted by the fact that he gave no figures for apparent diameters. He and his contemporaries had been brought up on the traditional scheme, and it would take more than general statements about the errors of the ancients to make his contemporaries abandon this. And when it came to planetary distances, the discussion in the *Dialogue* was conventional. Galileo picked on Scheiner's solar distance of 1208 e.r., a figure that had originated with Erasmus Reinhold (1511–53) and had been recalculated by Michael Mästlin (1550–1631), from whom Scheiner had taken it.

Galileo's contemporaries did little better on apparent diameters. Scheiner kept very close to the traditional figures, but his fellow Jesuit Johann Baptist Cysat made some estimates that were better. Cysat was in contact with Johannes Remus Quietanus, a correspondent of Johannes Kepler (1571–1630), and thus Kepler became aware of the estimates of both Cysat and Remus, a few of which would find their way into his *Epitome*. These figures were different from the traditional values, but they varied so much among themselves that they afforded Kepler great latitude in his harmonic speculations.

3. Kepler's speculations

Kepler's speculations on cosmic dimensions spanned four decades, beginning before the telescope was brought to bear on this issue. In his *Mysterium cosmographicum* (1596) he gave refined values for the relative heliocentric planetary distances, but made little headway in determining the absolute length of the Copernican astronomical unit, the solar distance. Over the next decade he attempted to use the eclipse diagram for this purpose, simplifying the procedure by proving that the sum of the horizontal parallaxes of the Sun and Moon equals the sum of the angular radii of the Sun and the shadow cone at the Moon's distance ($\angle DNC + \angle XNR = \angle MRN + \angle NCM$ in Figure 7.1). In *De stella nova* (1606) and *Astronomia nova* (1609) he chose solar distances between 700 and 2000 e.r. on the basis of these studies, but by about 1610 he had convinced himself that the eclipse method was inherently inaccurate for finding the solar distance, and henceforth he used this method only to find the width of the shadow cone for eclipse predictions. For the time being only a detailed examination of Tycho Brahe's measurements of the diurnal parallax of Mars held out any hope. In his *Astronomia nova* Kepler argued that these measurements showed that Mars's parallax was never greater than 4′, which put a limit of 2′ on the Sun's parallax, equivalent to a solar distance of more than 1700 e.r.

The telescope changed Kepler's ideas on cosmic dimensions only slightly. First, the new instrument held out hope that Aristarchus's venerable method of lunar dichotomy might now be more feasible, and Kepler urged his colleagues to attempt this method. Although several astronomers, from Remus Quietanus to Giambattista Riccioli (1598–1671) and even young John Flamsteed (1646–1719) believed this method to have promise, Kepler himself quickly lost faith in it: the telescope did not help in deciding the exact moment of dichotomy; in fact, it made it more difficult.

Kepler's final thoughts on cosmic dimensions, found in Bk IV (1620) of his *Epitome*, form a seamless cloth in which telescopic estimates of apparent diameters, analyses of Tycho Brahe's measurements of Mars's diurnal parallax, considerations of the problem of atmospheric refraction, principles of Aristotelian physics, and harmonic speculations are carefully woven together. He chose a solar parallax of 1′, which fitted better with his improved corrections for atmospheric refractions. The corresponding solar distance was 3469 e.r., and this meant that the lunar distance, 59 e.r., was very nearly the mean proportional between the Earth's radius and the solar distance. Kepler made the

volumes of the planets proportional to their heliocentric distances, their densities inversely proportional to the square-root of their distances, and the forces pushing them exactly equal. Further, he made the radius of the sphere of Saturn the mean proportional between the Sun's radius and the radius of the sphere of the stars, so arriving at a distance of 60 000 000 e.r. for the stars. Kepler's scheme of sizes and distances was not very influential as regards its details, although it was a very radical alternative to the traditional scheme and thus helped undermine it. His particular method of harmonic speculation was, however, adopted (in simplified form) by a number of astronomers in the seventeenth century.

4. Transits of Mercury and Venus

Where Galileo failed to drive home the surprising smallness of the apparent planetary diameters, Pierre Gassendi (1592–1655) succeeded. Kepler, who had long been interested in the possibility of transits, was able, on the basis of his *Rudolphine Tables* (1627), to predict transits of both Mercury and Venus within a month of each other in 1631. He therefore advised his colleagues to observe these transits, so that better measures of the apparent diameters of Mercury and Venus might be obtained (not, as we might think, to measure the parallaxes of these planets). The transit of Venus was visible only in America, but the Mercury transit was visible in Europe, and there were three observers who were not foiled by the weather or faulty instruments: Gassendi in Paris, Cysat in Ingolstadt, and Remus Quietanus in Ruffach. Of these, Gassendi's observation is the most important.

Gassendi made thorough preparations for the events, and in the case of Mercury he was rewarded with several brief glimpses of the planet on the Sun (Figure 7.2), as the clouds parted intermittently. When he first saw this black dot, Gassendi believed it to be a small sunspot, for he expected Mercury to be much larger. The tiny dot moved much too rapidly to be a sunspot however, and it was this motion that finally convinced Gassendi that he was observing Mercury on the Sun. But Mercury's apparent diameter was, according to his measurements, only about 20″, much smaller than he had expected on the basis of the traditional scheme of sizes. Gassendi shared his amazement with the learned world in a tract entitled *Mercurius in sole*

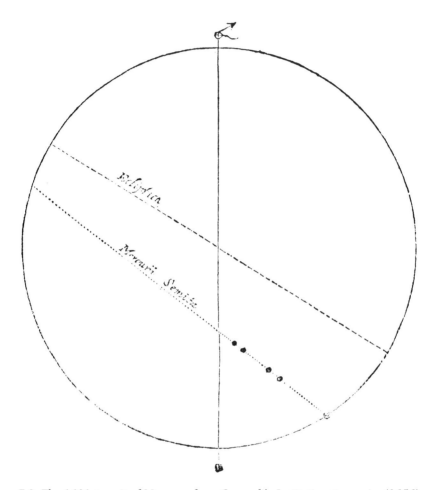

Discus Solis cum trajiciente Mercurio, prout intra obscuram Scenam se inverse in Circulo citra Telescopium objecto exhibuit.

7.2. The 1631 transit of Mercury, from Gassendi's *Institutio astronomica* (1656).

visus et Venus invisa (1632), written as an open letter to the German astronomer Wilhelm Schickard. Schickard had missed the transit because, misled by Kepler's forecast of the sizes of Mercury and Venus, he had used a camera obscura. In his reply, Schickard shared Gassendi's surprise at Mercury's "entirely paradoxical smallness", and (like Cysat, who had witnessed the transit) tried to explain it away by optical arguments. Remus had found Mercury's apparent diameter to be even somewhat smaller than had Gassendi, and he adopted his own measurement since it confirmed an *a priori* harmonic argument that gave all

the planets the same apparent diameter when seen from the Sun.

Important public support for Gassendi came from the Dutch astronomer Martinus Hortensius (1605–39), who had been interested in the problem of planetary sizes and distances for some time, but had been foiled by the weather in his attempt to observe the transit of Mercury. In *De Mercurio in sole viso* (1633) Hortensius accepted Gassendi's measurement and added to it his own figures for the apparent diameters of the other planets, based on telescopic estimates. Here for the first time was a complete list of apparent diameters (including those of the fixed stars), supporting Galileo's general statements on this issue in the *Dialogue*, which had been published the previous year. Galileo himself had not paid attention to Kepler's warning of the forthcoming transits.

If Kepler had been correct in predicting the Venus transit of 1631, he was unaware that these rare events come in pairs and so did not predict the transit of 1639. However, in provincial Lancashire, the young Englishman Jeremiah Horrocks (1618–41), a devoted Keplerian working in relative isolation, predicted a Venus transit for 1639 after he had made some improvements in Kepler's elements of Venus's orbit. Horrocks and his friend William Crabtree both managed to observe part of this transit, and Horrocks went on to write a book on the subject entitled *Venus in sole visa*.

Like his predecessors Kepler and Gassendi, Horrocks was attracted to transits such as that of Venus because of the opportunity to obtain an accurate planetary apparent diameter. He had also estimated the apparent diameters of the other planets as best he could, and in *Venus in sole visa* he tried to use this information to arrive at a scheme of sizes and distances. But he did so by combining these measurements with his own version of Keplerian harmonic speculation. Horrocks (see Chapter 10) was supremely confident of his own measurement of Venus's apparent diameter during the transit, namely about 76″. He also trusted Gassendi's measurement of 20″ for Mercury's apparent diameter during its transit. Knowing the relative heliocentric distances of both planets, he calculated how large their apparent diameters would be as seen from the Sun, and found that in both cases this was about 28″. This fact confirmed his earlier speculation that the planetary diameters were proportional to their distances from the Sun; and his estimates for the superior planets did not seriously conflict with this proportionality. If the Earth presented a disk 28″ in diameter to the Sun, then the horizontal solar parallax was 14″, corresponding to a solar distance of 15 000 e.r. (in round numbers). Horrocks now knew the size and distance of every body in the solar system. Unfortunately, because of his untimely death, his book did not see the light of day until 1662.

Estimating the diameter of a planet with a Dutch telescope was a difficult business. Hortensius and Horrocks used several methods. One was to determine the angular size of the field of one's telescope and then to estimate how much of the field was occupied by the disk of the planet. Another was to compare a planet in size to another planet or the Moon. Hortensius even tried to project the image of a planet onto a tablet and mark its size, and then compare this to the size of the projected image of the Sun the next day. Such estimates were obviously unreliable, especially for bright planets such as Venus. But a number of astronomers were now making such estimates, and slowly the traditional apparent sizes were being replaced by better telescopic estimates in the astronomical lore. In the early 1640s one observer, William Gascoigne (c. 1612–44), was making measurements of apparent planetary diameters with an astronomical telescope equipped with a micrometer; unfortunately, he died shortly afterwards in the English Civil War, and his work remained unknown until after the Restoration of the Monarchy in 1660. In 1649 Eustachio Divini published a copper engraving of a map of the Moon (Figure 7.3), based on instruments of his own construction that included a micrometer of gridiron design.

By the middle of the seventeenth century, then, the old schemes of sizes and distances had finally been discarded. Astronomers increasingly appreciated that the apparent diameters of the planets were much smaller than had been thought (even if their estimates were still often much too large), but as yet no satisfactory method for determining absolute distances had presented itself. In his *Almagestum novum* of 1651 Riccioli rejected the eclipse method and tried his hand at the traditional method of lunar dichotomy, using a new astronomical telescope whose field of view was large enough to show the entire Moon at one time. On the basis of several measurements he decided on a solar distance of about 7000 e.r. In his book he also reported the efforts of the Belgian astronomer Gottfried Wendelin (1580–1667), who, on the basis of lunar dichotomy and a Keplerian speculation that the planetary diameters are proportional to their heliocentric distances, arrived at a solar distance of at least 14 656 e.r., and a solar parallax of at most 15″. This value was, of course, unconvincing to those who did not believe in harmonic speculations, including Riccioli himself, and a consensus remained elusive.

7.3. The Moon when full, with other sketches of the Moon and planets, as depicted by Eustachio Divini (1649).

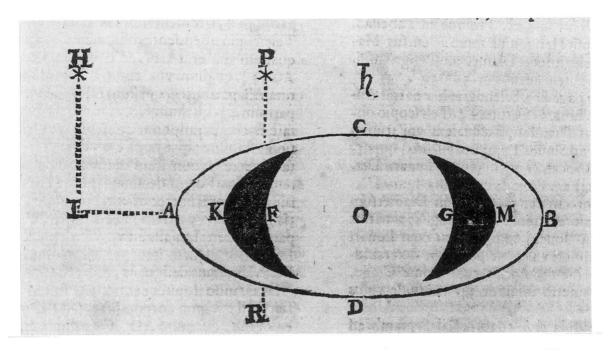

7.4. Saturn and its mysterious appendage, as measured by Riccioli. From his *Almagestum novum* (1651).

5. The micrometer

The astronomical telescope with its convex ocular began in the 1630s slowly to replace the Dutch telescope with its concave ocular, completely replacing it as the preferred research instrument by the middle of the century. Besides having a much larger field of view than the Dutch telescope, the astronomical telescope allows the introduction of a micrometer into the instrument. While Gascoigne's efforts in this direction remained unknown for the time being, those of Huygens did not.

Christiaan Huygens (1629–95) made his first research telecope in 1655 and immediately discovered a satellite of Saturn. He went on to unravel the mystery of Saturn's appearances (see Chapter 6), and in 1659 published *Systema Saturnium*, in which he set out his ring theory. In this work he also addressed himself to the important question of cosmic dimensions. Huygens had used a primitive micrometer to measure the apparent diameters of all planets except Mercury. His measurements were all in excess of the correct values by between 20% and 36%, and so much more uniform and accurate than those of his predecessors, and it is clear in retrospect that the problem of determining apparent diameters was now largely solved.

A solution to the problem of absolute distances was, however, as elusive as it had been since Kepler had destroyed the traditional consensus. Huygens examined the eclipse method and the method of lunar dichotomy and rejected both as hopeless. In their stead he proposed a Keplerian form of harmonic speculation, namely that the planetary apparent diameters are in proportion to their heliocentric distances. His own measurements, however, argued against such a relationship: while Venus was larger than Mercury, it was also larger than Mars, and while Jupiter was larger than Mars, it was also larger than Saturn. In the end Huygens simply chose a size for the Earth intermediate between Venus and Mars. Having found Venus to occupy $\frac{1}{84}$ of the Sun's diameter and Mars $\frac{1}{166}$, he took the Earth's diameter to be $\frac{1}{111}$ of the Sun's. This resulted in the unheard of solar distance of 25 086 e.r. and a solar parallax of 8.2″. Obviously the closeness of Huygens's solar parallax to the modern figure must not be taken as a measure of its precision! On the other hand, his measurements and calculations are clear evidence that the traditional system was now utterly discredited.

Huygens's micrometrical measurements were adopted by Thomas Streete (1622–89) in his

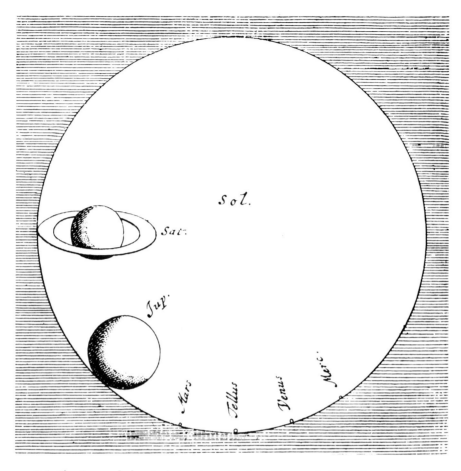

7.5. The Sun and planets drawn to scale, from Huygens's *Cosmotheoros* (1698).

Astronomia Carolina (1661), although Streete chose Horrocks's solar parallax of 15″ instead of Huygens's value. In his *Mercurius in sole visus* (1662) Johannes Hevelius (1611–87) presented his observations of the transit of Mercury of 1661 (and appended Horrocks's *Venus in sole visa*), but his measurements of Mercury and other planets were so excentric as to be without influence. When French astronomers developed Huygens's principle of the micrometer into full-fledged screw micrometers, and English astronomers resurrected the screw micrometer of Gascoigne, the measurement of apparent planetary diameters quickly became a science in which a progressive consensus can be seen. Flamsteed's apparent diameters, measured in the early 1670s, are reasonably close to present-day values, and much more uniform in their error margins than pre-micrometer esti-

mates. Residual errors, invariably still on the large side, must be ascribed predominantly to the poor optical quality of the anachromatic telescopes of the seventeenth century. By 1670, therefore, the problem of measuring apparent diameters had largely been solved. Astronomers now knew the relative distances of the planets with great accuracy, thanks to Kepler's third law, and therefore also the relative sizes of the planets, except for the Earth. But the measurement of that one absolute distance that would make all these measures absolute was still elusive. The final solution to this aspect of the problem was to come from an unexpected quarter.

6. Refraction and parallax

Although Ptolemy had made a detailed study of refraction in his *Optics*, and he and many of his successors were aware that light from heavenly bodies is refracted, no corrections for this effect were made to the measured positions of heavenly bodies until the sixteenth century. For nearly a millennium and a half, astronomers dutifully corrected the Sun's altitudes upwards for the effects of parallax (and their horizontal parallax was an order of magnitude too large at 3′), while the much larger error in the opposite direction caused by refraction (30′ at the horizon) was ignored. Bernhard Walther (1430–1504), the patron and student of Johannes Regiomontanus, was the first to draw attention to the Sun's upward displacement from the ecliptic as it approached the horizon. Tycho Brahe explored this problem in detail, even going so far as to send an assistant to Frombork to check by means of observations whether Copernicus had taken the effects of refraction into consideration.

Drawing up refraction tables was, however, a difficult business. Tycho based his parallax corrections to the altitudes of the Sun on the traditional horizontal solar parallax of about 3′, and as a result his refraction tables came out different for the Sun, Moon, and stars. Moreover, in the cases of the Sun and Moon Tycho's refraction corrections were zero above 45°, while for the stars they ceased at 20°. Because of his recognition of refraction, Tycho confidently 'corrected' the obliquity of the ecliptic from Copernicus's 23°28′ to 23°31½′. It remained for the seventeenth century to realize that this latter value was too high, and to reduce it gradually by 1½′ by removing the effects of Tycho's corrections for solar parallax.

Kepler was fully aware of the contradictions involved in having three different refraction tables for the Sun, Moon, and stars, and of having the corrections stop far short of the zenith. In his *Astronomiae pars optica* of 1604 (see Chapter 5) he was able to find a law of refraction that gave good results for altitudes above 10° and constructed a single table of refraction for all heavenly bodies. In his *Rudolphine Tables* (1627), however, he gave Tycho's refraction tables. Nevertheless, his successors now had two alternatives to choose from, and with the publication of Snell's law (by René Descartes in 1637), Kepler's approach became more attractive.

By the middle of the seventeenth century it had become clear to a number of astronomers that the interconnected corrections for solar parallax and refraction were together of fundamental importance for positional astronomy. If these corrections were faulty, then the measured solar declinations were in error. Inaccurate declinations in turn led to errors in fundamental parameters in solar theory, the obliquity of the ecliptic, the location of the equinoxes, and the solar eccentricity. Since the Sun is the centre of all planetary motions, these errors were propagated to all planetary theories, leading to faulty predictions. During the third quarter of the century solar theory was significantly improved thanks to new measuring instruments and a study of the relationship between refraction and solar parallax. The central figure in this campaign was Gian Domenico Cassini (Cassini I, 1625–1712).

7. The expedition to Cayenne

Cassini began his astronomical career in Bologna, where he was appointed professor of astronomy in 1650. One of his first research projects was solar theory. For this purpose he renovated the existing gnomon in the cathedral of San Petronio (in which a pin-hole image of the Sun passed across a meridian). Carefully plotting the Sun's elevations with this instrument, and comparing the latitude of Bologna derived from half the sum of the solstice declinations with the latitude of the city obtained by means of observations of the Pole Star, he found a discrepancy of several minutes of arc. He believed that this difference was due to faulty refraction and parallax corrections applied to the solar declinations. For the next several decades he pursued this subject.

Like most of his contemporaries, Cassini rejected the traditional solar parallax of 3′. Instead, he worked with Kepler's value of 1′. But when he employed this figure in his study of atmospheric refraction he ended up with three refraction tables, one for summer, one for winter, and one for spring and autumn. Only if he used a very small solar parallax – one of 12″ or less – could he reduce these three refraction tables to a single one. As an astronomer Cassini preferred the simplicity of a single refraction table, but as a product of his time he at first found a vanishingly small solar parallax, and the huge size of the solar system implied by this, daunting. Only very special measurements could answer the question as to which combination of parallax and refraction corrections was correct, and it was not until after his arrival in Paris, in 1669, that an opportunity for such measurements presented itself.

While in 1660 Paris was merely another city on the astronomical map of Europe, by 1670 it had become its foremost centre. Acting on the information in Huygens's *Systema Saturnium*, French astronomers were in the forefront of turning the telescope into a measuring instrument: installing a micrometer in the focal plane made possible the measuring of small distances in the telescope's field of view, and installing cross hairs in the focal plane made the telescope into a powerful and accurate sight with which to equip a measuring arc. At long last the vaunted accuracy of Tycho Brahe's measurements could be superseded. This astronomical work became centred on the new Academy of Sciences, founded in 1665. As a part of the Academy an observatory was designed and its construction begun, and Cassini was invited to come to Paris to take a leading part in the development of the observatory. In Paris he was able to combine the best telescopes (made by Giuseppe Campani) with the new measuring instruments in a lavish institutional setting backed by the wealth and power of the Sun King. His opportunity to solve the question of solar parallax and refraction came shortly after his arrival.

In their effort to establish a new precision astronomy on the basis of the accurate observations made possible with the new measuring instruments, the astronomers associated with the French Academy had come to appreciate the centrality of this problem of parallax and refraction. They had instituted a series of observations designed to solve the problem, but the effort lacked a clear focus. Cassini provided this focus. Comparing his proposed combination of parallax and refraction corrections with that of Tycho Brahe, he argued as follows. In the tropics, where the noon Sun is always high in the sky, Tycho's combination would use no refraction correction and a parallax correction based on his horizontal solar parallax of 3′. Cassini's combination, on the other hand, proposed refraction corrections continuing right up to the zenith and parallax corrections based on his much smaller horizontal parallax of 12″. While at European latitudes the Sun was not high enough to decide which of these alternatives is correct, an expedition to the tropics might solve the problem. The correct combination

of parallaxes and refraction would produce the same obliquity of the ecliptic near the equator and at European latitudes. Cassini was convinced that his combination was the correct one.

Since the Academy was an arm of the State, astronomical expeditions were feasible. Jean Picard (1620–82) had already measured the length of a degree in the plains of northern France, and he had been to Hven (see Chapter 9) to check the position of Tycho's observatory with respect to the Paris Observatory. Now a more ambitious expedition was desirable. And another factor made it urgent. Tycho Brahe had made many attempts to measure the diurnal parallax of Mars during the favourable opposition of 1582. Such oppositions, when Mars is at its perigee, occur about twice every fifteen years, and another one was to occur in 1672. If during this opposition Mars's position could be measured at Paris and at a spot very far removed from Paris, a comparison of these measurements might reveal the planet's parallax. On Cassini's and the Academy's recommendation an expedition under Jean Richer (1630–96) was sent to the French colony of Cayenne on the coast of South America, 5° north of the equator.

In the meantime, observers in Europe, now equipped with micrometers, telescopic sights, and pendulum clocks, would also attempt to measure the diurnal parallax of Mars; that is, in Copernican terms, the effect caused by the displacement of the observer in the course of the night as the Earth turns on its axis. Flamsteed in Derby, England, professed great faith in this method when he found that near opposition Mars would pass very close by several little stars in the constellation Aquarius. At this time, Mars and one or more of the stars could be brought into the field of his telescope, so that he would be able to measure their separations with his micrometer.

Although these micrometer measurements promised success, Flamsteed's results were not without contradictions. His first verdict was that Mars's parallax was not greater than 30″; he then stated that it was at most 15″, and he finally limited it to 25″, from which he derived a solar parallax of at most 10″. Cassini, using a pendulum clock to measure differences in right ascension between Mars and the little fixed stars, found figures for Mars's diurnal parallax that supported Flamsteed's final results. Mars's diurnal parallax, however, causes a displacement in right ascension of the order of a second of time, and Cassini's method was certainly not sufficiently sensitive to produce the required data. Knowing what he wanted Mars's parallax to be, Cassini was, however, able to choose the few measurements that would support his preconceived idea.

When Richer returned from Cayenne in 1673, it quickly became apparent that the expedition had been a great success. The solar declinations measured at Cayenne, when refined with Tycho's parallax and refraction corrections, produced a different obliquity of the ecliptic from the one found by measurements in Europe. Cassini's parallax and refraction corrections, on the other hand, resulted in the same obliquity of the ecliptic in both cases. Here was palpable proof that the Sun's parallax was vanishingly small – certainly less than 12″ – as Cassini had postulated. If the measurements of Mars did not produce the desired results, the expedition brought home an item of information that was entirely gratuitous: the amount by which the pendulum had to be adjusted in Cayenne indicated that the Earth has an equatorial bulge.

Richer's solar declination measurements were convincing, and the result was a change in the accepted value of the obliquity of the ecliptic from Tycho's 23° 31½′ to 23° 29′, with comparable adjustments to the locations of the equinoxes and the solar eccentricity. This resulted in a significant improvement in solar theory over Kepler's efforts, and a concomitant improvement in the theories of all the planets. Astronomers could now work with a single refraction table and try to improve it by taking into account temperature and pressure variations.

Cassini admitted that there could be an error of 2″ or 3″ in the measured parallax of Mars, which would result in an error of 1″ or more in the Sun's parallax. This is a serious underestimation of the errors involved in this procedure, for, in fact, he and his colleagues had not measured a parallax at all: Mars's parallax still lay within the error margin of the instruments. Yet, limits had now been imposed on the solar parallax: a solar parallax of 3′, which would imply that Mars's parallax would be about 8′ at favourable oppositions, was certainly ruled out by the accuracy of the new generation of measuring instruments. Cassini was entirely justified in making the Sun's horizontal solar parallax

less than the nominal 15″ error margin of measuring arcs with telescopic sights. He chose a value of 9.5″, which he extracted from the Mars data, a value that was supported by Flamsteed. The solar distance was, therefore, now at least 22 000 e.r., a figure larger than the radius of Ptolemy's entire universe!

In 1673 Cassini was Europe's foremost astronomer. Over the next decade Flamsteed joined him at the pinnacle of the profession. Their solar parallax of 9.5″ or 10″ was thus endowed with their combined authorities, and by the end of the century it had become the consensus value to which most influential scientists subscribed. Periodic attempts to measure the diurnal parallax of Mars during favourable oppositions merely confirmed the measurements of Flamsteed and Cassini. Yet, if these measurements were accurate, we might expect to see, in retrospect, an asymptotic approach to the currently accepted value of about 8.8″. In fact, successive attempts on the part of the best eighteenth-century astronomers produced figures that actually receded somewhat from this value, showing that the accuracy of these measurements is, at best, problematical. One astronomer who from the beginning to the end of his long and illustrious career refused to accept the consensus value of solar parallax was Edmond Halley (1656–1742).

8. Halley and the transits of Venus

Halley first attracted the attention of the astronomical world when he went on an expedition to the island of St Helena to map the southern skies, 1676–78. While there, he observed a transit of Mercury, but whereas others up to that time had made such observations to determine apparent diameters or to refine the elements of an orbit, Halley was the first to try to determine parallaxes by this method. The possibility of doing this had first been proposed by James Gregory in 1663 in *Optica promota*.

Upon his return to England, Halley found that (other than himself) only Jean Charles Gallet in Avignon had observed the transit, and a comparison of their observations produced a horizontal solar parallax for Mercury of 45″. Halley did not think well of this result, although he did not despise the method. The difficulty arose from the fact that Mercury's parallax was not much larger than the

Sun's. On the other hand, having assisted Flamsteed several times in making measurements of Mars's diurnal parallax, Halley was convinced that this approach to the problem was even less promising. In his *Catalogus stellarum Australium* of 1678 he concluded that transits of Venus would be the only occasions for accurately measuring a parallax. But, unfortunately, the next Venus transits would not occur until 1761 and 1769.

Halley returned to this subject on several occasions during his career, the most influential of which was when he published a paper in *Philosophical Transactions* of 1716. As his prestige increased – he became Astronomer Royal in 1719 – and the event approached, other astronomers took up the call. When Halley died in 1742 the astronomical community was confidently looking forward to the Venus transits of 1761 and 1769. And, indeed, Venus did not disappoint the learned world.

Although the transit of 1761 provided several surprises for the farflung network of observers, it and the 1769 transit did produce, for the first time in history, actual measurements of the parallaxes of Venus and the Sun. These measurements showed that the solar parallax found by Cassini and Flamsteed a century earlier was too large; but they did confirm roughly the new scale of the solar system that had come from their efforts.

Further reading

Dieter B. Herrmann, *Kosmische Weiten: Geschichte der Entfernungsmessung im Weltall* (Leipzig, 1977)

Albert Van Helden, The importance of the transit of Mercury of 1631, *Journal for the History of Astronomy*, vol. 7 (1976), 1–10

Albert Van Helden, *Measuring the Universe: Cosmic Dimensions from Aristarchus to Halley* (Chicago, 1985)

Harry Woolf, *The Transits of Venus: A Study of Eighteenth-century Science* (Princeton, NJ, 1959)

8

Selenography in the seventeenth century
EWEN A. WHITAKER

Pre-telescopic observations

Of all the heavenly bodies, the Moon is the only one to exhibit surface detail that is plainly visible to the unaided eye. Its pattern of light and dark spots had captured mankind's imagination since prehistoric times, as is evidenced by the ancient legends and folklore of many different cultures, where one often finds allusions to perceived images of familiar animals, human faces or inanimate objects in this pattern. It is rather surprising, therefore, that the only known naked-eye drawings dating from the pre-telescopic era are crude sketches (Figure 8.1) by Leonardo da Vinci (1452–1519) and a drawing (Figure 8.2) made by William Gilbert in about 1600. Gilbert (for whom see Chapter 4) bemoaned this lack of earlier images which, he reasoned, might have been useful for determining whether or not the spots had changed with the passage of time.

He provided names for thirteen islands, bays and other geographical features on his simple full-Moon sketch. Note particularly that he considered the darker areas to be land, the lighter water, a view that he discusses in the accompanying text, and one at first subscribed to by Johannes Kepler. Thus Gilbert may perhaps be thought of as the first selenographer, even though the 'map' and text were not published until 1651; by that time, they had been completely superseded by telescopic observations.

8.1(*a*). Sketches of the Moon by Leonardo da Vinci (enlarged).

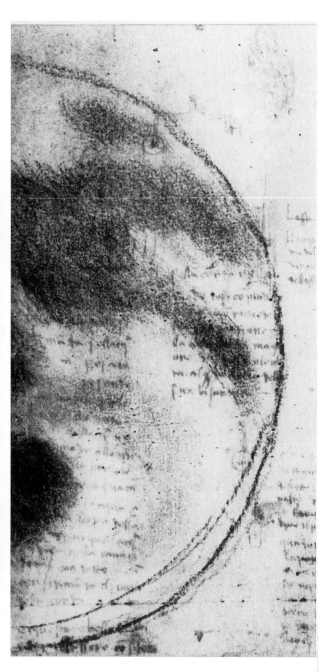

8.1(*b*). A further sketch of the Moon by Leonardo da Vinci (enlarged).

The first telescopic observations

It is not known who was the first to look at the Moon through one of the newly invented telescopes. Galileo Galilei (1564–1642) was the first to observe the Moon methodically (in 1609) and to interpret what he saw (see Chapter 6), but Thomas Harriot (*c*. 1560–1621), an astute mathematician, scientist and philosopher who is perhaps better known for his description of an expedition with Sir Walter Raleigh to Virginia in 1585, actually made the first known telescopic sketch of the Moon (Figure 8.3). The notes accompanying the sketch record the time and date as 9 p.m. on the evening of 26 July 1609 (OS) (that is, 5 August (NS)), a 6-power telescope being used on the 5-day-old Moon. This was four lunations before Galileo first pointed his own instrument towards that body. The sketch shows very little detail, possibly an indication of the generally poor quality of this particular "trunke", as Harriot termed it.

In any case, he apparently either paid little further attention to the Moon or at least recorded no more observations of it for about a year, although he did encourage his good friend Sir William Lower (*c*. 1570–1615) to observe the Moon frequently, having supplied him with a telescope. Observing from his home near Kidwelly in South Wales, Lower wrote this often-quoted but still charming description in a letter to Harriot dated 6 February 1610 (OS):

According as you wished I have observed the moone in all his changes. In the new I discover manifestlie the earthshine, a little before the dichotomie that spot which represents unto me the man in the moone (but without a head) is first to be seene. A little after neare the brimme of the gibbous parts towards the upper corner appeare luminous parts like starres, much brighter then the rest and the whole brimme along, lookes like unto the description of coasts, in the dutch bookes of voyages. In the full she appeares like a tarte that my cooke made me the last weeke. Here a vaine of bright stuffe, and there of darke, and so confusedlie al over. I must confesse I can see none of this without my cylinder.

Sometime in the late spring of 1610 Harriot apparently obtained a copy of Galileo's *Sidereus nuncius*. He must soon have apprised Lower of the general findings of Galileo since Lower replies, in a letter dated "the longest day of 1610":

Me thinkes my diligent Galileus hath done more in his threefold discoverie [concerning the Moon, Jupiter and

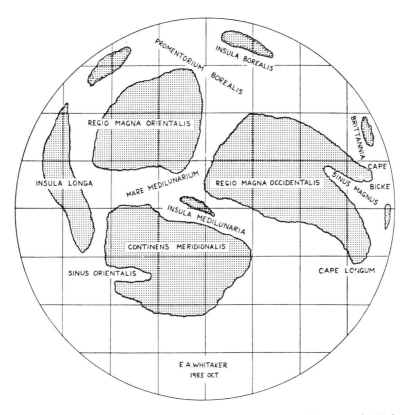

Labels on the figure:

PROMENTORIUM BOREALIS

INSULA BOREALIS

REGIO MAGNA ORIENTALIS

BRITTANNIA

INSULA LONGA

MARE MEDILUNARIUM

REGIO MAGNA OCCIDENTALIS

CAPE

SINUS MAGNUS

BICKE

INSULA MEDILUNARIA

CONTINENS MERIDIONALIS

SINUS ORIENTALIS

CAPE LONGUM

E. A. WHITAKER
1985 OCT.

8.2. William Gilbert's naked-eye image of the full Moon, *c.* 1600 (based on the original MS drawing, which is unsuitable for reproduction at this much-reduced scale).

the stars] than Magellane in openinge the streights to the South Sea or the dutchmen that were eaten by beares in Nova Zembla. I am sure with more ease and safetie to him selfe & more pleasure to mee. I am so affected with his newes as I wish sommer were past that I mighte observe the phenomenes also. in the moone I had formerlie observed a strange spottednesse al over, but had no conceite that anie parte thereof mighte be shadowes; since I have observed three degrees in the darke partes, of which the lighter sorte hath some resemblance of shadinesse but that they grow shorter or longer I cannot yet perceive.

The new revelations in Galileo's book undoubtedly rekindled Harriot's interest in telescopic observation in general because he resumed lunar observations in July of that year with renewed enthusiasm. His book of collected manuscript notes and drawings includes (*a*) 15 drawings of various lunar phases made during the following three lunations; (*b*) some rather more detailed sketches,

accompanied by descriptive notes, of selected lunar areas; (*c*) a quite carefully executed drawing of the full Moon annotated with seventy-two letter and number designations; (*d*) three attempts at determining the Sun's distance by the dichotomy/quadrature angle method (see Chapter 7); and (*e*) many pages of calculations dealing with the exact length of the lunation.

The lunar sketches contain nothing remarkable, even though he was then using different instruments (10-power and 20-power), but the annotated drawing of the full Moon (Figure 8.4), which was made and checked from a series of observations carried out during 1611 and 1612, is clearly the result of some patient effort. This has been referred to as the first lunar map, but it really deserves only a qualified 'first' since it does not include any of the topographic features of the Moon's surface. A page of notes referring to this drawing includes, in Harriot's own hand, the following observation: "1611 Decemb. 14th. ho. 8½

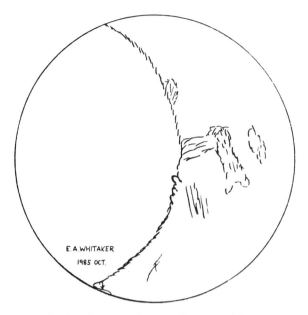

E.A.WHITAKER
1985 OCT.

8.3. The first known telescopic drawing of the Moon, made by Thomas Harriot on 26 July 1609 (OS), four months before Galileo first observed the Moon with his telescopes.

mane [i.e. 8.30 a.m.] I noted that the darker partes of 28, 26 were nerer the edge then is described." These numbers refer to the narrow dark marking nowadays named Sinus Roris and Mare Frigoris, which is situated near the north limb of the Moon. Calculation shows that maximum libration towards the north occurred only two days after this observation; Harriot was thus the first to observe the Moon's libration in latitude, although whether or not he considered this to be a true movement or just an effect of inaccurate drawing is not known.

Galileo's lunar observations

Although we now know that Harriot preceded Galileo in making telescopic observations of the Moon, Galileo remains the undisputed pioneer in the science of selenography. He first directed one of his home-made telescopes (the 20-power) towards a non-terrestrial object during the evening of 30 November 1609, the object being the four-day-old crescent Moon. The fact that he observed that body until almost moonset, describing the progress of sunrise on what has since been identified as a group of mountain crests in the Janssen area of the lower cusp, as well as making three separate sketches of

the Moon, shows that he immediately grasped the importance of what was being revealed. This was a time when confused notions about the true nature of the Moon were still held by most people – notions passed down from Antiquity. These included its being (*a*) a mirror, reflecting the terrestrial oceans and continents; (*b*) a polished, translucent crystalline sphere; (*c*) a body of condensed fire, etc.; and (*d*) a terrestrial type of spherical body with seas, mountains, valleys, plains, etc. Some of the more extreme ideas stemmed from the necessity of explaining such phenomena as the ashen light (by means of translucency or self-luminosity), visibility of the Moon during total eclipses (self-luminosity), and why the Moon does not fall to Earth (because fire rises and is luminous). Another factor that influenced thinking was that, unlike the Earth with its many imperfections, all objects in Heaven were Divine in nature, and therefore of necessity intrinsically perfect and spotless. This led to the idea that the dark markings were due to permanent clouds suspended between Earth and Moon. Galileo's telescope was showing him just how far from the truth most of these ideas were; the Moon's surface was by no means smooth, shiny or transparent – it was mostly very rough.

Galileo did not date any of his observations, but recent research has shown that he observed the Moon on the next three evenings (1, 2 and 3 December), making sketches on each occasion and noting the various changing appearances as the sunrise terminator advanced to the first quarter phase. He probably followed the progress of the entire lunation, as the weather allowed, but he made further sketches and gave descriptions only of the pre-sunrise observations of the waning phases of 18 and 19 December. A final sketch has recently been shown to date from 19 January 1610, again before sunrise. Eleven of Galileo's images of the lunar disk have been preserved; his little book *Sidereus nuncius*, which was published in March 1610, contains four copperplate engravings of different phases (see Figure 8.5 (*a*)), plus one repeated image, while a sheet preserved at Florence contains a group of seven brown-ink wash images (Figures 6.2 and 8.5 (*b*)).

One usually reads that Galileo's Moon drawings were very poorly executed, virtually none of the formations being recognizable. Part of this adverse criticism is a consequence of the limited print-run

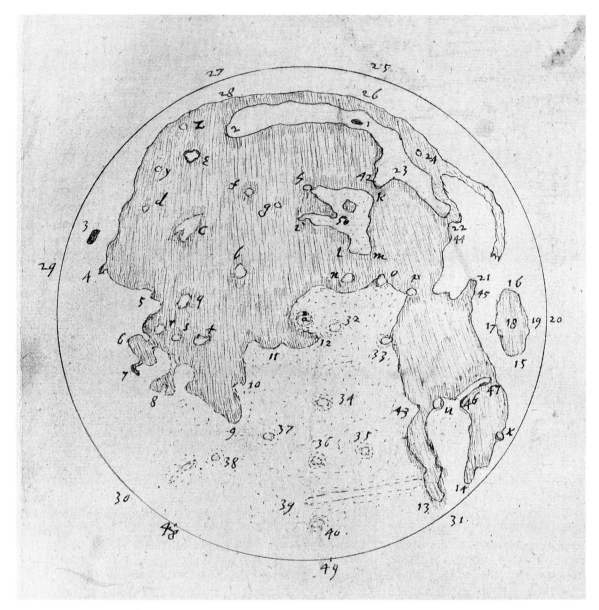

8.4. The first telescopic drawing of the full Moon, made by Harriot *c.* 1611. Since the numbers and letters were used only to identify points as an aid to their correct relative placement (Harriot's MS notes contain an identical image without the annotations), and no topographical features are shown, this does not quite qualify as a true map.

of the first edition of *Sidereus nuncius*, which led to hurried reprintings being made in several other European cities. These were illustrated with grossly inferior woodcuts, and later reprintings in the earlier editions of Galileo's *Opere* provide even poorer reproductions. The disparaging remarks of some recent critics may also be ascribed to their forming conclusions from a superficial perusal of

the images, since good-quality facsimiles and re-productions have been available for many decades.

Figures 8.5 (*a*) and (*b*) illustrate respectively one of the copperplate images from the first edition of *Sidereus nuncius* and one of the more interesting images from the Florence MS. Degraded recent photographs of the Moon at closely similar phases are included for comparison, with some common

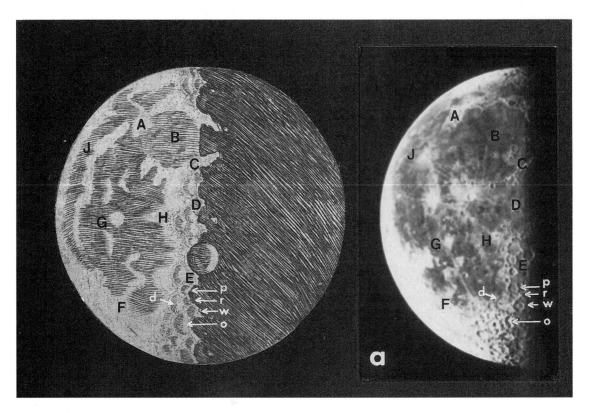

8.5. (*a*) Galileo's drawing of last quarter made on 18 December 1609, from the first edition of *Sidereus nuncius*, with a degraded photograph at a similar phase for comparison. *p* = Purbach crater, *r* = Regiomontanus crater, *w* = Walter crater, *o* = Orontius crater, *d* = Deslandres crater.

features noted. The super-large crater *E* has been identified as the 136 km diameter crater Albategnius; Galileo exaggerated its apparent size in order to illustrate more clearly his description of the progress of sunrise on this formation, which he compared with the mountain-ringed plain of Bohemia. The MS drawing shows very well the phenomenon of earthlight on the Moon, with the apparent dark region where it adjoins the terminator (an adjacency effect on the retina) nicely portrayed.

Galileo's descriptions of his lunar observations, and the conclusions he reached regarding the Moon and its surface features make fascinating reading. He knew that he had to write very convincingly if he was to bury two millennia of misconceptions. His conclusions may be summarized thus:

The Moon is a solid, opaque, spherical body with a rough surface; the ashen light is due to reflected earthlight

The roughness is mostly caused by hundreds of mountain-girt cavities and by circular mountain ranges enclosing fairly level tracts

The brighter areas are full of these formations, but very few occur in the darker areas

The darker areas are mostly smooth, and are lower than the brighter areas; some are bordered by lofty mountain ranges; none extends as far as the limb

The darker areas have lighter markings here and there; these cast no shadows and thus must be due to dissimilar materials

The cavities merge into the background at full Moon

The highest mountains attain altitudes of the order of 6000 metres

The Moon has an atmosphere (the only incorrect statement in the list)

In Galileo's capable hands, the impact of the telescope on man's knowledge of the Moon was clearly very great indeed. However, the story does not end there. Believing that a line joining the centres of the

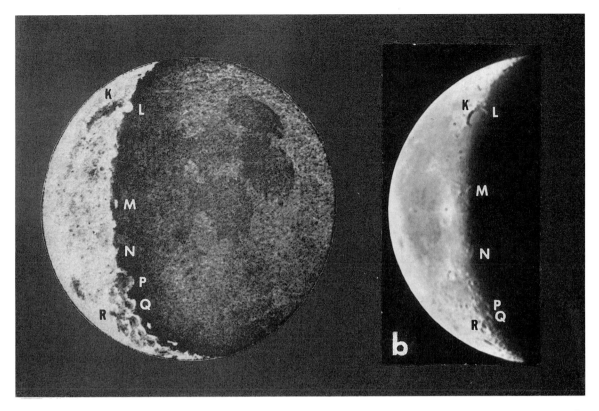

(*b*) Galileo's original drawing of the waning Moon, made on 19 January 1610, with a degraded photograph at a similar phase for comparison. Note the earthlight.

Earth and Moon passed through a fixed point on the lunar surface, he predicted an apparent side-to-side ('no') motion due to the observer's changing viewpoint from the surface of a rotating Earth. Observing the dark markings now named Grimaldi and Mare Crisium, he indeed noted that their apparent distances from the east and west limbs varied noticeably. This he announced in 1632. In actuality, the diurnal libration is far too small to be detected by plain viewing, so that Galileo had in fact observed the libration in longitude. Five years later he announced in a letter to Fulgenzio Micanzio that he was wrong in thinking that the Moon was fixed with respect to the Earth–Moon line – he had now detected both a nodding ('yes') motion that had a period of one month, that is, the libration in latitude, and also a pendulum-like swinging about the centre of the disk with a period of a year. He still thought that the motion in longitude was due to the changing viewpoint, and gave its period as one day. However, he may have fallen victim to the 'beat' effect here, since the crescent

phase is observed in the west, while two weeks later, when the libration in longitude has changed by half a cycle, the near-full Moon is observed in the east.

The final motion detected appears to be that caused by the $1\frac{1}{2}°$ tilt of the Moon's axis to the ecliptic. This has a period of about one year, but it is difficult to believe that Galileo actually saw the fairly subtle effects of this tilt. Whether he did or not, he certainly put the whole subject of lunar observation on a firm footing right at the outset.

Two new selenographical programmes

One might have expected that Galileo's revelations regarding the Moon would have sparked a flurry of observing activity, but this apparently was not the case. In fact, apart from one original drawing of the first quarter phase by Christoph Scheiner (1575–1650), published in 1614, and three very low-quality sketches appearing in books by Charles Malapert, Giuseppe Biancani and Christopher Borri (Figure 8.6), nothing was even attempted until the

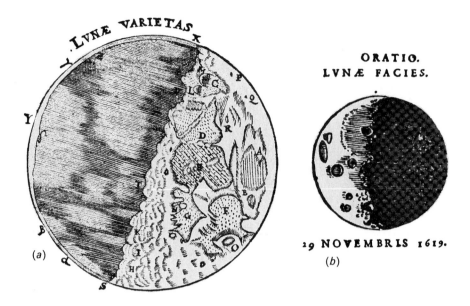

8.6. (*a*) Scheiner's drawing of first quarter with annotations, 1614.
(*b*) Malapert's image of last quarter, 1619.

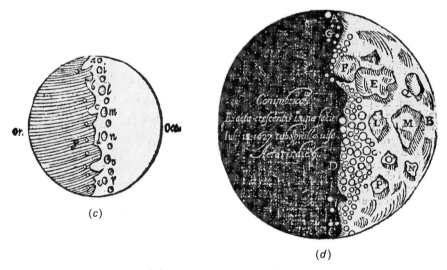

(*c*) Biancani's image of first quarter, 1620.
(*d*) Image of first quarter, made at Coimbra by Borri, July 1627.

mid-1620s when two quite independent programmes were conceived. The original aim in each case was to draw up an accurate map of the Moon, not so much for use as a starting point for lunar studies but rather as an aid in determing terrestrial longitudes!

(a) The Gassendi–Peiresc–Mellan program
Pierre Gassendi (1592–1655), a professor of mathematics at Paris and a leading astronomical thinker

and observer of his day, together with his friend Nicolas Claude Fabri de Peiresc (1580–1637), who was learned in many fields and an active promoter of astronomical observations, initiated one of these programmes. The basic aim was to obtain the longitude difference between Aix-en-Provence and Paris by noting the local times of the occultations and reappearances of small lunar spots during lunar eclipses, for which a good lunar map was needed. The artist and engraver Claud Mellan

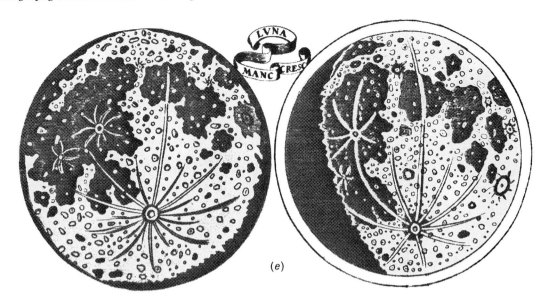

(*e*) Images made by Fontana in 1629 and 1630, used by Argoli and others.

(1598–1688) was finally engaged to undertake this task; using a telescope constructed from optical parts supplied by Galileo in 1634, he produced engravings of the full Moon and the two quarter phases (Figure 8.7). One does not need to be a lunar expert to see that these images are remarkably realistic in appearance, and that they represent a great improvement in both accuracy and content over the imagery of Harriot and Galileo – and, in fact, over all their predecessors. Many relatively small craters situated near the terminators in the quarter phase images can be seen, but more remarkable is the portrayal of several low ridges in Mare Imbrium, towards the top left corner of the first-quarter image. These are low-contrast, narrow features that do not immediately catch the attention, but Mellan obviously saw them with these optics by Galileo, which may give some indication of the quality of Galileo's craftsmanship. Gassendi notes that telescopes currently available in France were inferior.

Gassendi had already begun to work on compiling a nomenclature scheme by this time, in preparation for the proposed lunar map that would be based on these and future drawings, but Peiresc died in 1637 and the whole plan fell through. Only these three engravings were produced, apparently in very small numbers since they are not men-

tioned again until the latter part of the eighteenth century; they are exceedingly rare today.

(b) Van Langren: the founder of lunar cartography

The other programme for determining terrestrial longitudes stemmed originally from the offer by Philip III of Spain, of 6000 ducats to whoever could devise a practical and reasonably accurate scheme for solving this problem. Michael Florent Van Langren (Langrenus, *c.* 1600–75), member of a prominent Flemish globe and map-making family, and "Royal Mathematician and Cosmographer" to King Philip IV, conceived the idea of using the Moon's rotation as the celestial clock necessary to do this. By timing (in local time) the moments of sunrise or sunset on various identified lunar peaks and crater crests, one would not be confined to infrequent lunar eclipses for determinations of longitude – opportunities would be almost continuous. With an ephemeris giving the same events in standard time, longitudes at sea or in overseas lands might be determined on the spot.

Van Langren originally intended to construct a lunar globe, but opted later to prepare a flat map, which would be easier and cheaper to produce. He had, by 1643, already prepared thirty drawings of different phases, but procrastination, plus the death in 1634 of his patron Princess Isabella (his

Phasium Lunæ Icones, quos Anno Salutis 1634 et 1635 pingebat, ac sculp Agriis Sextis Claud. Mellan Gall. præsentibus ac Flagitantib. Illustrib. Viris Gassendo et Peyreschio.

8.7(*a*). Images of last quarter and full Moon, by Mellan for the Gassendi–Peiresc project, 1636.

source of funding), had delayed the final step – that of combining the drawings into a single map and supplying a suitable nomenclature for the various features.

Hearing of new Moon-mapping projects being undertaken by Juan Caramuel y Lobkowitz (1606–82) and Johannes Hevelius (1611–87), Van Langren was spurred into action and produced what is the first true map of the Moon; that is, it depicted not only the surface shadings as seen best at full Moon, but also a large number of topographical features (craters, mountain ranges, isolated peaks) which only become visible at other phases. As early as 1633 he had decided that "the names of illustrious men" would be used to "distinguish the mountains and the luminous and brilliant islands" of the lunar globe. In order to preserve his priority, Van Langren prepared a simpler, hand drawn and

coloured version of the map and provided it with forty-eight names to indicate the general scheme of feature depiction and nomenclature. The final engraved map (Figure 8.8) was published as a broadsheet in March 1645; only four copies are known to exist today.

A careful study of the map shows that it is more accurate than might be judged from the monotonous shading and somewhat stylized outlines of the darker areas. Craters are shaded as if illuminated by a morning Sun, a technique still used today on the best lunar maps; their positions and diameters are in general fairly trustworthy, although a small number of the portrayed features are not readily identifiable. In this respect the map can hold its own against those of the major figures that followed, Hevelius and Giambattista Riccioli (1598–1671). Even though his scheme for longitude de-

8.7(*b*). Image of the 8-day old Moon, by Mellan for the Gassendi–Peiresc project, 1636.

termination proved to be impractical because of the slowness with which peaks became illuminated or vanished, Van Langren deserves recognition for being the first to produce a serviceable lunar map, and for introducing the basic scheme of nomenclature that is used to this day.

A flurry of activity

The six-year period following the publication of Van Langren's map witnessed the appearance of no fewer than six Moon-related publications, ranging from single images to a bulky tome on the subject.

The first to appear was a rather curious image of full Moon by the Bohemian friar and optician Anton Maria Schyrlaeus (or Sirek) of Rheita (Rhaetia) (1597–1660), which was included in his large volume on optics and related subjects. It is presented south up, a result of Rheita's use of a positive eyepiece. It and its accompanying text did little to influence later work on the subject.

The following year (1646) saw the appearance of a small book devoted to telescopic observations of the planets as well as of the Moon, and also to microscopic views of small objects. It was published

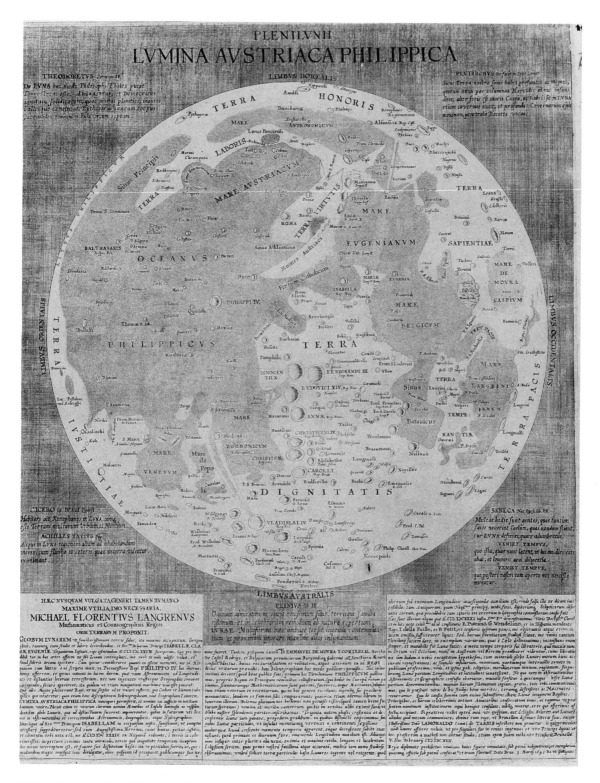

8.8. Van Langren's broadsheet (1645): the first true lunar map, including surface shadings, topographic features, and a viable nomenclature scheme.

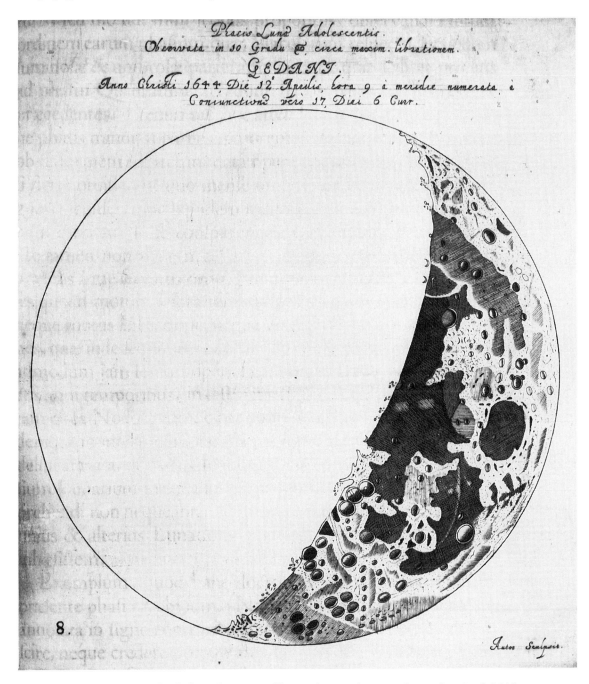

8.9. One of Hevelius's forty drawings of lunar phases. This was observed in April 1644.

by Francesco Fontana (1585–1656), a Neapolitan lawyer and keen amateur astronomer. The book contains twenty-seven small copperplate engravings of lunar phases plus one large image of the full Moon, each with a quaint description of the appearances noted ("shining pearls; gems; little fountains"). He was no great artist, and his Moon images have virtually no scientific value. Two images date from 1629 and 1630 and were used as illustrations by other authors on several occasions both prior and subsequent to 1646.

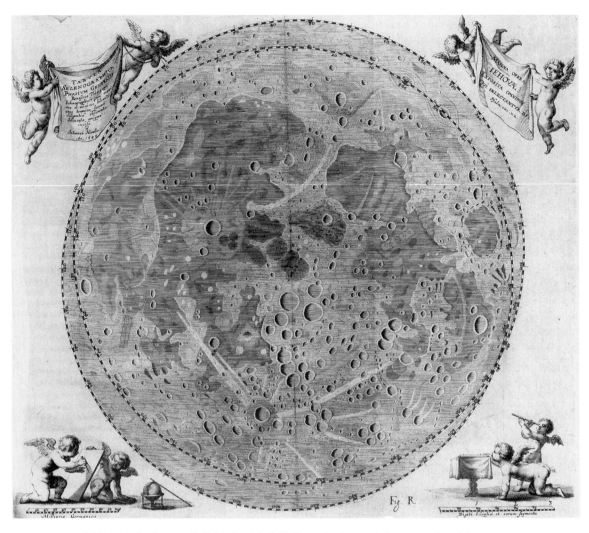

8.10. Hevelius's map *R* (1647), with full-Moon shadings and topographical features.

Hevelius and his *Selenographia*

The year 1647 saw the publication of Hevelius's *Selenographia*, a bulky and remarkably comprehensive tome devoted largely to the study of the Moon. It was the first published work of the wealthy Danzig brewer and amateur astronomer. He was well-educated, skilled in the construction and use of telescopes and other astronomical instruments, and an accomplished artist and engraver. He observed the Moon with a telescope or telescopes of his own construction from November 1643 to April 1645. He made drawings of forty different phases, from the slim waxing crescent, through full to the slimmest sliver of the waning Moon, each of which is presented as a copperplate engraving in the book (see Figure 8.9). The accompanying text gives descriptions of phenomena he noted, and his ideas and conclusions regarding what he saw. He also investigated the lunar librations, plotting on two of his maps the wanderings of the apparent centre of the disk and their relationship to the Moon's ecliptic longitude. He apparently also sought to confirm Galileo's "pendulum-like" motion, which manifests itself as a slight swing of the terminator line.

8.11. Hevelius's map *Q* (1647), with his system of nomenclature.

The best known products of this monumental investigation are the three large maps. One (*P*) is a full Moon image, the second (*R*) is a map with most of the shadings of the former plus the topography (Figure 8.10), while the third (*Q*) is for nomenclature (Figure 8.11). All are extremely attractively produced with elegant corner cartouches; each shows the limits of the combined librations in latitude and longitude at the time. Map *P* compares favourably with reality, the positions and proportions of the darker and lighter shadings being quite well portrayed. The topographic features in map *R* are rather less trustworthy, however; the percentage of non-existent craters is somewhat greater than in Van Langren's map, and a number of the objects depicted as craters are, in fact, peaks or ridges. Hevelius used Van Langren's convention for crater portrayal, but with reversed (sunset) illumination. The nomenclature map *Q* depicts the crater rims and mountain ranges as rows of 'anthills', then the normal convention in terrestrial maps. Inexplicably, Hevelius shows the bright rays as mountain ranges in this map, and even names some of them as such. Despite Van Langren's appeal to astronomers to accept his nomenclature without change – a forward-looking but unsuccessful attempt to try to prevent confusion in the future – Hevelius eschewed Van Langren's scheme of using the names of famous persons to identify the craters and mountains, fearing he might arouse jealousies amongst those whose names were omitted. Instead, he likened the Moon's face to a distorted image of the anciently known world, using classical geographical names for the markings and topographical features. Only ten of his 285 or so names are currently in use, and six of these have been moved to other features. *Selenographia* remained the standard work of reference on the subject for well over a century.

An interesting broadsheet was published in 1649 by Eustachio Divini (1610–1685), a skilled maker of telescopes and microscopes; it includes a quite detailed image of full Moon, a small image of the crescent Moon, and images of Jupiter and satellites, Saturn, and Venus. According to the accompanying text, Divini used telescopes of '24 palms' (about $5\frac{1}{2}$ m) and '16 palms' (about $3\frac{1}{2}$ m) focal length, the focus of the latter being provided with a reticle of fine threads to act as an aid in positioning the details correctly with respect to each other. However, the main image (Figure 7.3), supposedly made in March of that year, repeats so many idiosyncrasies of Hevelius's map *P* that Divini obviously started out with that map as a base and simply made additions and amendments as he noticed omissions or differences from what he observed.

Another portrayal of the full Moon was published in 1651 from an observation made in July 1650 by the Neapolitan Jesuit Gerolamo Sersale (Sirsalis, 1584–1654). Until a decade or so ago, the existence of this image was known only from references to it in Riccioli's 1651 work, *Almagestum novum*, but an engraved broadsheet containing it is now known to be in the library of the Naval Observatory at San Fernando, Cadiz and is reproduced for the first time in Figure 8.12. Comparison with a photograph of full Moon shows that its general accuracy is on a par with Hevelius's map *P*, but it is notably more complete and reliable with respect to the smaller details of the boundaries of the maria, and the smaller light and dark spots.

Riccioli, Grimaldi, and nomenclature

The six-year spate of publications devoted to lunar observations that began with Van Langren's map of 1645 culminated with work carried out in Bologna. Riccioli, a Jesuit and professor of philosophy,

theology and astronomy in that city, is best known for his monumental classic, *Almagestum novum*, published in 1651. His pupil Francesco Maria Grimaldi (1618–63), a fellow Jesuit who is better known for his discovery of the diffraction of light, prepared a new map of the Moon (Figure 8.13) for inclusion in this work. As the wording above the map notes, it was compiled by using the "best telescope" to observe the Moon at many different phases, and by partly confirming, partly correcting and augmenting the maps of Van Langren, Hevelius, Divini, Sirsalis and others. A careful study of the map shows that Grimaldi did indeed do as the heading says. The chief errors of the earlier maps have been removed, and most of the newly added detail can be identified with considerable certainty. Unfortunately, the map is aesthetically less pleasing than the Hevelius maps; its superior accuracy was not generally recognized, so that the Hevelius maps tended to be regarded as the more trustworthy.

Riccioli's contribution to this joint effort was the descriptive text and, more importantly, the nomenclature. Rejecting Hevelius's scheme of using classical geographical names, he restored Van Langren's plan of using the names of famous people, but confined his selection almost entirely to philosophers, scientists, mathematicians, astronomers, etc. He used sixty-three of Van Langren's names but allocated all but three (Pythagoras, Endymion and Langrenus) to different formations (see Figure 8.14). He distributed the names in a carefully and cleverly thought out scheme, grouping them by era, affiliation, nationality, philosophy and so forth, with the ancients near the top (north) and his contemporaries generally towards the bottom. He named the larger light and dark areas after various states of the climate and weather, and their effects on the land and humanity. This system of nomenclature was adopted almost exclusively in SW Europe right from the start, but elsewhere in Europe it was used in parallel with Hevelius's system for almost 150 years, when J.H. Schroeter's adoption of it ensured the final demise of all but ten of Hevelius's awkward names. The Riccioli system remains the basis of our current lunar nomenclature.

It is interesting that Van Langren, Hevelius and Riccioli all gave the darker areas watery designations such as Mare, Oceanus, Sinus, Lacus. The

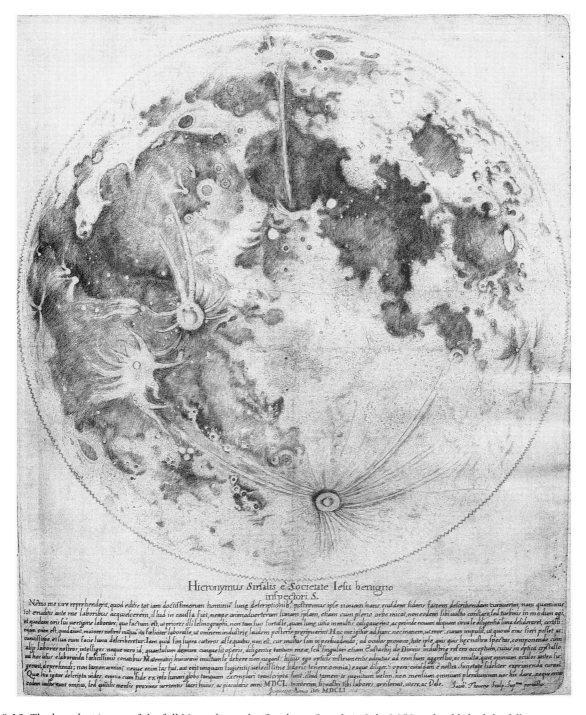

8.12. The long-lost image of the full Moon drawn by Gerolamo Sersale in July 1650 and published the following year.

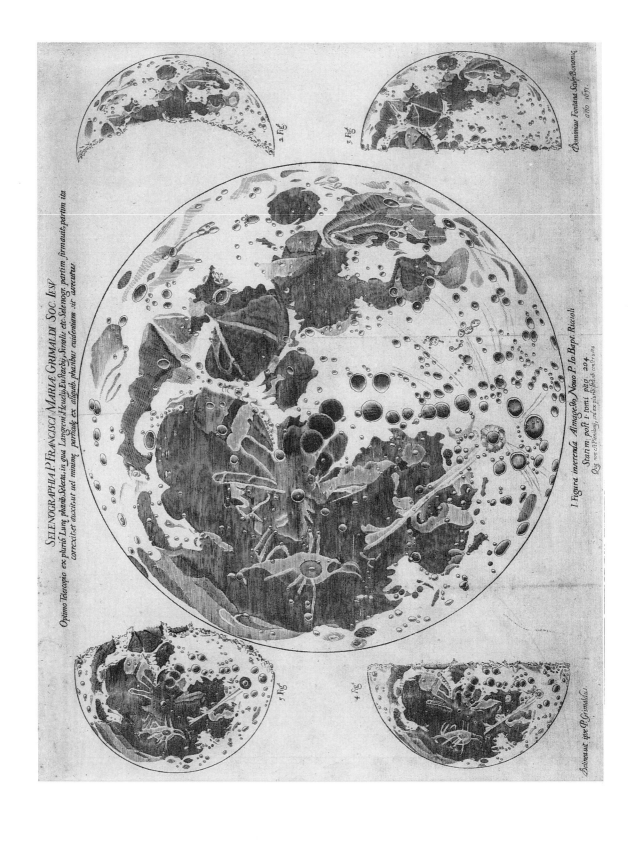

SELENOGRAPHIA P.FRANCISCI MARIÆ GRIMALDI SOC. IESU

Optimo Telescopio ex plurib. Lunæ phanib. Selecta, in qua Langreni, Hevelij, Eustachij, Sirsalis etc. Selenogr. partim firmauit, partim ita correxit: et vel minimæ particulæ ex aliquib. phasibus euidentiam ut videretur.

1. Figura incernula. Almagesto Nouo P. Io. Bapt. Riccioli

Statim post 1. tomi pag. 204.
Quæ non est Pliniani, sed ex plurib. phasib. construitur.

2. Fig.
3. Fig.
4. Fig.
5. Fig.

Dominicus Fontana Sculp.Bononiæ
anno 1651.

Delineauit ipse P. Grimaldus.

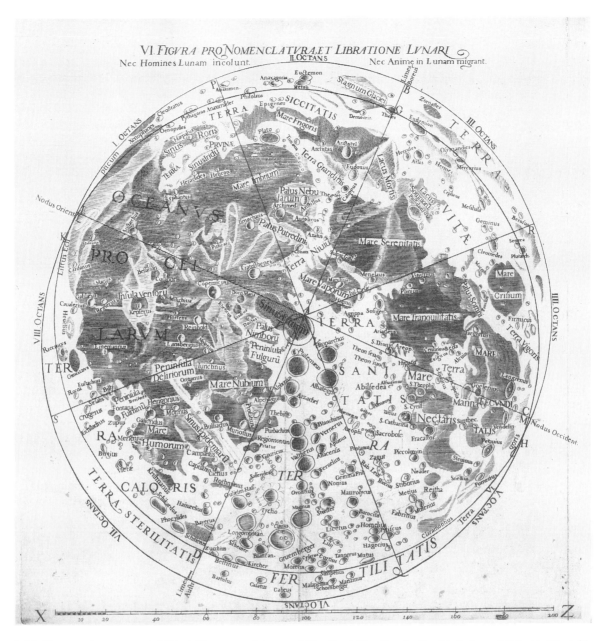

8.14. The same basic map with Riccioli's nomenclature and Hevelius's libration limits added, also published in Riccioli's *Almagestum novum* (1651). This scheme of names stood the test of time, and forms the core of modern lunar nomenclature.

8.13 (opposite). The map by Grimaldi, based on earlier maps and his own observations, and published in Riccioli's *Almagestum novum* (1651).

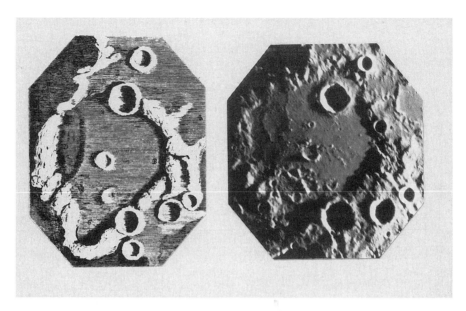

8.15. Hooke's drawing of the Hipparchus crater, made in 1664 with a 30 ft refractor, with a modern high-resolution photograph for comparison.

source of this categorization is usually stated to be Galileo, but nowhere does he actually say that he himself considers the dark spots to be stretches of water, and indeed in the First Day of the *Dialogue* (1632) he expressly denies that the Moon is made of earth and water. However, he was widely believed to hold this view and in the *Sidereus nuncius* he lays himself open to misinterpretation, saying ". . . so that if anyone wishes to revive the old opinion of the Pythagoreans: that the Moon is another Earth, so to say, the brighter portion may very fitly represent the surface of the land, and the darker the expanse of water".

Robert Hooke and selenology

There was, not unnaturally perhaps, a lull in lunar observation following the appearance of Hevelius's and Riccioli's landmark publications on the subject. However, in October 1664 Robert Hooke (1635–1703), the talented scientist, inventor, and curator of experiments at the newly formed Royal Society in London, made some pioneering observations with a telescope of 30 ft focal length. Drawing attention to the inadequate and stylized depictions of the crater Hipparchus on the maps in those earlier publications, he presents in his *Micrographia* (1665) a detailed drawing of that formation for comparison (Figure 8.15). Of more

interest and importance, however, is the text accompanying the drawing, in which Hooke speculates on the nature of the lunar surface and the origin of the craters: ". . . it is not improbable, but that the substance of the Moon may be very much like that of our Earth, that is, may consist of an earthy, sandy, or rocky substance, in several of its superficial parts, ..." He likens the craters to the temporary structures formed by dropping round bullets into a viscous mixture of pipeclay and water, but discounts this idea because he could not imagine where the source of such bodies might be in the case of the Moon. Lifting the bullets out again produced structures akin to the central peaks of larger craters. His second experiment was to heat dry alabaster powder in a pot; this released water vapour in the form of rising bubbles which left behind convincing replicas of lunar craters! He therefore speculated that the actual lunar craters might have been caused by volcanic or closely related phenomena. From considerations such as these, and also from the fact that the Moon is spherical apart from the proportionately minute topographical features, Hooke concluded that the Moon was substantially earthlike and possessed its own gravitational pull.

Two other selenographical efforts dating from the 1660s deserve mention. One is a rather strange

8.16. The 'Cassini large map' of 1679. Note the characteristic 'phi' marking in Mare Serenitatis, the 'volcano' in Petavius crater, and the lady's head at Prom. Heraclides.

looking map that was compiled from observations made in Modena on twelve consecutive clear nights in October 1662, the Moon's age ranging from $3\frac{1}{2}$ to $14\frac{1}{2}$ days. The observer and cartographer was Geminiano Montanari (1633–87), professor of mathematics at Bologna, and later of astronomy and meteorology at Padua. The artistic quality and positional accuracy of this map are very low, despite Montanari's use of a rotatable reticle of fine threads in the focal plane. Yet a surprising number

of craters and other features can be identified fairly easily. The map had no influence on subsequent work.

The second of these efforts was a globe of the Moon that was prepared by none other than Christopher Wren (1631–1723), then professor of astronomy at Oxford. Although Van Langren, one Matthias Hirzgarter (1574–1653), and Hevelius had mentioned the advantages of a lunar globe, none of them made one, so that Wren's is the first known example. It was based on a blank sphere "fixed on a pedestal of Lignum Vitae" made by Joseph Moxon, a leading maker at the time, and it represented not only the spots and various degrees of whiteness upon the surface of the Moon, "... but the hills, eminences and cavities of it moulded in solid work, which if turned to the light shewed all the phases of the Moon, with the several appearances that arise from the shadows of the hills and vales". Wren also apparently "measured the relative positions of the different formations" himself. The globe had been requested by the Royal Society, and was completed by the summer of 1661; it was subsequently placed among the curiosities of the King's Cabinet. Wren also worked on the problem of the Moon's librations, apparently making some sort of model to demonstrate them.

Cassini and La Hire

An early task undertaken by Gian Domenico Cassini (Cassini I, 1625–1712), the leading astronomer at the Paris Observatory during the last three decades of the seventeenth century, was the preparation of a detailed map of the Moon. In 1671 he had brought with him from Italy an objective between 8 and 10 cm in diameter and $5\frac{1}{2}$ m in focal length made by Giuseppe Campani, and soon obtained, from the same optician, another with twice those dimensions. Telescopes incorporating these lenses were the ones mainly used to make the observations, which commenced that year and continued until 1679. Cassini and two artist assistants, Jean Patigny and Sebastien Leclerc, produced some sixty drawings of the lunar surface during that period, some of them being detailed sketches of relatively small areas, others covering most of the visible disk. These drawings, which are still preserved in a large volume, scrapbook style, in the Observatory library, were used as the basis for the famous but quite rare 'Cassini large map'. The 22-inch diameter copperplate image was always assumed to have been engraved by the ageing Mellan, who prepared the three Gassendi images, but recent research shows that the actual engraver was Patigny.

The final map (Figure 8.16), some copies of which were available in 1679 or 1680, is quite dramatic in appearance, especially when seen at full scale. It gives the impression of portraying every detail of the lunar surface down to a size of about 15 km, but a careful comparison with photographs shows that much of this finer detail is illusory. The relative shapes, positions and dimensions of the larger formations leave something to be desired, but the map nevertheless contains much new verifiable detail, notably near the limb. Some points of special interest are the volcano-like mountain in the crater Petavius, the lady's head at Prom. Heraclides (Figure 8.17) and the characteristic 'phi' marking in Mare Serenitatis. The only recorded use of this map came over a century later, when Schroeter compared its portrayal of the Mare Crisium area with his own observations.

In earlier literature there is considerable confusion between this map, a smaller full-Moon image published in 1692 as an illustration accompanying an article by Cassini dealing with a forthcoming lunar eclipse, and two drawings by Philippe de la Hire (1640–1718), professor of mathematics and architecture in Paris: a 13 ft diameter image of the full Moon he made in 1686, and an engraved reduction, presumably of this, that appears in his astronomical tables first published in 1702. The 1692 map is usually referred to as "Cassini's map of the Moon", since it had a wide distribution in the Memoirs of the French Royal Academy of Sciences, whereas the large map apparently had a very limited original production. The author of this full-Moon image is not mentioned in the accompanying text, but it bears so many characteristics of the large map, such as the 'phi' and other markings in the maria, and the 23° tilt of the lunar axis, that obviously it was at least based on that map. The high-Sun markings appearing in this image that are not included in the large map can all be found in the original drawings. The main features are numbered 1–40 in the approximate order in which they are covered by the Earth's umbra, and an accompanying table identifies the features by their Riccioli names. This image was copied and recopied for over a century in textbooks, dictionaries, encyclopedias and the *Connaissance des temps* until the quality was so degraded that it was unrecognizable – except for the ubiquitous 'phi'.

Even though the lunar axis is rotated by 45°, the full-Moon image published in La Hire's tables (Figure 8.18) clearly shares many of the general characteristics of the previous image, yet it is everywhere different in regard to the smaller details, and is aesthetically more pleasing. This en-

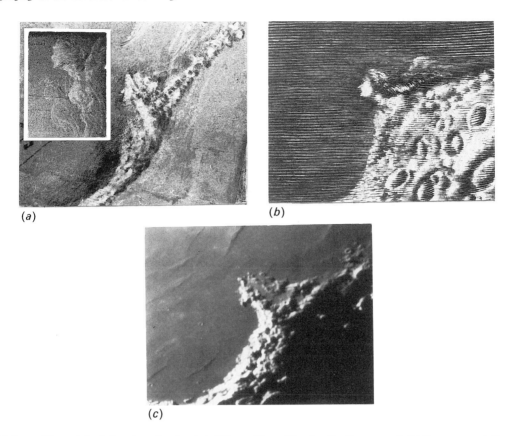

8.17. (*a*) Two of the original drawings by Cassini of Prom. Heraclides for the 'large map', (*b*) the representation in the 'large map', and (*c*) a modern photograph for comparison.

graving is no doubt a reduction of a 4 m diameter drawing (now lost) that La Hire made in 1686.

Cassini's best-known contribution to lunar studies is his formulation, in 1693, of the three laws that now bear his name. These deal with the orientation and motion of the lunar axis with respect to the ecliptic and the lunar orbit. The formulation of these laws could only have resulted from a lengthy period of careful quantitative observation of the lunar librations. Accurate determination of the coordinates of features on the Moon was impossible before these laws were formulated; but it would be over half a century before such determinations (by Tobias Mayer) would be accomplished for the first time.

Other seventeenth-century selenography

The only other known original observations from this period, apart from the inaccurate and fanciful full-Moon image by Georg-Christoph Eimmart

(1638–1705), engraver and amateur astronomer whose large collection of other lunar drawings may still be extant in a Silesian monastery, are the sporadic sketches by the Dutchman Christiaan Huygens (1629–95). Huygens apparently made very few lunar observations. At the close of 1658 he noted five bright craters ("round valleys") in Mare Fecunditatis (none is shown on then existing maps), and sketched a crater (Tycho?). His next recorded observations were not made until 1685 and 1686, when three separate 'discoveries' were logged. On the first occasion (11 April 1685), when "never before was the half-Moon seen better or more clearly", he observed and sketched "a brownish spot from which a ditch issued to a bright cavity *B*, thence further to *C*". This was clearly an observation of the Hyginus Rille and crater. In the spring of the following year he observed and sketched "an irregularly curved trench near a very bright sunken circle". Here he had discovered Schroeter's

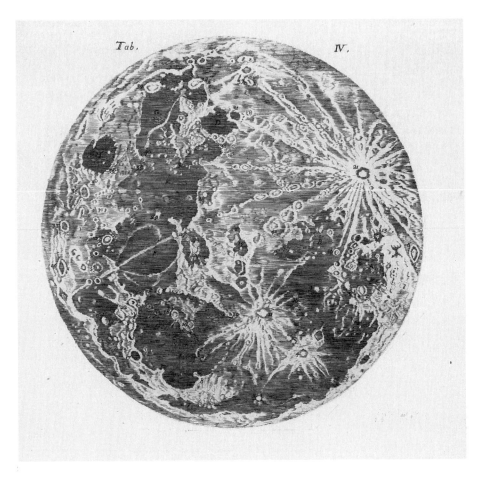

8.18. The image of the full Moon published by La Hire in 1702, and based on the 4 m diameter drawing made by him in 1686.

Valley over a century before that observer rediscovered it! At the end of May of the same year he observed the shadow of the Straight Wall (now Rupes Recta), commenting on its extreme straightness and noting its reduced width on the following night. He likened the appearance, which included the small group of hills to the south, to a sword. All three features were forgotten until the close of the eighteenth century, when Schroeter, with his reflecting telescopes, rediscovered them.

Conclusion

From our twentieth-century vantage point, the selenography discussed here may seem to be rather crude and naïve. At the time, however, it undoubtedly helped to sweep away mediaeval concepts concerning the Moon, and to reinforce the general astronomical renaissance of the period. Through most of the eighteenth century, the perceived excellence and breadth of Hevelius's *Selenographia* ensured its continuing reputation as the leading authority on the subject, although atlases and textbooks usually presented copies of the Hevelius and Riccioli maps side by side. As already noted, the Cassini full-Moon map was made rather widely available during this same period, for the ultimate purpose of improving terrestrial longitudes.

The main scientific value of the seventeenth-century selenography was probably in providing reference material and starting points for several major new selenographical projects undertaken between about 1790 and 1830. The most influential of these was that of Wilhelm Beer and Johann Heinrich Mädler, who adhered closely to the nomenclature of Riccioli, thereby providing the prime reference for our current lunar nomenclature. The scheme of using the names of famous people for designating surface features on the Moon, devised by Van Langren, has recently been extended to many of the other planets and satellites.

Further reading

Galileo Galilei, *The Starry Messenger*, in *Discoveries and Opinions of Galileo*, transl. by Stillman Drake (Garden City, NY, 1957)

P. Humbert, La première carte de la lune, *Revue des questions scientifiques*, vol. 20 (1931), 193–204 (for Gassendi, Peiresc and Mellan)

Z. Kopal and R.W. Carder, *Mapping of the Moon, Past and Present* (Dordrecht, 1974)

A. Paluzíe Borrell, Historia de la cartografía lunar, *Urania*, no. 266 (1967), 1–69

S.P. Rigaud, *Miscellaneous Works and Correspondence of the Rev. James Bradley*, Supplement (Oxford, 1833; reprinted New York, 1972) (for Harriot and Lower)

O. Van de Vyver, Lunar maps of the XVIIth century, *Vatican Observatory Publications*, vol. 1 (1971), 71–114

O. Van de Vyver, Original sources of some early lunar maps, *Journal for the History of Astronomy*, vol. 2 (1971), 86–97

T. Weimer, Carte de la lune de J.D. Cassini, *The Moon and the Planets*, vol. 20 (1979), 163–7

Ewen A. Whitaker, Galileo's lunar observations and the dating of the composition of *Sidereus nuncius*, *Journal for the History of Astronomy*, vol. 9 (1978), 155–69

The Galilean satellites of Jupiter from Galileo to Cassini, Römer and Bradley

SUZANNE DÉBARBAT and CURTIS WILSON

The discovery by Galileo Galilei (1546–1642) of the 'Medicean stars' in 1610 – four moons of Jupiter, and the first such objects to be discovered with the telescope – marks a new epoch in the history of science. Hitherto the universe had been equally accessible to all generations of enquirers. But in future, by using artificial aids to extend the powers of the human senses, scientists in general – and astronomers in particular – would be able in each generation to discover new information about the natural world. And from this new information would constantly arise new challenges to accepted theories and explanation.

The Galilean moons were also to play a key role in a controversy that had hitherto been 'academic' but that was now found to have far-reaching implications. The astronomers's perception of the universe is based upon the information that light brings us from the stars and planets. But does this light reach us instantaneously; or does it take time to traverse the intervening space, in which case we see celestial objects, not as they are when the light reaches us, but as they were when the light began its journey? The controversy was argued out in the later decades of the seventeenth century and brought to a conclusion early in the eighteenth, and much of the evidence came from eclipses of the Galilean moons.

The work of Galileo

On 7 January 1610 (NS) Galileo observed through his telescope three small stars very close to Jupiter, two to the east and one to the west of the planet, and forming with it a straight line. On the following night, there were three small stars to the west of Jupiter; on the 10th there were two to the east; and on the 12th there were once more two to the east and one to the west. Continued observations on 13–15 January led Galileo to the idea that these stars revolved in orbits about Jupiter, and were thus 'satellites' – to use the term that Johannes Kepler (1571–1630) later proposed for them.

In his *Sidereus nuncius* published that spring Galileo wrote that the period of the outermost satellite was about 15 days, but that he had not yet determined the periods of the others. Kepler in his *Narratio de Jovis satellitibus*, dated 11 September 1610 (though published in 1611), questioned whether the periods of the inner three would ever be known: how to distinguish them from each other? A lucky observation enabled Galileo to solve the puzzle: the configuration of the inner three on 10 December 1610 proved to be almost exactly the same as their configuration one week earlier, on 3 December, while in the same interval the fourth satellite had moved very nearly from greatest western to greatest eastern elongation. In fact, the periods of the first three satellites are almost as 1:2:4, the period of the third satellite being just a little over 7 days; the period of the fourth satellite is about 16.75 days.

By April 1611 Galileo had a fairly precise knowledge of the periods; he published these in 1612 at the beginning of his *Discorso intorno alle cose che stanno in su l'acqua o che in quella si muovono*, warning the reader, however, that they could still be in need of some correction. Compared with modern figures for the *mean* synodic periods, Galileo's values are very good indeed; the greatest error is a mere 0.05%. But although the motions of these satellites with respect to the stars are nearly uniform, their returns to the line from the Sun to Jupiter – their synodic periods – are not quite constant. As Galileo continued his observations, he became aware that perpetual tables were beyond his reach, and that it was necessary repeatedly to reconstitute the epochs from which the motions of the satellites were reckoned.

To predict with accuracy a conjunction or an occultation (the disappearance of a satellite behind the disk of Jupiter), Galileo found, it was necessary to take account of the 'prosthaphaeresis' of the Earth's orbit – the displacement of the observer due to the Earth's annual motion. It was with the aid of tables for the mean motions and tables of the prosthaphaeresis that Galileo, on 18 March 1612, first detected an eclipse of a satellite: at a moment when the fourth satellite should have been visible according to his tables, it was not to be seen, and he was able to conclude that it had passed into the cone of the shadow cast by Jupiter.

An eclipse, unlike a conjunction or an occultation, is a phenomenon of which the instant is independent of the place of observation. Galileo quickly recognized that the eclipses would make it possible to time the satellites with respect to their aphelia. Because of their clock-like recurrences, visible each time from a whole hemisphere of the Earth, they offered a means for the precise determination of differences in local time and, hence, in longitude. Already by September 1612 Galileo had presented a proposal to this purpose to the King of Spain, who since 1598 had been offering a handsome reward to anyone who could "discover the longitude".

The negotiations with Spain dragged on through two decades. At length, having failed to sell his idea to Spain, Galileo entered during his last years into negotiations with the States General of Holland. The repeated reconstitution of the epochs or 'radices' of the motions of the satellites, required if Galileo's tables of mean motions and prosthaphaereses were to serve for the accurate determination of longitude, became increasingly difficult for Galileo and finally impossible as failing sight turned to blindness. The negotiations ceased with his death.

Successors of Galileo to the mid-century

In 1614 Simon Mayr (Marius, 1573–1624), a Bavarian, published his *Mundus Jovialis* in Nuremberg (see Figure 9.1); he claimed therein that he had been observing the little stars near Jupiter during the course of December 1609, and by 29 December had concluded – because they remained close to Jupiter and Jupiter was then retrograde – that they were satellites of Jupiter. (As Galileo would later point out, Mayr was using Julian-style dates, so

that his 29 December is 8 January, Gregorian-style, hence one day after Galileo's first observation of the small stars.)

I do not recount these things [Mayr avers], as if I wished to rob Galileo of his reputation of having been the first among the Italians to discover these Jovial stars, not at all, but so that it may be understood that I found and observed these stars without help from others, by my own investigations, at almost the same time as, or somewhat sooner than Galileo first saw them in Italy.

The periods of the satellites as given by Mayr in his *Mundus Jovialis* are again very close to the modern values; the greatest deviation is only 0.03%. Indeed, in the case of the first two satellites, Mayr's figures are notably closer to the modern mean synodic periods than those given by Galileo in 1612. But Mayr was unaware that the epochs of the motions would repeatedly need to be reestablished; he imagined that the inequality of Jupiter's motion would have but negligible effect. Like Galileo, he took account of prosthaphaeresis, but attributes it to the motion of the Sun rather than to that of the Earth. He was in fact an adherent of the Tychonic system (see Chapter 1), which he claimed to have thought of before ever hearing of Tycho Brahe's description of it.

In his *The Assayer*, Galileo gives the following argument to show that Mayr had not been observing the satellites in 1610–11. According to Mayr, the satellites appear on a straight line through Jupiter parallel to the ecliptic only when they are at their maximum elongations from the planet; otherwise they depart from this line, southward when they are in the superior semi-circles of their orbits, northward when they are in the inferior semi-circles. Mayr did not realize that this configuration is temporary. But according to Galileo, the configuration in 1610 was different: the satellites were north of the line parallel to the ecliptic in their superior semi-circles, south in their inferior semi-circles. Not till 1612 did the configuration described by Mayr emerge, after the satellites had for a time appeared always in the line parallel to the ecliptic. According to Galileo, the change in the appearances was simply an effect of parallax, due to Jupiter's having moved from the southern to the northern side of the ecliptic; the plane of the orbits of the satellites, he held, was parallel to the ecliptic.

SIMON MARIVS GVNTZENH. MATHEMATICVS
ET MEDICVS ANNO M. DC. XIV. ÆTATIS XLII.

PERSPI CILIVM

JNVENTVM PROPRIVM EST: MVNDVS IOVIALIS, ET ORBIS
TERRÆ SECRETVM NOBILE, DANTE DEO.

9.1. The title-page of Simon Mayr's *Mundus Jovialis* (1614). This work contained the first tables of the mean periodic motions of Jupiter's four satellites, which we still know by the names Mayr gave them.

(Galileo is somewhat in error here, though less so than Mayr: the orbital plane of the satellites makes an angle of 3°7′ with the plane of Jupiter's orbit, which is in turn inclined to the ecliptic by an angle of about 1°19′.)

The plausible conclusion is that Mayr depended heavily on Galileo's prior announcements to identify the satellites and limit their periods. Of Mayr's attempt to persuade his readers that Galileo was not the first discoverer, the Jesuit Christoph

Scheiner (1573–1650) wrote, shortly after the appearance of Mayr's book in 1614, that it was "in vain and too late", and "most inopportune".

In France, the first observers of the Galilean satellites of Jupiter were Nicolas Claude Fabri de Peiresc (1580–1637) and, assisting him, Joseph Gaultier, on 24–25 November 1610 at Aix en Provence. Peiresc and Gaultier continued to record observations until June 1612. At Peiresc's request, Gaultier determined periods for the satellites, but his values were soon found to require correction. It was Peiresc's plan to publish observations and tables of the satellites' motions, but on learning that Galileo intended to do the same, he abandoned the project.

Like Galileo, Peiresc conceived the project of using observations of the satellites to determine differences in longitude. He sent Jean Lombard on an expedition to Marseilles, Malta, Cyprus, and Tripoli (Lebanon), with the object of recording in each place both the configuration of the satellites as observed and the local time. These observations, collated with those made at Aix, would permit calculation of the differences in longitude between the places. But Peiresc was not satisfied with the results. In 1612 he decided to leave the perfecting of the method to Galileo and Kepler.

Years later, Peiresc took up the problem of determining differences in longitude once again, this time, however, using eclipses of the Moon rather than observations of the Galilean satellites. On 20 January 1628, assisted by Gaultier and Pierre Gassendi (1592–1655), he observed an eclipse of the Moon in Aix, while Marin Mersenne and Claude Mydorge observed it in Paris, in order to obtain the difference in longitude between the two cities. During the last years of his life, from 1634 to 1637, he and Gassendi engaged in selenographical studies with a view to perfecting this method. Through Peiresc's encouragement and direction, the lunar eclipse of 28 August 1635 was observed in Rome, Naples, Aleppo, Cairo, Tunis, Aix, Digne, and Paris, mostly by observers trained by Peiresc, Gaultier, and Gassendi. Thus were obtained the first reasonably accurate values for the length of the Mediterranean – some 41° of longitude, replacing the 60° used in earlier navigational guides.

The possibility of using the Galilean satellites for the determination of longitude may not have been altogether abandoned by Peiresc and his colleagues: Gassendi continued observing the satellites over the years from 1624 to 1645. But the eclipses of the satellites could not yet be predicted accurately enough. Accurate prediction required that the two components in the synodic periods of the satellites – the sidereal motions of the satellites on the one hand and the circumsolar motion of Jupiter on the other – be calculated separately, with due account being taken of the unequal motion of Jupiter. It required also that the latitudinal motions

of the satellites be understood and reduced to calculation. For not all eclipses of the satellites are central in the sense of passing through the axis of the cone of Jupiter's shadow; indeed, the fourth satellite in its aphelion may have such latitude as to escape being eclipsed altogether for three or four years at a time. Not before the publication in 1668 of the *Ephemerides bononienses* of Gian Domenico Cassini (Cassini I, 1625–1712) were the means at hand for using observations of the satellites for the accurate determination of longitude.

The difficulty of the undertaking was not generally realized. Pierre Hérigone (Clément Cyriaque de Mangin, d. 1642) in his *Cursus mathematicus* (1634–37) announced "A new and easy method of finding longitudes both on the sea and on the land":

Let it be observed, with a good telescope, at what hour of local time one of the satellites of Jupiter arrives at the line of sight between our eye and Jupiter. Then find by astronomical tables at what hour of the day of observation the said satellite must be in the said line of conjunction. The difference in the time found by observation and that by the tables reduced into degrees and minutes . . . will be the difference in longitude between the place of observation and that for which the tables were constructed.

Hérigone is quite sure that the satellites can serve as an infallible clock, and that their periods about Jupiter are always equal. He speaks of the new method as "our invention", claiming to have thought of it two years before Galileo published the idea. He takes the periods of the satellites from Mayr's *Mundus Jovialis*.

New tables of the satellites, based on ten years of observations, were promised by Vincent Reinerius, a pupil of Galileo and mathematician at the University of Pisa, in a letter of 11 September 1647 to Giambattista Riccioli in Bologna. Two months later Riccioli learned of Reinerius's death, and despite his own efforts and those of the Grand Duke of Tuscany to recover the tables, they were never found.

The *Menologiae Iovis compendium seu ephemerides Medicaeorum* of Gioanbatista Odierna (or Hodierna, 1597–1660), mathematician to the Duke of Palermo, appeared in 1656. Odierna had observed the satellites assiduously from 1652 to 1655. Their synodic periods, he urged, should be determined with respect to their conjunctions not with Jupiter but with the mid-point of the planet's shadow; for Jupiter's motions are to be reckoned with respect to the Sun (Odierna implies that the Sun moves about a stable Earth, while the other planets move about the Sun). To this end he made a careful study of the latitudes of the satellites, and concluded that their orbital plane is invariable but not parallel to the ecliptic as Galileo had supposed. His values for the periods are within 0.02% of present-day values for the first three satellites; his value for the fourth errs by 0.04%. He provides ephemerides for the central moment of the eclipses of the satellites for the years from 1650 to 1682.

In the period between Galileo's discovery of the satellites and the middle of the century, it was several times suggested or claimed that the number of the satellites exceeded four. Scheiner, writing to Mark Welser in 1612 under the pseudonym Apelles on the subject of sunspots, said "it is suspected" that there are more than four, but a little later in his *Disquisitiones mathematicae* of 1614, published under his own name, recognized only four. The Capuchin monk Anton Maria Schyrlaeus de Rheita (1597–1660) claimed that on 29 December 1642 and at later times he had observed five additional satellites, to which he gave the name *Urbanoctavianes* after Pope Urban VIII. Gassendi, in refutation, argued that the putative satellites were fixed stars in the constellation of Aquarius. Schyrlaeus answered in his *Oculus Enoch et Eliae* of 1645 that the mutual distances of the stars changed, so that they could not be *fixae*; the fact that they had later disappeared altogether was not astonishing, since they had been newly generated, and so could perish.

Francesco Fontana in a series of observations from 1630 to 1646, reported in his *Novae coelestium terrestriumque rerum observationes* (1646), claimed that besides the Galilean four there were five other stars that accompanied Jupiter and whose mutual distances did not remain constant. Johannes Baptista Zupus, using one of the excellent telescopes made by Fontana, reached a similar conclusion in 1644. Riccioli, reviewing these claims in his *Almagestum novum* of 1651, concluded that the case was not yet clear: more evidence was required. Odierna in his *Ephemerides Medicaeorum* cited above denied that there were more satellites than four: Schyrlaeus and Fontana, he urged, were merely mistaking fixed stars for satellites.

The Galilean satellites and the system of the world

To Galileo, discovery of the Jovian satellites seemed a confirmation of the Copernican system: it showed that planets could have satellites, and so legitimated the idea that the Earth, about which the Moon revolves, could be a planet. That the prosthaphaereses due to the Earth's motion had to be taken into account in predicting the appearances of the satellites, seemed to Galileo a further confirmation of the Copernican arrangement. Others, however, found no difficulty in accommodating the Galilean satellites within the Tychonic system, for mere kinematic facts told little one way or the other. But Kepler introduced a new kind of argument, the dynamic; and the dynamic account of the motions of the satellites was to play a key role in the establishment of the Newtonian theory of gravity and so of the Copernican system it substantiates and implies.

Already in his *Dissertatio cum Nuncio sidereo* of May 1610, Kepler had boldly – without observational evidence for support – announced that Jupiter rotates on its axis, thereby causing the satellites to revolve in orbits about it. In his *Astronomia nova*, which had appeared in the previous year, he had argued that the circumsolar planets are pushed about in their orbits by an immaterial virtue issuing from the Sun, and turning with the rotation of the Sun; the solar rotation was shortly to be confirmed by the observation of sunspots. Observational evidence of the rotation of Jupiter was not to emerge for some years. Fontana in the mid-1640s hypothesized that the changing appearance of the bands on the planet's surface indicated rotation. The motion of a distinctly coloured spot at length enabled Cassini, in 1664, to establish the period of rotation.

But long before this, Kepler had taken another step along the road that would lead to a causal account. In 1619 he had discovered the law called his third: the squares of the periods of the planets are as the cubes of their mean solar distances. In his *Epitome astronomiae Copernicanae* he explained how a dynamics in which velocity varies directly as force and inversely as quantity of matter could account for this law, provided that the solar force and volumes and densities of the planets are certain monotonic functions of the distance from the Sun. In the same work he suggested that this law would apply to the satellites of Jupiter. Using the estimated distances of the satellites from Jupiter and their periods as provided in Mayr's *Mundus Jovialis*, he showed that the periods varied neither with the distances nor with the square of the distances, but with an intermediate proportion.

The unqualified claim that the periods of the satellites were in the sesquialterate proportion (that is, as the $\frac{3}{2}$ power) of their distances was made by Gottfried Wendelin in a letter to Riccioli, as the latter reports in his *Almagestum novum*. Wendelin had earlier, like Horrocks, arrived at the conclusion that the law applies so exactly to the circumsolar planets that their relative mean solar distances can be calculated from their periods, which are easier than the distances to ascertain with precision (see Chapter 10 below). Whether Wendelin's claim for the satellites was based merely on analogy or on an attempt at precise measurements of their distances from Jupiter is not indicated in Riccioli's report. Riccioli's own opinion of these Keplerian 'harmonies' was that they showed more cleverness than solid erudition or true doctrine.

Estimates of the relative distances of the satellites from Jupiter's centre in terms of the semi-diameter of Jupiter's disk were made by Galileo, Mayr, Schyrlaeus, and Odierna, always with considerable underestimation of the distances of the satellites, owing to the blurring in definition of Jupiter's disk in the early telescopes. Of these results, Schyrlaeus's were least accurate, Odierna's most accurate – good to within about 1% for the relative distances of the inner three satellites, but in error by 7% for the distance of the fourth. But a precise instrument for making the estimates was not yet available. The filar micrometer, brought into use in the 1660s, would make possible the precise verification of Kepler's third law for the Galilean satellites of Jupiter, and it was John Flamsteed (1646–1719) who would do the verifying.

Flamsteed's early astronomical education included the reading of Riccioli's *Almagestum novum*, and it is most likely in this source that he first encountered the idea of applying Kepler's third law to the Galilean satellites. Richard Towneley, who first supplied Flamsteed with a micrometer (in 1670), encouraged him to make a study of the distances and periods of the satellites; Towneley had already made his own study of the appearances, and wanted Flamsteed to compare his results with those of Cassini. So resulted Flamsteed's first letter to Cassini, dated 21 July 1673 (OS), in which Flamsteed makes several comparisons between Towneley's and his own measurements of the distances of the satellites and the cruder, premicrometer estimates in Cassini's *Ephemerides Bononienses* of 1668. Cassini acknowledged that Flamsteed's measurements were superior.

Flamsteed continued his study of the satellites, and by the time Isaac Newton (1642–1727) was engaged in writing his *Principia* could assure him (in a letter of 27 December 1684 (OS)) that the distances of the satellites from Jupiter's centre were "as exactly in sequialte proportion to [that is, as the $\frac{2}{3}$ power of] their periods as it is possible for our sences to determine". This fact was to become the first phenomenon cited in Bk III of the *Principia* (in the first edition it appears as Hypothesis V, after four 'hypotheses' that are regulative or suppositional rather than statements of phenomena). Flamsteed later told Newton, in a letter of 11 October 1694, that Cassini "allows [the satellites'] distances to be in sesquialter proportion to the periods of their revolutions. Though, to be thought a good Catholic, he says nothing of it, but conceals it . . .".

The first tables of Cassini

It was in 1650 that Cassini, then newly appointed professor of mathematics at the University of Bologna, took up the subject of the tables of the satellites of Jupiter. He examined Galileo's observations and those of Gassendi, which extended to 1645: he thus had at his disposal nearly three periods of the sidereal revolution of Jupiter. The compilation that he undertook convinced him that accurate prediction of the phenomena of the satellites could be achieved.

In 1652, using a copy of Torricelli's telescope that the Marquis Cornelis Malvasia had had made for him, Cassini began a series of observations that would continue for fifteen years. On the basis of these observations he drew up tables of mean motions, and thence deduced ephemerides predicting the phenomena to be observed. In 1664 in Rome, with the aid of the new telescopes of Giuseppe Campani and his own ephemerides, he was able for the first time to detect the shadows of the satellites, projected by the Sun onto the illuminated disk of Jupiter. Timing of the transits of the shadows made possible a new level of accuracy of prediction. Some contemporaries doubted the existence of these shadows. Partly to convince them, and partly to make known the quality of his researches, Cassini addressed an open letter, dated 22 July 1665, to the Abbé Falconieri, foretelling therein the principal phenomena for the months of August and September of the same year. Observation confirmed his predictions.

In March 1668 were published the *Ephemerides Bononienses Mediceorum syderum ex hypothesibus et tabulis Io. Dominici Cassini*. Cassini here takes the motions of the satellites to be circular and uniform, but in his introduction he remarks that the second and third satellites are subject to inequalities. Eventually, he believes, these will be accounted for by such theories as Giovanni Alfonso Borelli has developed, using the lunar inequalities as an analogy; in the interim Cassini applies empirical equations that he does not divulge.

Among other subtleties not taken into account in the tables is the fact that, when Jupiter is in quadrature to the Sun, about $\frac{1}{50}$ of the face directed toward the Earth is unilluminated, so that a satellite can be occulted before it reaches the illuminated disk. But the tables are sufficiently accurate, Cassini asserts, to serve as a basis for their own further refinement and to make possible good determinations of differences in terrestrial longitudes.

Their accuracy depends crucially on Cassini's discovery of the principles for determining the latitudes of the satellites: the constant inclination of their orbits to the plane of Jupiter's orbit, at an angle that Cassini takes to be double the inclination of Jupiter's orbit to the ecliptic (it is actually half a degree larger), with the node moving about 6° per sidereal revolution of Jupiter. For on the latitudes depend the exact times of eclipses. Following the tables Cassini gives ephemerides of the principal phenomena for the year 1668 (Figure 9.2). Throughout he uses for the satellites the names Pallas, Juno, Themis and Ceres. These names would be abandoned in the eighteenth century and replaced by Io, Europa, Ganymede and Callisto, the names proposed by Mayr.

Cassini's tables very quickly reached Britain: they are mentioned in *Philosophical Transactions* for 18 May 1668 (= 28 May NS). In France the *Journal des Sçavans* published a review of Cassini's work in the issue for 17 December 1668. The author (possibly Adrien Auzout (1622–91)) declares that "several observations have been made at the Bibliothèque du Roi to verify these ephemerides....

They have often been found to have greater accuracy than the author promised for them." There follows a description of eight observations, made between 7 October and 20 November, of the first, second and third satellites.

Cassini's tables and ephemerides of the satellites of Jupiter were to be of great importance for the future development of astronomy in France, for it was the quality of these tables that led to the invitation from Louis XIV to Cassini to come to France, to assist in the planning and operation of the new observatory, the first stone of which had been laid on 21 June 1667. Cassini arrived in Paris in 1669, and was installed in the Observatory some years later. He became the leading astronomer of France in his time, and the first of four Cassinis (the others all his direct descendants) who took leading roles in the astronomy of France. The post of Director General of the Paris Observatory was created for the third Cassini by Louis XV in 1771.

The observations of Picard and Cassini

On his arrival in Paris, Cassini was introduced to the king, and joined the Parisian group of astronomers, led by Jean Picard (1620–82) and Auzout, together with the Dutchman Christiaan Huygens (1629–95), who was in Paris from 1665 to 1681. One pressing need was to take advantage of the observations that Tycho Brahe had made at Uraniborg in the previous century, and for this it was necessary to reduce them to the meridian of Paris, by determining the difference in longitude between the two observatories. Cassini and Picard undertook to determine this difference by the method that Cassini had proposed in his *Ephemerides* of the Medicean stars.

Picard departed for Denmark in July 1671, while Cassini remained in Paris. Observations of the eclipses of the first satellite of Jupiter were made simultaneously in Paris and either at Uraniborg or in Copenhagen, the difference in longitude between the latter two places having already been accurately determined. Picard was assisted by a twenty-six-year-old Dane, Ole Christensen Römer (1644–1710), who was studying the observations made by Tycho.

The letters exchanged between Picard and Cassini bear witness to the care with which the operations were conducted. The project was completed in 1672 after eight months of work. The

Iannuarius. **1668.**

Configurationes Mediceorum.

Hora 7. P.M.

Dies

A 1		
2		
3		
4		
5	Primus in facie.	
6		
7		
A 8		
9		
10		
11	Secundus post ♃	
12		
13	Primus post ♃	
14		
A 15		
16	Tertius post ♃	

Con-

Iannuarius. **1668.**

Congressus cum Ioue, Eclipses, & maximæ digressiones Mediceorum.

Iupiter occidet nobis h. 8. o. db Occ.

Die 1. Hora 7.35. p. m. Maxima digressio orientalis primi. H. 12.42. maxima digressio orientalis secundi.

2 Hora 8.47. p.m. immersio inferior secundi, h. 11.13. eiusdem emersio. h. 12.44. ingressus ipsius vmbræ in Iouem, h. 15.10. egressus vmbræ.

3 Hora 7.20. maxima digressio occidentalis secundi.

4 Hora 8.43. immersio superior primi, hor. 11.7. eiusdem emersio. Hora 3.36. immersio superior secundi, h. 6.6. eiusdem emersio, h. 6.35. immersio in vmbram, hor. 8.24. emersio, h. 18.20. coniunctio inerior quarti.

5. Hora 5.47. immersio inferior primi, h. 7.4. ingressus ipsius vmbræ in Iouem, hora 8.11. emersio ex facie, h. 9.28. egressus vmbræ. H.10.24. immersio inferior tertij, h.13.20. ipsius emersio, h. 15.35. ingressus vmbræ in Iouem, h. 18.31. egr. vmbræ.

6. Hora 3.11. immersio superior primi, h. 5.35. emersio.

7. Hora 6.50. maxima digressio occidentalis tertij.

8. Hora 9.29. maxima digressio orientalis primi.

9. Hora 6.42.30. maxima digressio occidentalis primi. Hor. 11.20. immersio inferior secundi, h.13.46. ingressus ipsius vmbræ in Iouem, h. 11.46. emersio, h. 16.2. egressus vmbræ.

10. Hora 9.53. maxima digressio occidentalis secundi.

11. Hora 5.23.30. immersio superior secundi, h. 8.3.30. emersio, h. 8.16.30. immersio in vmbram, h. 10.56.30. emersio.

12. Hora 7.50.30. p. m. immersio inferior primi, hor. 8.11. eiusdem emersio h. 7.4. ingressus ipsius vmbræ in Iouem, hora 9.28. egressus. H. 9.29. maxima digressio orientalis secundi. H. 14. 8. immersio inferior tertij, h. 17.4. emersio, h. 19.35. ingressus ipsius vmbræ in Iouem.

13. Hora 5.29. p. m. immersio superior primi, h.9.11. emersio. H. 5. 15. coniunctio superior quarti.

16. Hora 4.22. immersio superior tertij, h.7.18. ipsius emersio ex Ioue, h.9.39. immersio in vmbram, h.12.35. ipsius emersio ex vmbra. H.8.38. maxima digressio occidentalis primi. H. 13.57. emersio inferior secundi, h. 16.23. emersio, h. 16.37. ingressus ipsius vmbræ in Iouem, h. 19.3. egressus eiusdem vmbræ.

A 2   Con-

9.2. Pages from Cassini's *Ephemerides Bononienses Mediceorum syderum* (Bologna, 1668).

instruments used included sectors and quadrants with sighting pins or telescopic sights, the latter equipped with the micrometers that Auzout and Picard had been using since the mid-1660s. The observers also employed aerial telescopes of great length: since the middle of the century (see Volume 3) opticians, among them Eustachio Divini and Campani in Rome, had been fabricating large objective lenses up to 12 cm in diameter and with focal lengths that could reach several dozen metres. The quality of the glass and the excellence of the polishing made possible important and much publicized discoveries (see Chapter 7 above).

Observation of an eclipse of a satellite of Jupiter was made by noting the clock-time when the phenomenon occurred. The clock, set to local mean time, had to be capable of keeping time with sufficient precision between resettings, which were usually undertaken by observations of the Sun at midday. Picard, Cassini and their collaborators used clocks with pendulums adjusted to periods of a second or a half-second; with such clocks it was possible to time events with a precision of a quarter of a second.

At the conclusion of the project, comparison of all the observational data collected led to a value for the difference in longitude between Uraniborg and Paris of 42 minutes 10 seconds (the modern value is 40 minutes 26 seconds). This degree of accuracy is remarkable since on maps of this period one can find longitudes in error by several minutes of time. The same method was used a little later by Cassini and Picard to rectify the longitude of several French cities, in particular those on the coast of Brittany, which were found to be closer to Paris than previously supposed by several dozen kilometres.

From their determination of the difference in longitude between Uraniborg and Paris, Cassini and Picard were able to reduce to the longitude of Paris not only Tycho's observations but also all the observations made during the course of the project of 1671–72. The latter brought to light some notable divergences between the observations and the predictions from Cassini's tables. These divergences could be as high as several minutes of time, and they were inexplicable, given the known precision of the measurements. Römer, who had been tempted away from Denmark by Picard and was now installed at the Paris Observatory, joined with Cassini in undertaking a detailed comparison between observations spread over nearly ten years and the corresponding predictions derived from Cassini's tables. Neither of them could admit the possibility of errors of such magnitude in either the tables or the observations.

The investigations of Römer and the discovery of the finite velocity of light

Owing to a series of unfortunate accidents, we lack the original documents that would enable us to trace with certainty the work of Römer on the satellites of Jupiter and his role in the discovery of the finite velocity of light. According to the later account of Bernard le Bouyer de Fontenelle (1657–1757), who in 1697 entered the Academy of Sciences as its Perpetual Secretary, Cassini on 22 August 1674 presented to the Academy a document in which he explained the inequality that had been found to exist in the immersions and emersions of the first satellite of Jupiter. The inequality "appeared to come from the fact that light takes some time to come from the satellite to us, and that it requires some ten or eleven minutes to traverse a space equal to the semidiameter of the terrestrial orbit". In the same paper Cassini apparently announced that an eclipse of the first satellite due on the following 16 November would happen ten minutes later than the time predicted from his tables – this at least is what we read in the *Histoire de l'Académie des Sciences* for the year 1707 (published in 1730).

The problem with this account is that in 1676 Cassini (as we shall see) was to offer a quite different explanation of the inequality and would strenuously challenge the finiteness of the velocity of light; and this has led one historian to suggest that

the document of 22 August 1674 was written by Römer rather than Cassini, and that Fontenelle was simply confused as to who had been responsible. This suggestion is supported by the fact that Cassini never challenged Römer's claim to priority in explaining the inequality by means of the velocity of light. In 1693, when presenting his new tables of the satellites, Cassini writes: "M. Römer explains very ingeniously one of these inequalities that he had observed in the first satellite over a period of several years, by the successive movement of light, which requires more time to come from Jupiter to the Earth when it is farther away from the Earth than when it is closer."

Some historians in the past, relying on Jean-Baptiste Du Hamel's *Regiae scientiarum academiae historia* (1698), and assuming that the original proposal of the finite velocity of light as an explanation of the anomaly was due to Cassini, have shifted the date of the proposal to 1675 or 1676; but the shift to either of these dates proves to be implausible. If we accept Fontenelle's account as referring correctly to Cassini, the conclusion to draw is that *Cassini proposed the finiteness of the velocity of light in 1674*, and later changed his mind.

Cassini's predictions of eclipses of the first satellite of Jupiter for August to November 1676 were published in *Journal des Sçavans* in August of that year. They included those of 9 November, predicted for 5 hours 27 minutes, and 16 November, predicted for 7 hours 21 minutes. In September Römer told the Academy that "if his supposition [that the velocity of light is finite] is true, the emersion of the first satellite that must occur on the following 16 November will happen 10 minutes later than it should according to the ordinary calculation [as in the tables of Cassini]". The records we have (see Figure 9.3) are in fact of the eclipse of 9 November. The emersion was observed by Picard at 5 hours 37 minutes 49 seconds, and it was later given by Römer as being timed at 5 hours 35 minutes 45 seconds, both times being of the order of ten minutes later than the prediction.

On 21 November – two years after Cassini's document cited by Fontenelle – Römer read a memoir to the Academy in which he undertook to show that light is not transmitted instantaneously; the record of the meeting shows that Römer was to confer with Cassini and Picard about publishing the memoir in the *Journal des Sçavans*. By this time

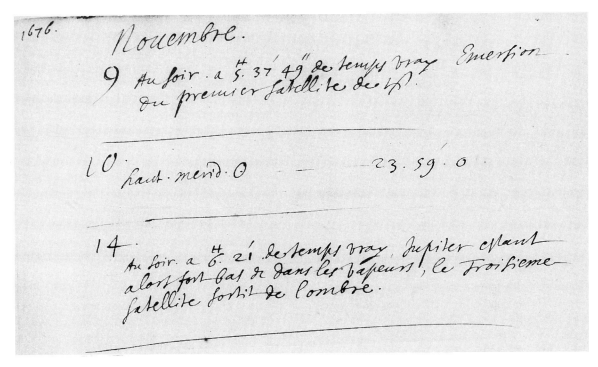

9.3. Picard's record of eclipses of Jupiter's satellites in November 1676.

Cassini had decided that the anomaly was due, not to the finite velocity of light, but to an actual inequality in the movement of the satellite. A week later "there was a discussion of the immersions and emersions of the first satellite of Jupiter. . . . It was thought relevant that M. Cassini should give in writing the arguments that he proposed, and that M. Römer should respond to them". The following week Cassini duly expounded his arguments to the Academy.

Two days later, on 7 December, the *Journal des Sçavans* carried Römer's memoir, "Démonstration touchant le mouvement de la lumière trouvé par M. Römer". There is no suggestion here that the explanation had been advanced earlier by Cassini, nor is there in the later account by Fontenelle of the activities of the Academy for that year: "[The discovery] follows from the observations of M. Römer. . . . If we deduce from it the distance that [light] traverses in a minute, and the distance it must traverse in order to cause a ten-minute delay in the phenomenon, we shall be terrified both by the immensity of the distances, and by the rapidity of the movement of light."

A modern scientist approaching seventeenth-century determinations of the time light takes to cross the Earth's orbit may be surprised to find astronomers of the period uninterested in the actual velocity of light. But this quantity could play little role in astronomy. Once it was realized that we see astronomical events, not when they actually occur, but some time later, it was important for purposes of prediction to know the difference in the time-lag for different positions of the Earth and of the object observed; but these could be defined in terms of fractions of the time-lag for the mean Earth–Sun distance (the 'astronomical unit'). Therefore, there was little incentive from astronomy to derive an actual velocity, even though the recent improvement in knowledge of the length of the astronomical unit made this feasible.

Römer, accordingly, makes little effort to determine the velocity of light. In the *Journal des Sçavans* for 7 December 1676, he "demonstrates that for a distance of about 3000 leagues, which is the approximate size of the diameter of the Earth, light requires less than a second of time. . . . This is because of the 22 [minutes that it requires] for the whole interval . . . double that from here to the Sun". Subsequent analyses of Römer's data by historians show that he might in fact have been expected to derive a more accurate value, in the region of 9 rather than 11 minutes for the retardation corresponding to the Earth–Sun distance. In the following year, in "Confirmatio doctrinae de mora luminis ex novis observationibus anni 1677", Römer obtained two new estimates for a distance of $1\frac{1}{4}$ astronomical units: 12 minutes from observations of eclipses and 14 minutes from observations of a spot on Jupiter.

Why did Cassini resist Römer's explanation? According to the historian J.E. Montucla, writing a century later, Cassini demanded: "Why does it happen that [the inequality] does not occur in the case of the other three satellites? Their eclipses should be subject to the same periodic accelerations and retardations as those of the first satellite, yet one observes nothing of the kind." Similarly, in two works published in 1693, *Les Hypothèses et les tables des satellites de Jupiter* and *De l'origine et du progrès de l'astronomie*, Cassini rejects Römer's interpretation of the inequality of the first satellite because Römer "did not examine whether this hypothesis fitted the other satellites, which should be subject to the same inequality of time". Cassini, as a rigorous scientist who has accumulated observations and trusts in their precision, needs convincing before he is willing to resort to a completely novel explanation.

In his Tables of 1693, Cassini introduced into the motion of each satellite an annual inequality, depending on the angular distance between the Sun and Jupiter, and rising at maximum to 2° in the motion of each satellite. In the case of the first satellite, the 2° corresponds to 14 minutes 10 seconds of time; and the inequality has the same effect in the prediction of eclipse phenomena as Römer's equation of light. But in the case of the other satellites, 2° of motion corresponds to longer times, since the angular motions of these satellites about Jupiter are slower. Römer's equation of light, by contrast, should be the same for all the satellites, since the time for light to come from Jupiter to us, for any given distance between the Earth and Jupiter, is the same for all. To the question whether Cassini's annual inequality fits the observations of the satellites other than the first, we shall return shortly. Undoubtedly, Cassini's investigations were complicated by the fact that the detection of the equation of light is more difficult in the outer satel-

lites, for which eclipses are less frequent. Moreover, Cassini had detected a number of further anomalies, as yet unexplained; and their presence made it difficult to isolate and measure the annual inequality. Confronted with these complicating factors, he concluded that the evidence for the equation of light was inconclusive.

The exchanges between Römer and Cassini concerning the inequality of the first satellite continued at meetings of the Academy throughout 1677 and well into 1678. Both men read to the assembly letters they had written to Huygens on the subject; Huygens sided with Römer, though in writing to Jean Baptiste Colbert, the Minister of Louis XIV, he tactfully hinted that final proof was still wanting: "I have learned recently with much joy of the beautiful discovery made by M. Römer, to demonstrate that light in spreading from a source requires time, and even to measure this time. It is a very important discovery, in the confirmation of which the Royal Observatory will be worthily employed." In his great *Traité de la lumière*, written about this time but not published until 1690, Huygens presents Römer's discovery as an established truth. For his part, as far as the surviving records show, Römer made his final observation of the first satellite in January 1678, after which he turned his attention to other tasks, first in France and then in Denmark.

The retardation of light in England

In England Römer's discovery found a powerful advocate in Flamsteed, the first Astronomer Royal. Flamsteed became convinced of the truth of Römer's claim when he was able to discuss it with him face-to-face on the occasion of Römer's visit to England in 1679. He introduced the equation of light into his Gresham College lectures in astronomy in 1681 and made use of it in correcting tables predicting the eclipses of Jupiter's satellites. In 1684, in a letter to Newton, Flamsteed told him that in the tables "I use [the satellites'] motion altogeather aequable only allowing Roemers aequation of light, without which allowance the error of my tables would be above 10 minutes of time".

His words fell on receptive ears. Earlier, in 1675, in writing a long letter to Henry Oldenburg on the colours of thin films, Newton had remarked that "it's possible light it self may not be so swift as some are apt to think, for notwithstanding any argument that I know yet to the contrary it may be an houre or two, if not more in moveing from the sunn to us". But what Newton learned from Flamsteed in 1684 changed his mind. In his *Opticks* of 1704, in Prop. XI of Part III of the Second Book, he wrote:

Light is propagated from luminous Bodies in time, and spends about seven or eight minutes of an hour in passing from the Sun to the Earth. *This was observed first by Romer, and then by others, by means of the Eclipses of the Satellites of Jupiter. For these Eclipses, when the Earth is between the Sun and Jupiter, happen about seven or eight minutes sooner than they ought to do by the Tables, and when the Earth is beyond the Sun they happen about seven or eight minutes later than they ought to do . . . .*

Meanwhile, in the *Philosophical Transactions* for November–December 1694, Edmond Halley (*c.* 1656–1743) published "Monsieur Cassini his New and Exact Tables for the Eclipses of the First Satellite of Jupiter, reduced to the Julian Stile, and Meridian of London". Applying these tables to three observations of the first satellite, one by Flamsteed, one by himself, and one by Cassini, Halley found the errors to be – 1 minute 21 seconds, – 3 minutes 27 seconds, and + 0 minute 9 seconds. He then commented:

After this manner I have compared these Tables with many good and certain Observations, and scarce ever find them err above three or four Minutes of Time; which proceeds, as may well be conjectured, from some small Eccentricity in its Motion, and from the Oval Figure of Jupiter's Body . . . . But we may hope future Observations may shew how to divide those compounded causes of Error, and correct them; which Errors are exceeding small in comparison of the short time that the Satellites have been discovered, and argue the Skill and Diligence of the deservedly Famous Author of these Tables.

There is one feature of Cassini's tables, however, with which Halley is in sharp disagreement: it is the form of Cassini's second or annual inequality, dependent on the distance of the Sun from Jupiter:

The distribution of this Inequality he makes wholly to depend on the Angle at the Sun between the Earth and Jupiter, without any regard to the Excentricity of Jupiter, (who is sometimes $\frac{1}{2}$ a Semidiameter of the Earth's Orb farther from the Sun than at other times) which would occasion a much greater difference than the Inequality of Jupiter and the Earth's Motion, both of which are accounted for in these Tables with great Skill and Address. But what is most strange, he affirms that the same Inequality of two Degrees in the Motion, is likewise found in the other Satellites, requiring a much greater time, as about two Hours in the Fourth Satellite: which if it appeared by Observation, would overthrow Monsieur Roemer's Hypothesis entirely. Yet I doubt not herein to make it demonstratively plain, that the Hypothesis of the progressive Motion of Light is found in all the Satellites of Jupiter to be necessary, and that it is the same in all; there being nothing near so great an Annual Inequality as Monsieur Cassini supposes in their Motions. . . .

Halley then proceeds to show for a pair of observations of the third satellite in 1672, and again for a pair of observations of the fourth satellite in 1682, that Cassini's annual equation is much too large, while the "equation of light", quantitatively the same as Cassini's annual equation for the first satellite, is almost exactly right in the case of the first pair of observations, and within near range of the observational result in the case of the second pair. Halley's argument might seem sufficient to lead to a rejection of Cassini's notion of an actual annual inequality in the motion of each satellite; but we find Cassini's table for this inequality repeated in the *Tables astronomiques* published by his son in 1740.

The work of Bradley

When Halley's own *Tabulae astronomicae* finally appeared posthumously in 1749, they included tables of the satellites of Jupiter drawn up in 1719 by James Bradley (1693–1762), who was to succeed Halley as Astronomer Royal in 1742. Bradley had compared past observations of the satellites over a period of some four revolutions of Jupiter, and had detected a number of inequalities hitherto unidentified. His tables included Römer's equation of light, with the time for light to traverse the diameter of the Earth's orbit put at 13 minutes 53 seconds, somewhat less than the present-day value (16 minutes 38 seconds). The orbit of the fourth satellite, he found, was definitely elliptical, with an eccentricity close to the orbital eccentricity of Venus; in his tables he therefore made use of the table of equations of Venus to compute the position of this satellite. The first three satellites, it was also clear, were subject to notable inequalities, especially the second. Some eccentricity might be involved in their motions, but Bradley did not believe that all of the inequality could come from this source; for the satellites in their digressions did not exhibit easily detectable eccentricities. The inequalities in the case of these satellites, Bradley conjectured, likely came from their mutual attractions. The periods of the errors corresponded very nearly to the time in which the three interior satellites returned to the same configuration with respect to one another and with respect to the axis of Jupiter's shadow, namely 437 days. Since the satellites after this period were not in the same place in the ecliptic, the errors, Bradley found, could vary a little from one 437-day period to another. To analyse apart the compounded causes of the inequalities, he concluded, would be very difficult on the basis of observation alone. Indeed, it turned out to require the sophisticated mathematics of a Lagrange and a Laplace.

It was not this close study of the satellites of Jupiter, however, but rather another and quite different investigation, that was to lead to the definitive confirmation of Römer's discovery.

In the 1670s Robert Hooke (1635–1703) had attempted to detect the annual parallax of a star, that is, the apparent change in position of the star caused by the Earth's displacement in its motion about the Sun (see Volume 3). For his series of observations Hooke chose Gamma Draconis, a star that culminates near the zenith of Gresham College where Hooke was observing; thus the effects of refraction would be reduced to a minimum. His observations appeared to show variations in the positions of Gamma Draconis, and he attributed these to annual parallax; but his results were not generally regarded as conclusive. In 1725 Samuel Molyneux (1689–1728), a fellow of the Royal Society and a student of astronomy and optics, undertook to verify Hooke's claims, and for the programme of systematic observations that he envisaged, enlisted the help of Bradley.

By late 1727 or early 1728 Bradley had discovered a periodic oscillation of amplitude 20″ in the position of Gamma Draconis. He then examined the behaviour of other stars that passed close to the zenith in London, and found a pattern which he

eventually was able to explain, in terms of "the Aberration of Light". The parallactic displacement claimed by Hooke, Bradley found, did not occur: each of the stars did indeed change position during the course of the year, but not in the direction required by parallax, namely, a direction opposite to the Earth's change in position. Instead, each star was displaced in the momentary direction of the Earth's motion about the Sun. The amount of this displacement, according to Bradley's interpretation, was given by the angle whose tangent is the quotient of the velocity of the observer by the velocity of light. Knowing this angle to be at maximum 20″, and knowing the angular velocity of the Earth about the Sun, Bradley calculated that light travels 10 210 times faster than the Earth in its orbit about the Sun, within 20% of the correct value. Knowing the angular velocity of the Earth about the Sun, Bradley deduced with similar accuracy the time for light to reach us from the Sun: first 8 minutes 12 seconds, then 8 minutes 13 seconds when different stars were used for the calculation.

Bradley's result, published in *Philosophical Transactions* for 1729, was in agreement, to within the limits of the observational errors, with the earlier determinations of "the retardation of light". Since Bradley's discovery concerned a stellar phenomenon, completely independent of the eclipses of the satellites of Jupiter, there could no longer be any doubt of the fact that light has a finite velocity. The hypothesis of Römer thus became a confirmed reality some fifty years after it was first enunciated. Bradley's observations, like the observations of the satellites of Jupiter, had a systematic character that permitted the secure detection of the abnormal phenomenon within a sequence of observations of known precision.

In navigation the satellites of Jupiter did not play the role that their discoverer had envisaged for them: it was geodesy that was able to make use of them in the determination of differences in terrestrial longitude. As for Römer's discovery, it came about through the efforts of those who applied themselves to observations of the satellites, and to the construction of theories and tables of their motions. The difference the discovery made was greater than might at first appear: if it had not already been established that light had a finite velocity, would Bradley have been able to comprehend the nature of the phenomenon that he discovered in 1727? The discovery of aberration, a brilliant confirmation of the Copernican hypothesis of the motion of the Earth about the Sun, owes much to the satellites of Jupiter.

Further reading

Seymour L. Chapin, Astronomical activities of Nicolas Claude Fabri de Peiresc, *Isis*, vol. 48 (1957), 13–29

I.B. Cohen, Römer and the first determination of the velocity of light, *Isis*, vol. 31 (1940), 327–79

Stillman Drake, Galileo and satellite prediction, *Journal for the History of Astronomy*, vol. 10 (1979), 75–95

Stillman Drake, *Galileo at Work* (Chicago, 1978)

Derek Howse, *Greenwich Time and the Discovery of the Longitude* (Oxford, 1980), chap. 1

Pierre Humbert, *Un Amateur: Peiresc* (Paris, 1933)

René Taton (ed.), *Roemer et la vitesse de la lumière* (Paris, 1978)

Albert Van Helden, Roemer's speed of light, *Journal for the History of Astronomy*, vol. 14 (1983), 137–41

PART III

Planetary, lunar and cometary theories between Kepler and Newton

Predictive astronomy in the century after Kepler
CURTIS WILSON

In the century following the publication of Johannes Kepler's *Rudolphine Tables* (1627), improvement in the accuracy of astronomical theories and tables depended in large measure on the adoption, gradual and successive in some instances, abrupt and wholesale in others, of six Keplerian innovations:

(1) Replacement of the Mean Sun (which for all earlier astronomers had been the chief point of reference for the planetary orbits and motions) by the true Sun. Thus the apsidal lines, previously taken as passing through either the Earth or the Mean Sun, were now taken as passing through the true Sun. Similarly, the eccentricities of the orbits were now to be measured from the true Sun; and conjunctions, oppositions, and (in the case of the inner planets) greatest elongations were to be determined in relation to the true Sun.

(2) Adoption, as a corollary to this, of the postulate that each orbital plane shall pass through the true Sun, at a constant inclination to the plane of the ecliptic. The earlier, Ptolemaic and Copernican theories of the planetary latitudes had involved oscillating epicycles and deferents. Kepler found this complication incredible, and succeeded in removing it at one stroke. The planetary latitudes derived from the new postulate admitted of repeated and persuasive verification.

(3) Bisection of the eccentricity of the Earth's orbit. Ptolemy's theories for Venus and the superior planets had already incorporated bisection of the eccentricity; thus the equant point, or centre about which the planet's motion was assumed to be angularly uniform, was placed twice as far from the Earth as the centre of the planet's eccentric or deferent circle. Nicholas Copernicus, while rejecting the equant device as improper, introduced an epicycle mechanism which produced almost exactly the same paths and motions in these planets.

Kepler was the first to apply the bisection to solar theory, that is, the theory of the Earth's motion. This meant that – to express the postulate in heliocentric terms – the centre of the Earth's orbit was taken to be only half as far from the Sun as in earlier theories, the remainder of the inequality, as determined by the traditional methods, being accounted for by the assumption that the Earth's motion along its path was not only in appearance but also in reality non-uniform, slower at aphelion and faster at perihelion. This innovation was consistent with Kepler's belief in the Sun as the cause and source of the orbital motions of the planets. The bisected eccentricity in solar theory was verified initially from Tycho Brahe's observations of Mars and Venus, then in observations of apparent solar diameters at apogee and perigee.

(4) Replacement of the eccentric circle of each planet by an elliptical orbit, with the Sun at one focus. Because of the level of unavoidable error in observations of position, and the near circularity of the orbits, the departure from circularity could be detected observationally only in the orbits of Mars and Mercury, where the eccentricity was exceptionally large; and even in these cases the choice between ellipse and other oval shapes was, as far as the observations could show, a matter for conjecture. Kepler, of course, had reasons for his choice: a causal account which led both to the elliptical orbit and to the planet's motion on that ellipse, with close agreement between the theoretical prediction and observation in the particular case of Mars. Ironically for Kepler, it was the ellipse that caught on most quickly and widely, as a simple curve that could be produced by a combination of circular motions, and as able to account for the bisection of the eccentricity and for the observably oval paths of Mars and Mercury; while Kepler's rationale of the planetary motions was all but universally rejected.

The Keplerian ellipse became emblematic of the new astronomy.

(5) Determination of the motion of the planet on its orbit by the rule of constant areal velocity: the areas swept out by the radius vector from Sun to planet are proportional to the times. This rule, as indicated under (4) above, was a consequence of Kepler's causal account of planetary motion; as such, and because it did not admit of a direct calculation leading from time to heliocentric longitude, it was mostly ignored, and various geometrical devices were introduced to replace it. To be successful, however, all such calculational schemes had to approximate closely to the areal rule, or to be more specific, they had closely to agree with the 14 Tychonic observations of the heliocentric longitudes of Mars against which Kepler had tested his areal rule, and which remained the touchstone for planetary theories until, late in the century, new and more accurate observations became available.

(6) Adoption of Kepler's 'third law', the rule according to which the periods of the planets are as the $\frac{3}{2}$ powers of their mean distances from the Sun. This rule was first put forward in Kepler's *Harmonice mundi* of 1619, but Kepler did not employ it in the construction of his tables. Yet if deemed exact, it offered a way of improving the accuracy of the tables by permitting computation of the less easily ascertainable mean solar distances from the more precisely known planetary periods. It was Jeremiah Horrocks (1618–41) who first stated that such a use of the rule should be made, and the London Excise Office clerk Thomas Streete (1622–89) who in 1661 first so used it in published tables. The resulting improvement in accuracy was particularly noticeable in the cases of Mercury and Venus. In the case of Jupiter and Saturn the rule is less useful because of mutual long-term perturbations which appear to alter the periods, and which led Edmond Halley (*c*. 1656–1743) in his tables to postulate a secular acceleration in Jupiter and a secular deceleration in Saturn.

In the century following the publication of Kepler's *Rudolphine Tables*, there were two other developments that would eventually be of the first importance for astronomical accuracy: the appearance of Isaac Newton's *Principia* (1687), and a revolution in the instruments of observation, with the introduction of the filar micrometer (invented by William Gascoigne about 1640, and made widely known by Richard Towneley, Adrien Auzout, and Jean Picard in the 1660s) and the pendulum clock (invented and improved by Christiaan Huygens in the period 1655–60), and the application of telescopic sights to graduated arcs (accomplished by Picard in 1668). But during the period here under consideration, Newton's theory of gravitation played no role in the construction of planetary tables, unless it was to underline for John Flamsteed (1646–1719) and Halley the importance of approximating the areal rule very closely in the calculation of equations of centre. For a long time, doubt as to whether the planetary aphelia were mobile or immobile with respect to the stars – their immobility suggesting that mutual perturbations were negligible – led Newtonian theorists to neglect the problem of planetary perturbations. The first perturbation of a planet to be calculated from the inverse-square law and included in tables was the lunar perturbation of the Earth's motion, incorporated by Leonhard Euler in his solar tables of 1746. As for the Moon's motions, Newton's attempt at a lunar theory was not successful. Efforts to base lunar tables on the Newtonian rules were made by R. Wright and by Charles Leadbetter in the 1730s, without achieving clear superiority over tables constructed by Flamsteed on the basis of a theory due to Horrocks. Halley's lunar tables, based on the Newtonian theory, were not published until 1749; they were accompanied by Halley's tabulation of the discrepancies between tables and observation over an eighteen-year period from 1722 to 1739. The errors frequently came to 4' or 5', and were sometimes as high as 7' or 8', so that David Gregory's earlier claim (1702) that the errors would not exceed 2' was evidently unfounded. The first successful lunar tables to be founded on a systematic calculation of perturbations were those published by Tobias Mayer in 1753. Only in the case of cometary theory did Newton's efforts quickly prove of decisive importance for prediction.

On the other hand, the revolution in instruments of observation, while leading to much improved values of solar and lunar parallax, did not immediately produce the sharp improvement in the accuracy of tables that one might have expected, but rather turned up new anomalies requiring to be explained. The positions of planets were determined in relation to fixed stars, and, from about 1670 onwards, the stars were showing themselves to be other than fixed: Picard, Robert Hooke, Flamsteed, and others were detecting in them movements amounting to sizable fractions of a minute of arc. In a number of cases the movements were mistakenly attributed to stellar parallax. Where systematic (and genuine) they were in fact due to two different causes, the aberration of light and the nutation of the Earth's axis, the first of these producing an apparent annual motion of each star in an ellipse with major axis equal to about 41″ of arc, and the second causing an oscillation in the Earth's longitude, with amplitude 17.25″ either way from the mean position, and a variation of more than 9″ either way in the obliquity of the ecliptic. It was in the autumn of 1728 that James Bradley first saw how to account for the phenomena caused by aberration; he was suspecting the second effect a few years later, but did not formally announce it till 1748, after he had traced its full 18.6-year cycle. As long as ignorance of these deviations endured, positional astronomy had willynilly to be limited to an accuracy measured in minutes rather than seconds of arc. The first astronomers to incorporate aberration and nutation in their astronomical tables were N.-L. de Lacaille and Tobias Mayer, in the 1750s.

But while astronomy in the century after Kepler did not yet know how to calculate perturbations or that allowance must be made for aberration or nutation, advances were nevertheless made in dealing with astronomical refraction, in estimating solar and lunar parallax, in understanding Kepler's discoveries and what they entailed, and in drawing from these developments significant improvements in the numerical parameters from which astronomical tables are derived. During the earlier part of our period astronomers were mostly astrologers, with widely different viewpoints and aims; the principal points of difference and controversy were whether the Sun or the Earth is the static centre of the world, how accurate were ancient observations, especially those of Ptolemy, and whether Kepler's physical causes and harmonic speculations ought to be admitted into astronomical theory. At once the sanest and the most innovative of the early successors to Kepler was Horrocks; unfortunately, however, his ideas and discoveries remained largely unknown for some twenty years

after his death, until in part made public in the 1660s and 1670s. It is at this latter epoch that the scientific revolution truly begins to come into its own. With the foundation of the Academy of Sciences in Paris in 1666, and of the Royal Observatory at Greenwich in 1675, and with the appointments of Picard, Auzout, Huygens, and Gian Domenico Cassini (Cassini I) to the former and of Flamsteed to the latter, astronomy took a long step forward along the path of professionalization, and a stronger emphasis was placed on the possible services of astronomy to navigation and geography. The development of astronomical tables during our period can be regarded as culminating in the *Tables astronomiques* of Jacques Cassini (Cassini II), published in 1740 but deriving in significant measure from the labours of Cassini I in the previous century, and in Halley's *Tabulae astronomicae*, resting largely on the observations of Flamsteed, and essentially complete by 1717, but first published posthumously in 1749.

From Kepler to Horrocks: Astronomy in the 1630s

Kepler's own *Admonitio ad astronomos rerumque coelestium studiosos*, published in 1629 (and reissued the next year), is an effective introduction to the problems and the controversies of astronomy in the 1630s. Here he is predicting, for late 1631, transits of Mercury and Venus across the face of the Sun, praying for clear skies on the designated days, and calling on all astronomers to watch for and observe these events with care.

Kepler's concern is twofold. In the first place, observation of such transits would for the first time permit the heliocentric longitudes of the inner planets to be determined with accuracy: they had never before been observed when in line with the Earth and the Sun. Yet Kepler is still more interested in a second question: what are the diameters of these planets? The question of sizes and densities of the planets was of central importance in Kepler's causal account of the planetary motions. In this account, as we have seen (in Chapter 5 above), the velocities, masses, volumes, and densities of the planets all turn out to be monotonic functions of the mean solar distances. In his *Epitome astronomiae Copernicanae* and *Rudolphine Tables*, accepting the telescopic observations and conclusions of his friend Remus Quietanus, Kepler took the volumes of the planets to vary directly as their mean solar distances. From this and his assumptions about dynamics, it followed that the masses are as the square roots, and the densities inversely as the square roots, of the distances.

Such 'harmonies' could seem irrelevant to an exact positional astronomy, but in fact the assumed proportionality between planetary volumes and mean solar distances impinged on a crucial astronomical constant, the horizontal parallax of the Sun. This constant is involved in the correction of all observed altitudes of the Sun, and it was on such observations that all solar theories, up to the 1690s, were based. Through much of the seventeenth century, an erroneous solar parallax (see Chapter 7) was one of the principal causes of error in solar theories, the other being mistaken refractions; and from these sources error was propagated into the theories of all other planets.

The assumption about planetary volumes was subject to observational refutation, and by the late 1620s, indeed, Remus was finding that his original hypothesis could not stand. In its stead he suggested that the apparent diameters of all the planets could be supposed the same as seen from the Sun, and equal to about 34″ to 36″. This would reduce the horizontal parallax to 17″. Kepler in his *Admonitio* reports but does not adopt this proposal; more evidence, he says, is needed. An observed transit of Venus, he opines, would by determining the size of this planet help to resolve the question; unfortunately the transit he has predicted for 6 December 1631 (NS) will, according to his tables, occur after sunset and so be visible only in America.

Kepler died in 1630, but the first of the transits he had predicted, that of Mercury, was observed (see Chapter 7) by Pierre Gassendi (1592–1655): "I have found him, I have seen him where no one has ever seen him before!" he exclaims in his open letter to Wilhelm Schickard, announcing the event. Kepler's prediction proved to be in error by 13′ in longitude, 1′ 5″ in latitude, 5 hours 49.5 minutes in time. What impressed Gassendi above all was the tininess of Mercury, its apparent diameter being scarcely $\frac{1}{90}$th of that of the Sun. Even Remus's second proposal would not quite suffice here, and Gassendi is inclined to think we should simply acknowledge our ignorance of the harmonic structure Kepler had believed in and sought

to know. Epicurus, Gassendi urges, was right: the stars are smaller than they seem. Mercury, esteemed thrice-great on Earth, proves thrice-small in the heavens. The astrologers' claim of Mercurial influence for good or ill in human lives is shown, in Gassendi's opinion, to be absurd.

Gassendi's letter was published in 1632, as also was a response from Schickard, professor of Hebrew at Tübingen, friend of the late Kepler, and himself a competent astronomer who appreciated the Keplerian innovations. Gassendi's observation, Schickard points out, shows the *Rudolphine Tables* to be far better than others; in predicting the egress, they erred by but 14′ 24″ in longitude, while the Ptolemaic tables were in error by 4° 25′, the Copernican by 5°, and Longomontanus's *Astronomia Danica* (1622) by 7° 13′ – a matter of days rather than hours. Gassendi's observation will make possible still further improvement in the Keplerian parameters. With regard to the apparent diameter of Mercury, Schickard thinks there may have been some deception in the observation, a reduction in the planet's apparent size somehow caused by the light of the Sun.

Martinus Hortensius's *De Mercurio in sole viso* was a response to both Gassendi's and Schickard's letters. Hortensius (1605–39) had been a disciple of and co-worker with the Belgian astronomer Philip van Lansberge (1561–1632), assisting him in his determination of astronomical constants and construction of astronomical tables. Van Lansberge was a Copernican, and like Copernicus assumed the exactitude of all the ancient observations, and attempted to devise a theory that would fit all of them. In this project he claimed to have succeeded; thus his tables, first published in 1632, bore the proud title: *Tabulae motuum coelestium perpetuae: ex omnium temporum observationibus constructae, temporumque omnium observationibus consentientes.* Van Lansberge's claim involved him inevitably in a polemic against the 'Tychonici', Tycho, Longomontanus, and Kepler; and one of the chief points at issue was the treatment of Ptolemy's alleged observations of equinoxes and of the obliquity of the ecliptic. It was in seeking to accommodate these observations that Copernicus had been led to introduce a 1717-year inequality in the precession of the equinoxes, and a 3434-year oscillation in the obliquity of the ecliptic. Tycho had called the original observations into doubt, and Longomontanus and Kepler, although not dismissing altogether the possibility of an inequality in the precession and the obliquity, treated these matters as uncertain, and incapable of being decided until a longer sequence of modern observations had been accumulated. Kepler openly acknowledged that positions of the stars and planets taken out of the *Rudolphine Tables* differed in longitude by more than a degree from Ptolemy's observations and tables. To van Lansberge and Hortensius, such an admission, and the profession of doubt as to the exactness of Ptolemy's observations, constituted a dereliction of the astronomer's task. In a preface to van Lansberge's *Commentationes in motum terrae diurnum & annuum* (1630), Hortensius discoursed at length on the errors of the Tychonici.

Van Lansberge in his *Uranometria*, published in 1630, claimed to be the first to make correct use of the 'diagram of Hipparchus' of Ptolemy's *Almagest* (V, 15) for the determination of solar parallax; he thereby obtained 2′ 13″ for this constant, and new sizes and distances for all the celestial bodies. He and Hortensius claimed the solar diameters at perigee and apogee to be 36′ and 33′ 34″, a great deal larger than the values found by Tycho and Kepler (the correct values in 1600 were 32′ 32″ and 31′ 27″). The *Tabulae perpetuae* were based on geometrical theories differing little from those of Copernicus's *De revolutionibus*. The tables were accompanied by a *Thesaurus* giving a wealth of observations, ancient and modern, and ostensibly showing their exact agreement with van Lansberge's tables and theories.

Hortensius in his *De Mercurio in sole viso* of 1634 has to acknowledge that the Keplerian tables predict the transit of Mercury of 1631 more accurately than any other tables. The Ptolemaic, Copernican, and Danish tables differ from the observation so much that "no medicine would suffice to heal the wound". The Lansbergian calculation of the event, he allows, does not exhibit the very truth, but it approaches the Keplerian prediction; "therefore, after all the highly praiseworthy industry of the old man [van Lansberge], there no doubt remains something to be corrected in his theory of Mercury, although in all other observations the celestial truth is given accurately enough by his tables". The discrepancy between Gassendi's observation and the prediction from van Lansberge's tables,

Hortensius finds, is 1° 8′ in longitude, 17′ in latitude.

But Hortensius passes quickly over the discrepancy to other topics. He argues persuasively that, contrary to Schickard's supposition, Mercury was as small as Gassendi saw it, and the rays of the Sun could not significantly reduce the size of the projected shadow. While agreeing with Gassendi that judicial astrology is uncertain and vain, he by no means concedes that smallness entails impotency; he is sure that Mercury, like the Sun and Moon and the other planets, has some influence in the sublunary world. Finally, he outdoes Gassendi in holding Kepler's harmonic speculations up to ridicule. For a long time, Hortensius says, he has been telling astronomers with loud voice that those speculations are worthless. Posterity will laugh at the labour wasted on such idle dreams. Only observations and geometrical demonstrations are to be trusted.

A single observation, of course, can hardly be taken to refute an entire astronomy, including its underlying assumptions as well as its numerical parameters. But it was on the basis of the transit of Mercury of 1631, with additional observations of Mercury by Gassendi in 1632–34, and also observations of Mars, that Noël Durret, cosmographer to the king of France, claims to have abandoned the Lansbergian tables and adopted the Keplerian. In 1635 Durret had published his *Nouvelle théorie des planètes . . . avec les tables Richeliennes et Parisiennes*, dedicated to Cardinal Richelieu. Both the theoretical text and the tables of this work are largely direct translation from van Lansberge, except that Durret has recalculated the epochs of the tables for the longitude of Paris, and insists in opposition to van Lansberge and all Copernicans on the immobility of the Earth. Then in 1639 Durret published a *Supplementum tabularum Richelienarum . . . cum brevi planetarum theoria ex Kepleri sententia*, and in 1641 a set of ephemerides for fifteen years, of which the first six years were derived from Lansberge, and the last nine from Kepler. In the prefaces to both works he insists upon the greater accuracy of the Keplerian tables, citing as proof the observations just mentioned. From observations made by his acquaintance Valesius Virgilia, however, he finds these tables less successful with Jupiter and Saturn than with the other planets. Durret takes pride in having simplified the form of the *Rudolphine Tables*, but he evinces no interest in the physical hypotheses underlying them. He wrote rather extensively on astrological prognostication in medicine and meteorology. Jean-Baptiste Morin (1583–1656), another astrologer and ardent anti-Copernican, but a more highly competent astronomer, published another simplified version of the *Rudolphine Tables* in 1650, in the preface to which he stated that Durret's work was so filled with calculational errors as to be a disgrace.

Van Lansberge's tables, republished in 1653, and again in 1663 as part of his *Opera omnia*, continued in use into the 1660s. On the basis of them Francisco Montebruni in Bologna calculated ephemerides for the years 1645–60, and the Marquis Cornelis Malvasia for the years 1661–67, although Malvasia realised that van Lansberge's tables were less accurate than could be wished. Growing recognition of the unreliability of van Lansberge's tables was fostered by the publication in 1640 of a duodecimo volume, Johannes Phocylides Holwarda's *Dissertatio astronomica*, which includes a 167-page *Examen astronomiae Lansbergianae*. In instance after instance, Holwarda shows that van Lansberge has 'fudged' – neglected a correction that elsewhere he insists on, altered an observational result, or allowed a calculational error to slip by, in order to make the agreement between his theory and observation appear exact. The inequality of the equinoxes that van Lansberge took over from Copernicus with minor modification, Holwarda demonstrates, is untenable: the length of the tropical year as determined from observations of equinoxes by Hipparchus, Albategnius, Bernhard Walther, and Tycho – although not those that Ptolemy claimed to have made – has remained constant at 365 days, 5 hours, and 49 minutes (the actual decrease of 0.53 seconds per century was not detectable in the observations at Holwarda's disposal). Although allowing that the obliquity of the ecliptic and eccentricity of the Earth's orbit may be undergoing long-term oscillations as van Lansberge and Copernicus had claimed, Holwarda denies the exactitude of van Lansberge's parameters for these oscillations, proposes three different hypotheses of his own to accommodate some of the better known ancient and modern observations, and warns that this matter cannot be settled definitively until a lengthy sequence of more exact observations has been collected.

Holwarda repeatedly contrasts the enormous industry, sincerity, and learning of "optimus Keplerus" with van Lansberge's boasting and dishonesty, but his opinions are not uniformly Keplerian. He is unwilling, for instance, to follow Kepler in ascribing discrepancies between observation and prediction of solar or lunar eclipses to unaccountable physical influences causing fluctuations in the diurnal or annual motions of the Earth. Against van Lansberge and Hortensius he defends Kepler's bisection of the eccentricity in the Earth's orbit – he has verified it in observations of Saturn and Jupiter as Kepler confirmed it in the Tychonic observations of Mars; but unlike Kepler he does not find the eccentricity exactly bisected in the superior planets (probably he is here proceeding along the lines of Kepler's '*vicarious hypothesis*', without realizing the limitations of the latter). On the basis of observations made with an 18 ft quadrant, and in opposition to both Tycho and van Lansberge, Holwarda suggests that refractions go at least to altitudes of 51° (Tycho had held that refractions of the stars cease above 20°, and those of the Sun above 45° of altitude). Whereas van Lansberge claimed to be the first modern to understand and follow strictly the 'diagram of Hipparchus', Holwarda shows that van Lansberge's procedure remains inconsistent with the diagram, and that his value for the solar parallax, 2′ 13″, is worthless; and Holwarda himself inclines with Kepler toward reducing this constant to 1′. In his *Epitome astronomiae reformatae* of 1642, Holwarda again subjects van Lansberge's astronomy to devastating criticism, and in the posthumous *Philosophia naturalis seu physica vetus-nova* of 1651 he shows himself a partisan of Kepler's elliptical astronomy and celestial physics, with which he combines an atomistic philosophy.

A still more powerful attack on van Lansberge's astronomy, and a staunch defence of Keplerian tables and theories, came from a young man we have already heralded as one of the greatest astronomers in the century after Kepler: Jeremiah Horrocks. Of Horrocks's writings, the *Venus in sole visa* was published in 1662; other papers, letters and observations were published under the editorship of John Wallis (1616–1703) in the *Opera posthuma* (editions in 1672 and 1673, with what is essentially a reissue of the latter in 1678). To Horrocks are owing major improvements in solar, lunar, and planetary parameters, and in lunar theory. Although his influence was exerted primarily in the 1660s and later, his observational and theoretical work was accomplished in the brief period from 1636 until his death on 3 January 1640–41 (OS). It is in the post-Keplerian but pre-Cartesian context of those short years that it is rightly to be understood.

Horrocks first applied himself to astronomical studies in about 1633, while a student at Emmanuel College, Cambridge. Put onto the astronomical works of van Lansberge by a complimentary reference in the writings of Henry Gellibrand, he accepted the pronouncements of the Belgian astronomer for a time. In 1636, after returning to his native Toxteth, a Lancaster village now swallowed up in the eastern suburbs of Liverpool, he came to know and correspond with another amateur of astronomy, the draper William Crabtree of Broughton, near Manchester; it was Crabtree, on the basis of his own observations, who led Horrocks to doubt the Lansbergian claims to a perfected astronomy and to begin the study of Kepler's works. The conversion was soon complete; and it was as one of "nos Kepleriani" that Horrocks henceforth wrote, although recognizing that the numerical constants of Kepler's tables required some revision, and that Kepler's celestial physics contained implausibilities that should and could be removed.

Like Holwarda, Horrocks demonstrates that van Lansberge has been inconsistent in his use of the diagram of Hipparchus, and that his value for the mean horizontal solar parallax, 2′ 13″, is without real foundation. But Horrocks goes beyond Holwarda in proving that, because of unavoidable observational error, the Hipparchic method for solar parallax is doomed to fail. The method (see Chapter 7) requires that we know the apparent semi-diameter of the Sun, the horizontal parallax of the Moon, and the apparent width of the Earth's shadow where the Moon crosses it; the last of these is found in turn from the apparent diameter of the Moon and the extent and duration of the lunar eclipse. The unavoidable observational error involved in these determinations, Horrocks shows, far exceeds the horizontal solar parallax that is sought. Even with respect to the apparent diameters of the luminaries, more accurately ascertainable than the other numbers, van Lansberge is grossly in error: Horrocks employed a number of observational procedures and always found that

the perigeal and apogeal diameters differ by no more than about 1′.

In rejecting the Hipparchic method, Horrocks is following the indication left by Kepler, but in reducing solar parallax, he goes further than Kepler, arriving at a value of 14″ or 15″. His arguments for this reduction are as follows:

(1) The parallax of Mars when in opposition to the Sun is between two and three times that of the Sun, and yet Kepler found it to be observationally undetectable. Also, without making any correction for parallax in the observations, he found the nodes of the Martian orbit to be precisely on a line intersecting the Sun. In comparisons of the apparent diurnal motion of Mars with its true motion as calculated from the theory, Kepler never found the parallax of Mars to be more than $1\frac{1}{2}′$, sometimes found 1′, and often found less. Therefore the parallax of the Sun must be less than 1′.

(2) Horrocks's own observations of Venus lead to a similar but even stronger conclusion. When Venus is near inferior conjunction with the Sun, its parallax has to be nearly quadruple that of the Sun, and yet, comparing the position of Venus with star positions at different altitudes of the planet as it rises when a morning star or descends when an evening star, Horrocks has found its parallax to be observationally undetectable. The conclusion is that the solar parallax must be of the order of $\frac{1}{4}′$ or less. This finding, of course, presupposes accurate elements for the orbits of both Venus and the Earth, whereby to calculate theoretical geocentric coordinates of Venus for comparison with its observed positions as subject to parallax and refraction. Horrocks's next point concerns the obtaining of such elements.

(3) The chief error in Kepler's solar and planetary theories, Horrocks has come to believe, stems from an exaggerated solar parallax; for this leads to an exaggerated eccentricity in the solar (or terrestrial) orbit, and the latter in turn produces distortions in the orbital elements of the other planets. Let us see how this is so.

Tycho had determined the meridian altitude of the equator at Uraniborg by finding the altitude of the pole from circumpolar stars, whose altitudes he believed he could assume to be unaffected by refraction or parallax. His result for the altitude of the pole was 55° 54′ 40″, whence the meridian altitude of the equator is 34° 5′ 20″. To find the times of the spring and autumn equinoxes – these being essential to the determination of the elements of solar theory – it is necessary to know when the Sun is exactly on the equator; and for this purpose Tycho interpolated between noonday altitudes of the Sun on successive days, after first correcting the observed altitudes for refraction and parallax. Now Tycho's value for the solar parallax at an altitude of 34° was much too large (2′ 29″ instead of about 7″), and his value for the solar refraction too small (44″ instead of 1′ 25″); hence he took the equinox to be occurring when the Sun appeared about 1′ 45″ below the celestial equator, whereas in fact it occurred when the Sun appeared 1′ 18″ above it. The difference in altitude corresponds to a motion in longitude of over 7′; and since the Sun's equation of centre was near maximum value at the equinoxes, Tycho's value for this constant, namely 2° 3′ 15″, is too large by a little more than 7′.

By reducing solar parallax, Horrocks reduced the error in the Tychonic solar theory by more than half. He obtained an equation of centre of 1° 59′ 18″, corresponding to an eccentricity of 0.017 35 (earlier on he gave the eccentricity as 0.017 30, implying an equation of centre some 21″ smaller). The computations leading to this value do not appear to be among his extant papers; we do not know what correction he would have used for refraction. He states that Tycho's refractions are too large because of Tycho's excessive values for parallax (the discrepancy between observed altitude and altitude calculated from solar theory is refraction minus parallax, so that the parallax has to be added to the discrepancy found observationally to obtain the refraction by itself). In fact Tycho's values for refraction are too large up to an altitude of about 28°, and thereafter too small. Horrocks seems to have supposed that the refraction at an altitude of about 35° was such as almost exactly to balance the parallax: at least this assumption results in the solar eccentricity and equation of centre he found. It is a pity that neither Horrocks nor any one else chose to use the table of refractions Kepler had worked out on the basis of theory in his *Astronomiae pars optica* of 1604; this lists refractions all the way to the zenith, and above 15° of altitude errs by only about 10″. With such corrections for refraction Horrocks would have obtained an equation of centre differing from the true by no more than 35″. Nevertheless, with his revised equation of centre differing from the true by

+ 3′, Horrocks was in possession of a more accurate solar theory than had ever been achieved before.

The improvement in solar theory leads to improved planetary theories. In the case of the superior planets, the theories are based primarily on observations of the planet when in opposition to the Sun, but the exact moment of opposition must be calculated from the solar theory, and any error in the solar equation of centre is reflected into the aphelion and equation of centre of the planetary theory. Thus Kepler, because of his faulty solar theory, obtained a theory of Mars in which the aphelion was too far advanced by about a third of a degree, and the eccentricity was too small (0.092 65 instead of 0.093 03). Horrocks in a letter to Crabtree of 19 January 1638 proposed retracting the aphelion by 10′ and increasing the eccentricity to 0.092 92, a correction which reduces the error in the maximum equation of centre from 2′ 37″ to 45″. Such improvements would manifest themselves in observations of the planet, not when it was in opposition, but at other times. Horrocks made many such observations, comparing the planets with star positions taken from Tycho's star catalogue; and some testing of his results against Tuckerman's *Planetary, Lunar and Solar Positions, A.D.2 to A.D.1649* shows them to be accurate to within 2′ – and frequently to within less than 1′ – of arc.

The planet Venus presented a different problem, since up to this time it had never been observed when in line with the Sun, and its orbital elements had been derived chiefly from observations of maximum elongations. We should pause to admire Horrocks's insightful 'fiddling' with the planetary parameters. Mean heliocentric longitude, eccentricity, radius vector, node, inclination, all these along with the elements of solar theory are involved in each observed position of Venus, and all admit of adjustment. During the first eight months of 1638 Horrocks manages to adjust these constants so that theory agrees with his own and Tycho's observations to within the limits of the observational error. One conclusion he draws is that Kepler's value for the mean radius vector is too large, and that this constant should be reduced so as to coincide exactly with the value obtained from the periods of the Earth and Venus by means of Kepler's third law. It is with Horrocks that the practice of obtaining the mean solar distances in this manner originates; it was to be adopted by Streete and passed on to Flamsteed, Halley, and J.-J. L. de Lalande. It leads to improved accuracy for the planets Mercury, Venus, and Mars, but is less useful for Jupiter and Saturn. As Horrocks himself discovered (letters to Crabtree of 29 September 1638 and 14 September 1639), the mean motions of these latter planets did not appear entirely constant.

A further result of Horrocks's refinement of the orbital elements of Venus was his prediction and observation of the transit of this planet across the face of the Sun on 24 November 1639 (OS) – the first such ever observed. The observation permitted still further refinement of the orbital elements.

(4) Horrocks asserts that eclipses of the Sun and Moon are best computed if the solar parallax is assumed to be zero, or else very small.

(5) Horrocks's further reasonings on solar parallax have to do with the sizes of the planets – the question so important to Kepler. In the Copernican system, embraced (so Horrocks tells us) by all astronomers worthy of the name, the Earth is one of the primary planets. The other planets, according to Horrocks, all have a semidiameter as seen from the Sun of about 15″ or less; Horrocks's observation of Venus on the Sun, for instance, gave 14″ for the semidiameter, Gassendi's observation of Mercury yielded about the same value, and despite some uncertainties in his telescopic observations of the sizes of Mars, Jupiter, and Saturn, Horrocks believes each of these would have about the same angular size as seen from the Sun. (In fact, the figure for Jupiter is 19″, and that for Mars 3″; Horrocks's other values need revision downward, although closer to the truth than those of any of his contemporaries.) What follower of the most noble Pythagorean philosophy, Horrocks asks, will doubt that our planet has a similar measure? If the solar parallax of the Earth were 1′ as Kepler postulated, the Earth would be larger than Mars, an inferior planet larger than a superior, which seemed to Horrocks improbable. Keplerian physical speculations support the smaller size: the Sun which moves the planets remains immobile, and therefore must be immensely greater than all of them combined.

Horrocks's revisions of the Keplerian parameters for Venus and the Earth are given in his *Venus in sole visa*; for the other planets, we have only the indications in the correspondence, showing that he

had improved the elements for Mars, was struggling with the difficulties presented by Jupiter and Saturn, and had begun to work on the elements of Mercury. But it is particularly for his lunar theory that Horrocks became known in the seventeenth century. We postpone our description of it to a later section, but here observe that, as the theory involves a variable eccentricity, Horrocks stood in need of a method of rapidly calculating equations of centre for different eccentricities and mean anomalies.

Procedures for this purpose are found in his unpublished *Philosophicall Exercises*, in part identical with computational techniques given in Bonaventura Cavalieri's *Directorium generale uranometricum* (Bologna, 1632 – a book which does not, however, appear to have been known directly to Horrocks). Cavalieri (*c.* 1598–1647) applies trigonometry and the logarithms of Henry Briggs to the solution of mathematical problems, including the famous 'Kepler's problem'. In Figure 10.1 *AOP* is half of an eccentric circle, *C* its centre, *A* its aphelion, and *CS* the eccentricity. If the eccentric anomaly *E* at some moment is $\angle ACF$, then according to Kepler the mean anomaly *M* corresponding to it is given by the ratio of the area *AFS* to the whole area of the eccentric circle. It is easy enough, given the eccentricity and a particular value of the eccentric anomaly, to compute the mean anomaly, and this is the course Kepler takes in calculating his tables. But he and every other Keplerian astronomer would have preferred to proceed in the reverse direction: given *M*, which represents the portion of the planet's period that has elapsed since it last passed aphelion, to find *E*. Cavalieri proposed the following procedure: if area *ADC* (or angle *ACD*) represents the mean anomaly, connect *S* and *D*, and draw *CF* parallel to *SD*; angle *ACF* will be the eccentric anomaly very nearly. The approximation here rests on the near equality of the areas *AFS* and *ADC*; and Cavalieri goes on to show how partially to correct for the difference, which is represented by the area of the segment formed by the chord *DF* and the arc *DF*.

Horrocks gives no such correction, but his procedure is otherwise equivalent, although his geometrical construction looks different because he uses an epicycle (see Figure 10.2). He recognizes that he is in effect replacing Kepler's equation $M - E = e \sin E$ by $\sin(M - E) = e \sin E$. However, he

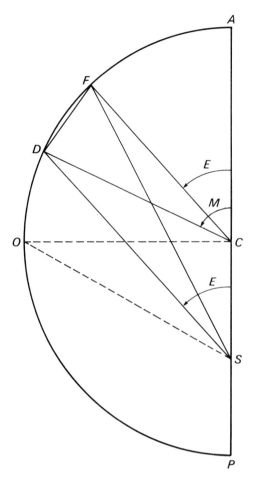

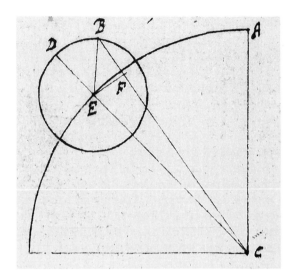

10.2. Horrocks's procedure for determining the eccentric anomaly (E) corresponding to a given mean anomaly ($M = \angle ACD$): with centre at intersection of CD with the eccentric circle, construct an epicycle with radius equal to the eccentricity; draw EB parallel to CA, and connect B to C; then, very nearly, $E = \angle ACB$.

10.1. Cavalieri's procedure for determining the eccentric anomaly (E) corresponding to a given mean anomaly (M): on the diameter of semi-circle with centre at C, lay off CS equal to the eccentricity; with $M = \angle ACD$, draw SD; then, very nearly, $E = \angle ASD$.

underestimates the magnitude of the error, which is proportional to $e^3/6$, and is a maximum when $E = 90°$; thus, for Mars and Mercury, the visible planets with the highest values of e, it amounts in heliocentric longitude to 28″ and 4′ 58″ respectively. Like Cavalieri, Horrocks employs the tangent rule for finding E when M is given; thus in triangle CEB of Figure 10.2, (corresponding to triangle SDC of Figure 10.1)

$$\frac{\tan \frac{1}{2}(E - \angle ECB)}{\tan \frac{1}{2}M} = \frac{1-e}{1+e} = \cot^2(45° + \tfrac{1}{2}\epsilon),$$

where $\epsilon = \arcsin e$ is very nearly the maximum value of $\angle ECB$. Also like Cavalieri, Horrocks solves

this equation logarithmically. But whereas Cavalieri leaves the ellipticity of the orbit out of account, Horrocks gives the formula for passing from eccentric anomaly to true anomaly in the ellipse:

$$\tan \tfrac{1}{2}v = \sqrt{\left(\frac{1-e}{1+e}\right)} \tan \tfrac{1}{2}E = \cot(45° + \tfrac{1}{2}\epsilon) \tan \tfrac{1}{2}E,$$

where v stands for true anomaly. This formula was not original with Horrocks; for after stating it logarithmically he adds: "Inquire ye demonstration". It is in fact easily derivable from Euclid VI.3 and properties of the ellipse.

Another of Horrocks's concerns was the replacement of the "harsh and unlikely conceits, and meer shifts" of Kepler's "magnetical philosophy", whereby Kepler sought to account for the ellipticity and eccentricity of the planetary orbits. Horrocks objects, for instance, that the Sun would turn the "friendly side" of the planet to itself and so cause the latter to unite with itself (the same objection is made by Ismaël Boulliau and later by Newton). He also observes that the Earth's magnetism is oriented in the wrong direction to bring about the approach to and recession from the Sun that Kepler is attempting to explain. In a move that will later have great importance in the thought of Hooke, Horrocks turns to the conical pendulum as his model.

The planet is attracted to the Sun just as the pendulum bob tends to its lowest possible place. In analogy with the circular motion of the hand that makes the pendulum move in an oval orbit rather than in a straight line, Horrocks supposes an influence deriving from the rotation of the Sun to be required to move the planet out of the line of apsides, in which it would otherwise librate. In the conical pendulum the apsidal line advances as in the planetary orbits, but the path is without eccentricity, and in order to introduce it Horrocks imagines in each planet an "affection to its aphelion", a natural desire "to rest in ye same place where ye hand of nature hath at first placed it". When the conical pendulum was once more put forward as a model of elliptical planetary motion, by Hooke in the 1660s, the inspiration no doubt derived from the papers of Horrocks.

Horrocks's extraordinary achievement in the improvement of predictive astronomy, truncated by early death though it was, remained unmatched and largely unknown during the middle decades of the seventeenth century. Before closing our account of the immediately post-Keplerian years, however, we should mention the work of an older astronomer whose findings were at some points, and in particular with respect to solar parallax, close to those of Horrocks: Gottfried Wendelin (1580–1667), "the Hipparchus of the Belgians" as Pierre Dupuy (Puteanus) called him, and the teacher of Gassendi. From the early years of the century Wendelin made solar and lunar observations, and studied ancient eclipse, equinox, and solstice records. In 1626 he published his *Loxias seu de obliquitate solis diatriba*, seeking to establish that the obliquity of the ecliptic undergoes an oscillation of 30' to either side of a mean value of 24°, with a period of 9840 years, the minimum obliquity to be reached in AD1860. Wendelin's penchant for neat schemes is evident enough here, but the hypothesis at least recognizes the decrease in the obliquity during the past 1000 years, which Horrocks also recognized but which was doubted by Kepler and rejected by Gassendi, Flamsteed, and most other seventeenth-century astronomers. Just as important were the corrections for parallax and refraction on the basis of which Wendelin defended his

hypothesis. From telescopic observation of lunar dichotomies in 1625, using the method of Aristarchus, he found the Moon when dichotomized to be at least 89° from the Sun, and concluded thence that the solar parallax could not exceed 1'. Given this correction in the solar parallax, he inferred (in opposition to Tycho) that the refractions of the Sun would not differ from those of the Moon and stars; none of these refractions, he thought, was detectable above an altitude of 18°.

In a letter to Gassendi of May 1635 (published in 1658), Wendelin claimed that the solar parallax should be still further reduced to 14" or 15", basing his remarks on observations of the diameters of the planets, from which like Horrocks he concluded that, as seen from the Sun, their diameters are no more than 28" to 30". In a later writing (his *Luminarcani arcanorum caelestium lampas* . . . of 1643), he reduced the eccentricity of the Earth's orbit to 0.017 45, with a maximum equation of centre of 2° 0' (*praecisissime*, Wendelin claimed). Moreover, like Horrocks, Wendelin improved on the Rudolphine values of the mean solar distances of Mercury and Venus: 0.387 11 and 0.723 43 as compared with Kepler's 0.388 08 and 0.724 13 (the correct values being 0.387 10 and 0.723 33); presumably he was using Kepler's third law to arrive at these results. Wendelin's propensity to insist on the exactitude of quantitative propositions of a speculative kind is especially evident in the *Luminarcani . . . lampas*, in which he announced that the equation of time to be used in calculation of solar and lunar eclipses is zero, and that the motions of the aphelia of the planets vary – exactly! – as the sixth roots of their eccentricities. Neither claim can stand. But Wendelin's reduction of the solar parallax, of which he informed Giambattista Riccioli (1598–1671) by letter in 1647, became widely known through Riccioli's *Almagestum novum* of 1651, and no doubt stimulated Riccioli's own endeavour to determine solar parallax from lunar dichotomies, from which he obtained a value of 28". The necessity of reducing the ancient value was thus recognized in Bologna when Cassini I began his observational work there on refractions and the solar elements in the 1650s.

But astronomers in mid-century were by no means singlemindedly engaged in the revision of Keplerian parameters. Looming at least as large was a different concern, the replacement of

Keplerian hypotheses and calculational procedures judged "a-geometrical" by hypotheses and
procedures of strict mathematical exactness. This
concern was to be brought into prominence by
Ismaël Boulliau (1605–94).

Planetary theory from Boulliau to Streete and Mercator

Writing to Kepler on 20 January 1607, David
Fabricius urged that the principle of uniform circular motion should not be abandoned, and added:
"you can excuse the ellipse by another small
circle". Copernicus had set down uniform circular
motion as a fundamental postulate, believing that
the perpetuity of the celestial motions would not
otherwise be accounted for; Tycho agreed, and the
idea persisted. It was hardly surprising, then, that a
later proponent of the principle, having recognized
the superior accuracy of the Keplerian tables,
should undertake to arrive at essentially Keplerian
results by an unKeplerian path, using epicyclic or
other combinations of uniform circular motions.

Already in a letter to Gassendi of 1633 Boulliau
was objecting to the magnetic mechanism hypothesized by Kepler to account for the eccentricity of
the planetary orbits:

*Indeed astronomy and geometry have always rejected,
as mere dreams, these hypotheses and physical equations of motion, which exclude geometrical demonstrations. The axiom that the celestial motions are circular
or composed of circles must stand, and therefore I reject
and repudiate his ellipses unless he should suppose them
to be described by means of a Copernican or Tychonic
epicycle. For it is easily demonstrated that an ellipse is
described by means of a small circle with diameter equal
to the difference between the [transverse and conjugate]
axes of the ellipse, if the planet moves on the epicycle
with double the speed with which the centre of the
epicycle moves on its large orbit . . . .*

By the time he composed his *Astronomia philolaïca* (1645), Boulliau had imagined another way
of deriving elliptical orbits from uniform circular
motions, one which he believed had the merit of
proving for the first time, "from the general circumstances of planetary motion", that these orbits are
elliptical. The argument, briefly summarized, goes
as follows:

(1) The revolution of a planet about the Sun

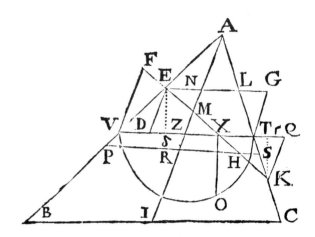

10.3. The scalene cone from which Ismaël
Boulliau derived the elliptical path and
(supposedly) the motion of a planet in his
Astronomia Philolaïca of 1645.

always occupies the same time, and always occurs
in the same way along the same re-entrant line, as
the observations of all ages attest. Boulliau concludes that, as the motion accords with the *leges
circulorum*, it must be composed of circular motions. Such motion, according to Boulliau, follows
from the inner nature of the planet.

(2) The equality is conjoined with inequality: as
observed from the Sun, the planet is seen to move at
different speeds in different parts of its orbit, slower
at aphelion, faster at perihelion. About half this
inequality is due to the fact that the Sun is off-
centre, and the other half to an actual change in
speed along the planet's path. Since the equal motion remains unaffected, Boulliau concludes that it
takes place not on a single circle, but on many
circles, indeed on an infinite number in a continuous series, increasing in size continuously from
aphelion to perihelion, while the angular rate of
motion on all these circles remains the same. The
path of the planet must be inclined to these circles
so that it passes continuously from one to another.
According to Boulliau, the circles must lie, not in a
plane (he gives no reason for this conclusion), but
in a solid surface, such as a sphere, cylinder, ellip-
soid, or cone.

(3) In 90° of mean motion from aphelion, the
planet's increase in speed is half the total increase
in speed in its passsage from aphelion to perihelion.
Considering plane sections of the various solids of

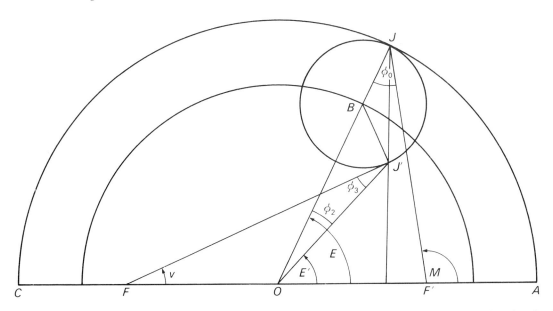

10.4. Diagram for Boulliau's derivation of true anomaly (v) from mean anomaly (M). The derivation involves, besides the angles shown, an angle ϕ_1 that cannot be represented in the same diagram.

revolution, Boulliau argues that only a conic section is able to meet this requirement. *Via ergo planetae est elliptica.*

To duplicate the observed motion he uses (as indicated in Figure 10.3) a scalene cone whose cutting plane EK makes with its axis AI the same angle $\angle AME = \angle AIC$ as its base BC; whence the triangle MXZ and all those similar to it are isosceles. It can readily be shown that the point M of intersection is a focus of the ellipse of major axis EK cut out of the cone.

According to Boulliau, the mean motion takes place about the axis AI in the circles parallel to the base. At the time of writing the *Astronomia philolaïca*, Boulliau failed to recognize that this implies an equivalent uniform angular motion about the (non-solar) focus M; the hypothesis is thus equivalent to the empty-focus equant that Kepler's "Uranian friend" Albert Curz had proposed for the Moon, and to which Kepler refers in the *Rudolphine Tables*. The equation of centre in this hypothesis, to the third power of the eccentricity, is given by

$$M - v = 2e \sin M - e^2 \sin 2M + \tfrac{2}{3} e^3 \sin 3M, (1)$$

where M is mean anomaly measured from aphelion, and v is the true anomaly. The corresponding expression derived from the Keplerian hypothesis is

$$M - v = 2e \sin M - \tfrac{5}{4} e^2 \sin 2M$$
$$+ \tfrac{1}{12} e^3 (13 \sin 3M - 3 \sin M) \ldots \quad (2)$$

and the difference $[v_{\text{Boulliau}} - v_{\text{Kepler}}]$ is

$$-\frac{e^2}{4} \sin 2M + \frac{e^3}{12} (5 \sin 3M - 3 \sin M) + O(e^4).$$

Here the largest term, $-\tfrac{1}{4}e^2 \sin 2M$, gives for Mars in the octants of anomaly a discrepancy of about $7\tfrac{1}{2}'$. Boulliau, as we shall see, avoids so large an error.

In doing so, he cannot, of course, be following the strict implications of his theory, but he gives no sign of being aware of the inconsistency. His procedure for calculating equations of centre, as described for instance in Chap. 4 of Bk II, consists of the following steps:

(1) Given an angle of mean anomaly M, Boulliau computes an angle ϕ_0 such that $\sin \phi_0 = e \sin M$. He calls the angle $E = M - \phi_0$ the "first equated anomaly". The relations between M, ϕ_0, and E are correctly represented in Figure 10.4.

(2) He next computes the "equation of the elliptic epicycle", angle $\phi_2 = \angle JOJ' = E - E'$ in Figure 10.4. Angle E', the "coequated anomaly", is given by

$\tan^{-1}(\sqrt{(1-e^2)}\tan E)$, and angle $E-E'$ approximately by $\frac{1}{4}e^2\sin 2E$. To compute this equation Boulliau makes use of the epicyclic production of the ellipse: the epicycle has a radius $BJ=\frac{1}{4}e^2$ when the semi-major axis of the ellipse or OA is 1, and the radius of the deferent or OB is $1-\frac{1}{4}e^2$; the angle JBJ', traversed clockwise by the planet on the epicycle, is twice the angle BOA traversed counterclockwise by the centre of the epicycle from the upper apse of the deferent. Boulliau thus obtains ϕ_2 from solving triangle OBJ' of which he knows OB, BJ', and the external angle JBJ'.

(3) From E' Boulliau then computes ϕ_1, the "equation of the equant circles", by the relation $\sin \phi_1=e\sin E'$. The angle ϕ_1 is not represented in Figure 10.4.

(4) The third equation to be computed is the "optical equation", angle $\phi_3=\angle FJ'O$ in Figure 10.4. Here $\sin\phi_3=e\sin E'/(1+e\cos E)$.

(5) Finally, the total equation is formed from the sum $\phi_1+\phi_2+\phi_3$, and this is subtracted from the mean anomaly M to obtain the true anomaly v_B.

We note that J' in the diagram does *not* represent the true position of the planet on its orbit: for that to be the case the total equation would have to be $\phi_0+\phi_2+\phi_3$. Hence v in the diagram is not the same as the true anomaly v_B, and ϕ_3 is not the true optical equation. Thus the individual steps of Boulliau's computational procedure fail to add up to a geometrically representable theory. But the results are better than those derivable from the simple elliptic theory of Figure 10.3, and far better than those derivable from the theory that Figure 10.4 appears to represent. A series expansion of the equation of centre implied by Boulliau's procedure, to the third power of the eccentricity, yields

$$M-v_B=2e\sin M-\frac{5e^2}{4}\sin 2M+\frac{1}{2}e^3\sin M$$
$$-\frac{5e^3}{3}\sin^3 M\ldots, \qquad (3)$$

which differs from the Keplerian equation by $v_B-v_K=\frac{5}{2}e^3\sin M-\frac{8}{3}e^3\sin^3 M$. By this formula the maximum error in heliocentric longitude occurs at about 34° from aphelion; for Mars it would be about $2\frac{1}{2}'$. Actually Boulliau's value for e in the Martian orbit, namely 0.092 39, is smaller than Kepler's (which at 0.092 65 is itself too small). Comparison of Kepler's and Boulliau's tables shows differences in the equations of centre for Mars going

to $-2\frac{3}{4}'$ in the first and fourth quadrants of anomaly and to $+2\frac{3}{4}'$ in the second and third. The error in geocentric longitude near the time of opposition can rise to nearly three times as much. Late in the day, after his book had already gone to press, Boulliau became concerned about these discrepancies, and had an extra leaf inserted into the *Astronomia Philolaïca*, giving the Rudolphine equations of centre, and explaining how by his own procedure these same equations might be derived, with an increase of the eccentricity to 0.101 92, the Sun's distance from the centre of the ellipse remaining fixed at 0.092 39.

Besides differing from Kepler in his way of computing equations of centre, Boulliau also differed as to procedure for determining the eccentricity and aphelion of an orbit from observations. The approximative procedure used for this determination by Ptolemy and Copernicus had been sharply criticized by François Viète in an appendix to his *Apollonius Gallus* (1600), which draws upon his still unpublished *Ad harmonicon coeleste*. Yet more complicatedly, Kepler's procedure in Chapter 16 of the *Astronomia nova* involved double false position and successive approximations. With a view to avoiding such a-geometrical procedures, Boulliau turned to a *problema* proposed for the purpose by Viète: given three points on a circle, to find a diameter such that the perpendiculars dropped from the points onto the diameter cut off intercepts having to one another a given ratio. In Boulliau's procedure, the circle is centred at the centre of the ellipse, and has a radius equal to the eccentricity sought; thus in Figure 10.5 the three points A, B, C correspond to three heliocentric longitudes of a superior planet. The angles at the centre, $\angle AXB$, $\angle BXC$, $\angle CXA$, are neither the angles of mean motion nor the angles of observed motion, but an average of the two; Boulliau believes this to be very nearly what would be observed from the centre of the ellipse. Viète's *problema* is then employed to locate a diameter PQ such that the intercepts GH, HI, IG formed by normals dropped from A, B, C onto PQ are in the ratios of the sines of the half-differences between the intervals of mean motion and intervals of apparent motion; in symbols, with M representing mean anomaly and v true anomaly as measured from a line of apsides through X at right angles to PQ,

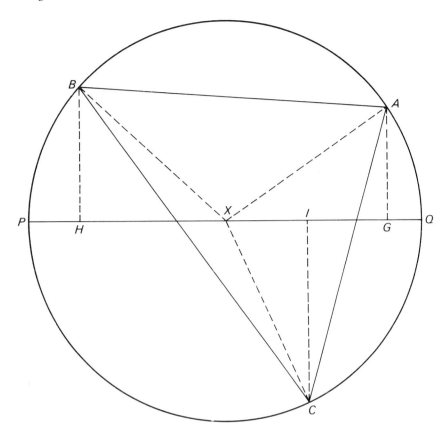

10.5. Boulliau's procedure (using a *problema* proposed by Viète) for determining the eccentricity and aphelion of a planet from three observations.

$$e \left[\sin \frac{M_B + v_B}{2} - \sin \frac{M_A + v_A}{2} \right]$$

$$= \sin \left[\frac{(M_B - M_A) - (v_B - v_A)}{2} \right]. \qquad (4)$$

Equation (4) may be viewed as an approximation, good to the order of e^2, to the assumption that for each true anomaly v and the corresponding mean anomaly M,

$$e \sin \frac{M+v}{2} = \sin \frac{M-v}{2}. \qquad (5)$$

Boulliau allows that his method is not quite exact, but urges that it is less laborious than Kepler's procedure. Equation (5), we note, is strictly implied by Boulliau's original theory, the theory of Figure 10.3 which, unbeknownst to Boulliau, was equivalent to the theory of Curz.

In the case of the inner planets, Boulliau depends on maximum elongations from the Sun to fix the size, eccentricity, and aphelion of the orbit. Like Kepler, he proceeds as if the line of sight in such observations were, to a good enough approximation, perpendicular to the line from point of contact to centre of the orbit. He manages to improve on Kepler's values for the aphelia, and (with the aid of Gassendi's observation) on Kepler's mean longitude for Mercury, but in the other parameters he does no better than Kepler and often worse. He is emphatic in insisting that the motions of Mercury cannot be made subject to numerical law unless the elliptical orbit is adopted, with the faster motion near perihelion, and the slower near aphelion: "whoever attempts to twist the paths of the planets into other orbits wastes time and trouble, and rolls the stone of Sisyphus."

In the two decades immediately following its publication, the *Astronomia Philolaïca* no doubt served to promote the wider acceptance of elliptical astronomy, and at the same time it stimulated the search for mathematical procedures more direct than Kepler's. In the longer run, the Philolaic tables were judged to be too inaccurate. When the Academy of Sciences was formed in Paris in 1666, Boulliau was excluded from membership; and Picard and Huygens, among the members, insisted that Kepler's *Rudolphine Tables*, although needing revision, had not yet been improved upon. Boulliau was an inexact and unreliable observer, and for the early members of the Paris Academy, a new level of observational precision was a basic objective. Huygens opposed Boulliau's exaggerated value of solar parallax (2′ 21″) and of the solar equation of centre (2° 2′ 41″), as well as Boulliau's faulty equation of time (on which more below). Both Huygens and Picard, we know, were familiar with Horrocks's *Venus in sole visa* and knew that solar parallax must be reduced. They and their colleagues looked to a renovation of astronomy that would begin systematically with first things: find north, find the height of the celestial pole, find the obliquity.

During the late 1640s and 1650s, however, Boulliau's astronomy received considerable attention, particularly in England. In the earliest known letter of Jeremy Shakerley (1626–c. 1655), addressed on 26 January 1647/48 to the astrologer William Lilly, we read that the Philolaic tables of Bullialdus "are holden for the best and exactest now extant". In his *Synopsis compendiaria* of 1651 Shakerley computes eclipses from "the latest and best Astronomical tables of Bullialdus, but reduced to the Meridian of London". Then in 1653, after Shakerley had left England for India (where on 23 October 1641 (OS) he observed a transit of Mercury, which he had predicted in the *Synopsis compendiaria* on the basis of both Kepler's and Boulliau's tables (but believing the latter to be more accurate), there appeared his *Tabulae Britannicae*. Here the planetary and lunar numbers were simply those of the Philolaic tables, recalculated for the meridian of London and for the Julian calendar. The title page misleadingly states that the tables are calculated "from the Hypothesis of Bullialdus and the observations of Mr Horrox": no observations of Horrocks are used, but the precepts accompanying the tables make use of Horrocks's logarithmic procedures of calculation. Shakerley had had access to Horrocks's papers

while living at the Towneley estate in Lancashire, and he learned much from them, including Horrocks's analysis of the diagram of Hipparchus and his method of observing transits. But he did not penetrate deeply enough to see the crucial importance of Horrocks's reduction in solar parallax and consequently in the solar equation of centre.

Another version of Boulliau's tables, again calculated for the meridian of London, appeared in 1657, the *Astronomia Britannica* of John Newton (1622–78). But of greater interest are the astronomical tables devised by the largely self-taught Vincent Wing (1619–68) – the *Harmonicon coeleste* of 1651, the *Astronomia instaurata* of 1656, and the *Astronomia Britannica* of 1669 – all of which show the influence of Boulliau, while incorporating mathematical innovations and innumerable confirmatory calculations due to Wing. Wing's earliest book on astronomy had been the *Urania practica*, written with William Leybourne and published in 1649. This work had been roundly attacked by Shakerley in *The Anatomy of Urania Practica* (1649): it was, said Shakerley, a hotchpotch of theory made up from unreliable authors like Andrea Argoli and van Lansberge, who had not yet accepted the principles of elliptical astronomy. In particular, Shakerley criticized the work for assuming with Copernicus and van Lansberge a trepidation or inequality in the precession of the equinoxes (Shakerley draws his refutation from Holwarda and Boulliau), for employing a faulty lunar theory and equation of time, and for failing to fulfil the requirements of the diagram of Hipparchus in the calculation of eclipses. Wing and Leybourne responded with their *Ens fictum Shakerlaei: or the Annihilation of Mr Jeremii Shakerley, his in-artificiall Anatomy of Urania Practica . . .* (1649), accusing Shakerley of being "popular, and a great enemy of antiquity", that is, of failing to recognize the authority of ancient observations, and for the rest of having misunderstood whereof he wrote.

Nevertheless, Wing himself appears to have been sufficiently influenced to turn to the study of Boulliau and Kepler and Tycho, and in his *Harmonicon coeleste* of 1651 we find him rejecting the inequality of the precession of the equinoxes, using the diagram of Hipparchus, accepting elliptical astronomy, and presenting us with an

. . . accurate, facile, and brief Theory of the Planets,

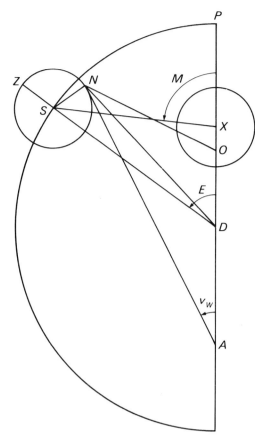

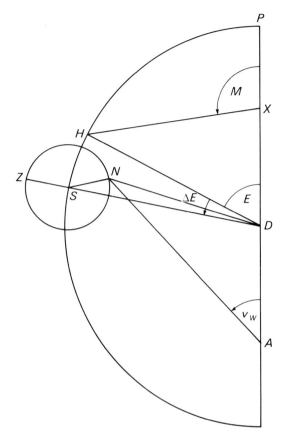

10.6. Vincent Wing's procedure, in his *Harmonicon coeleste* of 1651, for deriving true anomaly (v_W) from mean anomaly (M).

10.7. Wing's improved procedure, in his *Astronomia instaurata* of 1656 and his *Astronomia Britannica* of 1669, for deriving true anomaly (v_W) from mean anomaly (M).

newly devised by the Authour, wherein is plainly and succinctly delivered . . . how to calculate the Motions of all the Planets Trigonometrically, wherein I much dissent from all other Authours that have treated hereof in other Languages, and have delivered the same more methodically for practice, than any hath done before me . . . .

Wing's new procedure is in fact a modification of Boulliau's. In Figure 10.6 the ellipse is produced by an epicycle of radius $\frac{1}{4}e^2$ moving on a circle of radius $1 - \frac{1}{4}e^2$; M is the mean anomaly, and E the "equated anomaly", determined by the relation $\sin(M - E) = e \sin M/(1 - \frac{1}{4}e^2)$. The angle ZSN of epicyclic motion is $2E$. The eccentricity $DX = AD = e$ is varied by subtracting a sinusoidal term $XO = \frac{1}{4}e^2 \sin 2E$, and the total equation of centre is given by $\angle OND + \angle DNA$. The resulting true anomaly v_w can be shown to differ from the Keplerian value by

$$v_K - v_W = \tfrac{1}{4}e^2\sin 2M - \tfrac{1}{4}e^2\sin M \sin 2M - 2e^3\sin M + \tfrac{8}{3}e^3\sin^3 M - \tfrac{1}{2}e^3\sin^4 M.$$

In the case of Mars, this error rises to 5' in the second quadrant of anomaly.

By the time Wing published his *Astronomia instaurata* in 1656, he had detected the error in this theory by comparing it with acronychal observations of Mars. Moreover, he had found a way of eliminating most of this error; it consisted in adding to the angle E a correction term equal to $k \sin 2E$, where k was to be determined empirically. The value of k should be about $\frac{1}{2}e^2$; in the case of Mars, Wing in his calculation takes it to be $14'$ $55'' \approx \frac{1}{2}e^2 + \frac{2}{3}e^4$. The new theory, which is also that of the *Astronomia Britannica* of 1669, is represented in Figure 10.7. Once again the radius DS of the deferent is $\frac{1}{2}(1 + \sqrt{(1 - e^2)}) \approx 1 - \frac{1}{4}e^2 - \frac{1}{16}e^4$, so that the radius SN of the epicycle is $\frac{1}{4}e^2 + \frac{1}{16}e^4$, while, with $\angle PDH = E$, $\angle HDS = (\frac{1}{2}e^2 + \frac{2}{3}e^4) \sin 2E$; and the maximum departure from the measure of mean anomaly M by the areal rule is reduced to

$$\tfrac{1}{12}e^3 \sin 3M - \tfrac{1}{24}e^4(11 \sin 2M - 5 \sin 4M),$$

or for Mars at most about 20″. (Wing himself finds a discrepancy of 43″, but his numerical computation is not everywhere exact.)

Meanwhile, the mathematical errors in Boulliau's *Astronomia Philolaïca* were exposed in 1653 in the *Inquisitio brevis in Ismaelis Bullialdi astronomiae Philolaïcae fundamenta* of Seth Ward (1617–89), Savilian professor of astronomy at Oxford. Paul Neile it was who first put Ward onto the fact that all was not well in Boulliau's procedures. After showing that Boulliau's conical construction of the planet's motion (as in Figure 10.3 above) implied equal angular motion in the plane of the ellipse about the empty focus, Ward had no difficulty in proving that Boulliau's actual computation of equations of centre was incompatible with this construction. Methods for calculating the equations correctly, he pointed out, were not lacking, for instance the method proposed by Neile and indicated in Figure 10.8. Here F is the superior focus, and ∠ *AFP* = M is the mean anomaly; the line *FP* has been extended to Q so that *FQ* is equal to the major axis *AR*, whence *PQ* = *PS*. It follows that the equation of centre ∠ *FPS* is double ∠ *FQS*, the latter being determinable in triangle *FQS* since the two sides *FQ*, *FS* are known, along with the included angle. Ward also showed that Boulliau's attempted derivation of his procedure for finding the eccentricity and aphelion from three observations was a *non sequitur*, and complained that the method itself was "a-geometrical", that is, inexact.

Ward's *Astronomia geometrica* of 1656, dedicated to Boulliau among others, is based on the assumption that the "simple elliptical hypothesis", with superior focus as equant point, is true. Kepler and Boulliau had failed to recognize its truth, he believes, only because they were too much burdened by their own erudition and industry. Proceeding on this hypothesis, Ward proposes ways of determining aphelion and eccentricity from observations "geometrically", without approximation. Nowhere does he face the question whether the hypothesis is empirically correct. As for the proposed procedures, we must conclude with J.-B.J. Delambre that they are impractical, chiefly because dependent on observations that it would be difficult or impossible to obtain with sufficient accuracy.

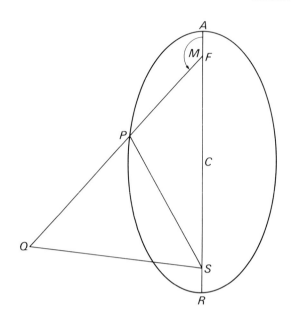

10.8. The procedure of Paul Neile and Seth Ward for deriving true anomaly from mean anomaly.

Another proponent of the simple elliptical hypothesis was Emile-François de Pagan, Comte de Merveilles, who in 1657 published *La théorie des planètes du Comte de Pagan. où tous les orbes celestes sont geometriquement ordonnez, contre le sentiment des astronomes.* For the construction of tables Pagan employs the Rudolphine values of the aphelia, mean motions, and maximum equations of centre, but calculates equations of centre for other points of the orbit on the assumption of the superior focus as an equant point. He lists the maximum discrepancies between Kepler's tables and his own, noting that they occur in the octants of anomaly, and asserting, without proof, that they show Kepler's tables to be wrong, his own correct.

Boulliau responded to Seth Ward's attack in his *Astronomiae Philolaïcae fundamenta clarius explicata et asserta* of 1657. He begins by acknowledging an error in his previous publication: the error which had led him to put Kepler's table of equations of centre for Mars in place of his own. Seth Ward, although he had noted this substitution, had not shown how the matter could be set right. The truth is that the mean motion about the axis of the cone, or as Ward makes it, about the empty focus, is subject to a small inequality, owing to "the translation of the planet's motion from circle to ellipse". In

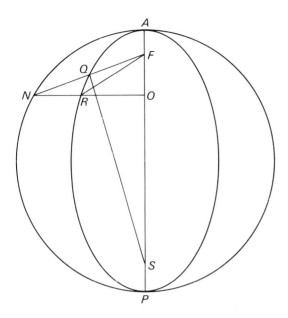

10.9. Boulliau's modified procedure, in his *Astronomiae Philolaïcae* of 1657, for deriving true anomaly from mean anomaly.

Figure 10.9, where F and S are foci of the ellipse $AQRP$, let $\angle AFR$ be the mean anomaly. To find the corresponding position of the planet on the ellipse, drop RO perpendicular to the axis AP, and extend it backwards to N on the circumscribing circle; connect N and F, and the point Q where this line intersects the ellipse will be the location of the planet. In a lengthy and obscure course of reasoning and explication that we shall not reproduce here, Boulliau seeks to show that this procedure is logically implicated in or at least consonant with his original conical construction and derivation of the elliptical orbit. He then compares the consequences of the new procedure with 28 Tychonic observations of Mars given in Kepler's *Astronomia nova*; the average discrepancy is 2' 4", the maximum 5' 33" – results a little better than those obtained by Kepler himself. In fact, the error in Boulliau's new procedure can be shown to be

$$v_{B'} - v_K = \tfrac{2}{3}e^3 \sin^3 M,$$

which in the case of Mars yields a maximum value of about 1' 51", six times the error in Wing's final procedure and four times that in the procedure of Cavalieri and Horrocks. In a final *Monitum*, Boulliau criticizes the impracticality of Seth Ward's proposals, in his *Astronomia geometrica*, for deter-

mining aphelia and eccentricities.

Boulliau's procedure for calculating equations of centre was made use of in Streete's *Astronomia Carolina* (1661), where, however, it is attributed to Streete's friend Robert Anderson. But in some respects Streete's planetary tables are rather better than Wing's or indeed any other tables published earlier; as Flamsteed will write in 1669: "I esteem Mr Streete's numbers the exactest of any extant." For Mercury, the Earth, and Mars, Streete's chosen values of the orbital parameters are clearly superior, as can be seen from Table 10.1, in which the eccentricities, aphelia, and mean solar distances used by Kepler, Boulliau, Wing, and Streete are compared with Simon Newcomb's values for these elements in 1600.

What underlay Streete's improvements in orbital elements were two Horrocksian propositions, gleaned in all likelihood from his reading of a manuscript copy of Horrocks's *Venus in sole visa*: (1) the solar parallax being reduced to about $\tfrac{1}{4}'$, the eccentricity of the Earth's orbit is to be reduced considerably, and the eccentricities and aphelia of the other planets readjusted as a consequence; (2) the mean solar distances of the planets are to be calculated from their periods by way of Kepler's third law. Adoption of the second proposition was particularly helpful in the cases of Mercury and Venus, where it eliminated errors of 3' or 4' in the *parallaxis orbis* of earlier tables.

To determine the eccentricity and aphelion of a planet from observations, Streete adapts a procedure given in 1637 by the Paris mathematician Pierre Hérigone (in his *Cursus mathematicus*) for a circular orbit with uniform motion about the centre, to the simple elliptic theory, in the process making use of Paul Neile's proposition. Thus in Figure 10.10, a, b, and c represent the directions from S (the Sun) of three observed positions of the planet; $abcg$ is a circle with radius equal to the major axis of the orbital ellipse, and F is the superior focus of the ellipse, the distance and direction of which from S are to be found. From the fact that the angles SaF, SbF, ScF are the halves of the corresponding equations of centre, it can be shown that $\angle aFb, \angle bFc, \angle aFc$ are the averages of the mean and apparent motions in the corresponding intervals, and so determinable. By means of auxiliary lines such as bF extended to g, cg, and ag, and by solutions of triangles, it is then possible to find FS in

Table 10.1 *Orbital elements of the planets adopted by seventeenth-century authors, compared with Newcomb's values for 1600*

	Eccentricity	Aphelion		Mean distance
Mercury				
Newcomb (for 1600)	0.205 55	251°14′ 9″		0.387 10
Kepler (K−N)	+0.004 50	+	1°35′49″	+0.000 98
Boulliau (B−N)	+0.004 52	+	23′38″	−0.001 25
Wing 1651 (W−N)	+0.004 85	+	1°34′ 6″	−0.000 70
Wing 1669 (W−N)	+0.004 84	−	7′54″	−0.001 10
Streete (S−N)	+0.000 34	−	31″	0
Venus				
Newcomb (for 1600)	0.006 97	305°55′51″		0.723 33
Kepler (K−N)	−0.000 05	−	4°41′29″	+0.000 80
Boulliau (B−N)	+0.000 87	−	32′46″	+0.000 65
Wing 1651 (W−N)	−0.000 05	+	32′41″	−0.000 26
Wing 1669 (W−N)	+0.000 36	−	6°55′41″	+0.000 74
Streete (S−N)	+0.000 18	−	3°22′50″	0
Earth				
Newcomb (for 1600)	0.016 88	276° 4′ 2″		1.000 00
Kepler (K−N)	+0.001 12	−	19′54″	0
Boulliau (B−N)	+0.000 96	−	28′38″	0
Wing 1651 (W−N)	+0.000 99	−	20′34″	0
Wing 1669 (W−N)	+0.001 00	−	20′34″	0
Streete (S−N)	+0.000 44	+	21′26″	0
Mars				
Newcomb (for 1600)	0.093 04	148°41′58″		1.523 69
Kepler (K−N)	−0.000 39	+	17′56″	−0.000 19
Boulliau (B−N)	−0.000 65	+	17′54″	−0.000 19
Wing 1651 (W−N)	−0.000 55	+	18′ 2″	+0.001 31
Wing 1669 (W−N)	−0.000 39	+	15′ 4″	−0.000 02
Streete (S−N)	−0.000 50	+	6′24″	0
Jupiter				
Newcomb (for 1600)	0.047 84	187°53′27″		5.202 7
Kepler (K−N)	+0.000 38	−	1° 1′26″	−0.002 7
Boulliau (B−N)	+0.000 72	+	7′55″	+0.010 5
Wing 1651 (W−N)	+0.000 46	+	53″	+0.020 5
Wing 1669 (W−N)	+0.000 01	−	8′20″	+0.013 3
Streete (S−N)	+0.000 32	−	27′25″	−0.001 6
Saturn				
Newcomb (for 1600)	0.056 93	265°13′24″		9.546
Kepler (K−N)	0	−	15′48″	−0.036
Boulliau (B−N)	+0.000 81	+	46′22″	−0.004
Wing 1651 (W−N)	+0.000 56	+	46′36″	−0.013
Wing 1669 (W−N)	+0.000 56	+	56′36″	−0.013
Streete (S−N)	+0.000 42	+	53′49″	−0.008

terms of the radius of the circle, and $\angle aSF$, which gives the position of the aphelion. Streete recognizes that the method is not quite accurate, as being dependent on the simple elliptic theory.

Undermining the accuracy of earlier planetary tables, Streete claimed, were the "unnatural Translations and never to be proved Motions of the Aphelions and Nodes of the Primary Planets"; and accordingly he declared the aphelia and nodes to be stationary with respect to the fixed stars. This assertion, as well as Streete's adoption of Horrocksian values for solar parallax and for the Sun's equation

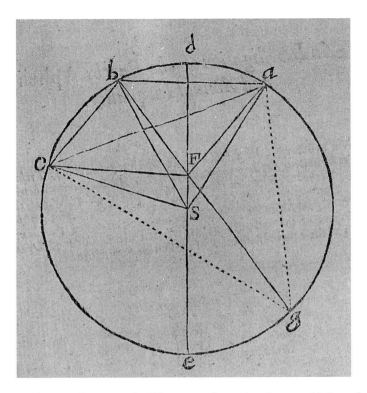

10.10. Streete's adaptation of a procedure given by Hérigone to determine the eccentricity and aphelion of a superior planet from observations, from his *Astronomia Carolina* (1661). *S* is the Sun, and *F* is the second focus of the planetary ellipse, which the procedure is designed to locate.

of centre, was disputed bitterly by Wing in his *Examen astronomiae Carolinae* of 1665; and Streete responded with argument and asperity in his *Examen examinatum* of 1667. If we give Streete the victory in the matter of parallax, we must take it away again in the matter of the aphelia and nodes; on which his position was nevertheless influential, for it no doubt underlay Newton's reference to the "quiescence of the aphelia" (*Principia*, Bk III, Prop. 2), and it served as a basis for Euler's initial belief, held into the mid-1740s, that the mutual planetary perturbations were negligible. Streete's planetary tables were longer in use than Boulliau's or Wing's. They were republished by Nicholas Greenwood in his *Astronomia Anglicana* of 1689, with minor alterations in the mean motions and epochs of the Sun and Saturn; in a Latin version using Greenwood's emended numbers by the Nuremberg mathematician Johann Gabriel Doppelmayr in 1705; in their original form by Halley in 1710 and again in 1716, and by Newton's successor in Cambridge, William Whiston, in his *Praelectiones astronomicae* (1707; English editions, 1715, 1728).

A different way of deriving equations of centre, essentially a superimposition of Kepler's vicarious hypothesis upon the elliptical orbit, was proposed by a Danish mathematician resident in London, Nicholaus Mercator (*c.* 1619–1687), in his *Hypothesis astronomica nova* of 1664. Thus in Figure 10.11 the distance between the foci *s* and *f* is divided in the 'divine proportion' or 'extreme and mean ratio' at *M*, so that $Mf:sM::sM:sf$ ($=\frac{1}{2}(\sqrt{5}-1)=0.618$) and with *M* as centre and radius equal to the semi-major axis of the ellipse, a circle is drawn. For a given angle of mean anomaly $M=\angle afd$, the angle of coequated anomaly, $\angle asd$, is determined by treating point *f* as an equant point in the circle; the place of the planet, however, is not at *d* on the circle but at *e* on the ellipse. The error here (to the order of e^4) is $-0.014e^2\sin 2M$, which for Mars ($e=0.093\,03$) amounts at maximum, when $M=97°\ 11'$, to some $2'\ 8''$. The accuracy of the procedure would be marginally improved were the ratio *sM:sf* increased to 0.621, yielding, at $M=93°$

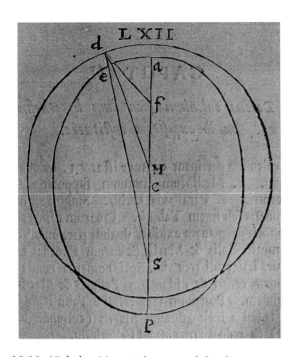

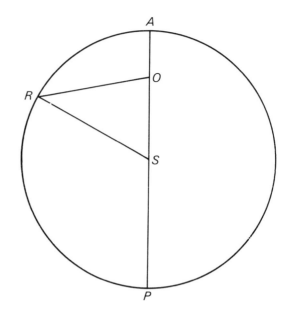

10.11. Nicholas Mercator's proposal for determining true anomaly from mean anomaly. The diagram is taken from his *Institutionum astronomicarum libri duo* (1676). Except for its use of lower-case letters, it is essentially identical with the diagram from Mercator's *Hypothesis astronomica nova* (1664).

10.12. A consequence of the simple elliptic hypothesis on which Cassini I depended in his procedure for determining eccentricity and aphelion: if S is the Sun, O the centre of the planetary ellipse, $v_R = \angle\, ASR =$ true anomaly, and ARP is a circle about S with radius equal to the semi-major axis of the ellipse, then $\angle\, AOR = \frac{1}{2}(M_R + v_R)$.

58′, a maximum discrepancy of 2′ 7″. But Mercator is extremely pleased with being able to use the divine proportion here as well as with the trinitarian 'signature' which he sees in the three points s, M, and f. The planetary ellipse is moreover reflected in the erect stature of man: when the arms are extended so that the finger tips are at middle height, the figure fits into an ellipse with foci at knees and heart (the latter corresponding to the Sun), perihelion at head and aphelion at feet. The hypothesis is an answer to those who babble about merely accidental gyration of vortices and fail to discern the mind of the Creator in his creation. Mercator compares his hypothesis with 14 observed oppositions of Mars from 1582 to 1610, and with 28 other observations of Mars from 1582 to 1595; the differences between observation and hypothesis in the first group average to 2′ 17″, and in the second to 1′ 45″. In the same publication Mercator espouses Streete's position on the quiescence of the aphelia and nodes, and on the derivation of

mean solar distances by means of Kepler's third law.

It was probably Mercator, as much as anyone, who gave the *coup de grâce* to the simple elliptic hypothesis. In the *Journal des Sçavans* for September 1669 there appeared a description of a "new, direct, and geometrical method of finding the apogees, eccentricities, and anomalies of movement of the planets", proposed by Cassini I. Apparently Cassini had discovered the method years earlier, for he refers to the invention of such a method in a letter to Gassendi of 21 October 1653. He assumes the simple elliptic hypothesis, or more directly, the following consequence of that hypothesis (see Figure 10.12): if S is the focus where the eye (or Sun or Earth) is located, O the centre of the ellipse, ARP a circle with radius equal to the semi-major axis of the ellipse, and if for a given observation the true anomaly is $v_R = \angle\, ASR$ and the mean anomaly M_R, then $\angle\, AOR = \frac{1}{2}(M_R + v_R)$. Cassini's actual procedure is indicated in Figure 10.13, where $AB = v_A - v_B$, $DF = M_A - M_B$, $BC = v_B - v_C$, $DE = M_B - M_C$: only the *differences* in true and mean anomaly being observationally determinable. The

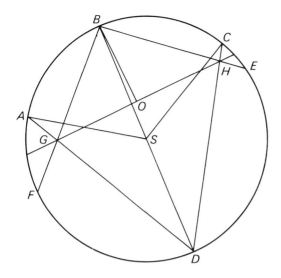

10.13. The procedure of Cassini I for determining eccentricity and aphelion of a planet from observations.

intersections *G* and *H* fix the line *GH* on which point *O*, the centre of the ellipse, is found by dropping a perpendicular from *B*. The fact that many observations could be brought to bear in determining the position of the line *GH* was, in Cassini's view, a great advantage; but the statistical averaging would be merely graphical.

Mercator's response to Cassini's proposal was published in the *Philosophical Transactions* for 25 March 1670, and the gist of it was that, as Kepler and Boulliau had already maintained, the simple elliptic hypothesis would not do, being incompatible with Tycho's observations of Mars. The error of that hypothesis in the octants of anomaly, amounting for Mars to nearly $7\frac{1}{2}'$, was corrected almost exactly, Mercator believed, by Boulliau's emended hypothesis of 1657, as well as by the hypothesis he himself had published in 1664. Mercator was to repeat his account of this matter in the second book of his *Institutionum astronomicarum libri duo* of 1676; an Englished version of which was published as Bk IX of Leybourne's *Cursus mathematicus* of 1690.

In July 1670 John Collins wrote to James Gregory (1638–75): "you likewise have that Transaction wherein Mercator glanceth against Cassini, who being incensed intends a reply ....". Gregory, in an appendix to his *Optica promota* of 1663, had proved several theorems on the determination of aphelia and eccentricities, assuming the simple

elliptic hypothesis, but also (p. 125) expressing uncertainty as to its truth. In a letter to Collins of 23 November 1670 he announced a method "whereby I turn a geometrick problems . . . into an infinite series, and among others Kepler's problem (to divide a semi-circle in a given ratio by a straight line through a point on the diameter) I resolve with ease". The series solution, which gives Kepler's eccentric anomaly in terms of the mean anomaly, was based on Christopher Wren's solution of the same problem by way of a prolate cycloid, first published in an appendix to Wallis's *De cycloide* of 1659, and reworked and published by Newton – without mention of Wren's name – as Prop. 31 of Bk I of the *Principia*. Newton also had read the account of Cassini's proposal, and no doubt Mercator's response; and he also, by 1676, had obtained series solutions to Kepler's problem. It did not mean that Kepler's area rule from being hypothesis had become empirical law, only that it or something fairly close to it was true.

As for Cassini's reply, it is unknown, but around 1690 he invented the *cassinoid*, with the aim of obtaining a possible orbit for the planets in which the superior focus would serve as equant point. The hypothesis was first mentioned in 1691 by Jacques Ozanam in his *Dictionnaire mathématique*, and was described by Cassini in 1693 in his treatise "Sur l'origine et le progrès de l'astronomie". In the cassinoid the product of the distances from any point to the two foci is a constant. Cassini by this supposition sought to make the increment in angle at the superior focus proportional to the increment in area swept out by the radius vector about the lower focus. David Gregory in his *Astronomiae physicae et geometricae elementa* (1702; English edition, 1726), Bk III, Prop. 8, attempted to demonstrate, in opposition to Cassini, that in curves that are everywhere concave inward the equant property cannot hold for one point and Kepler's areal rule hold for another; but his supposed proof is mistaken.

Still another sequel to the clash between Mercator and Cassini was an essay proffering a "Direct and geometrical method for the investigation of the aphelia, eccentricities, and proportions of the orbits of the primary planets, without supposing equality of angular motion about the second focus of the ellipse . . .", which appeared in the *Philosophical Transactions* for 25 September 1676. This was the

10.14. Facing pages from Vincent Wing's *An Ephemerides of the Coelestiall Motions for VII Years*. An ephemeris is a table showing the predicted positions of a heavenly body for every day during a given period. These pages give longitudes and latitudes of the Sun, Moon, and planets on the left, and lunar and planetary aspects on the right, for the month of October 1654. Thus the triangle in the first row on the right under Jupiter tells us that on 2 October the Moon was in trine with (or 120° from) Jupiter – a favourable aspect.

first publication of the young Edmond Halley (1656–1742). The method made use of the fact that the distances of two given points on the ellipse from a focus differ by the same amount as their distances from the other focus; it employed three empirically determined positions on the ellipse and proceeded to locate the second focus by the intersection of two constructed hyperbolas. Newton in Lem. XVI of Bk I of his *Principia* was later to show that the recourse to hyperbolas was unnecessary. The method was free of theoretical difficulties, but because it required the exact observational determination of three solar distances with the corresponding heliocentric longitudes, cannot be regarded as eminently practical.

Improvements in solar and planetary parameters: from Cassini I and Flamsteed to Cassini II and Halley

To improve the accuracy of solar and planetary parameters beyond the point reached by Streete, a necessary requirement was the adoption of an improved table of astronomical refractions. The Tychonic and Lansbergian tables hitherto used were derived empirically on the basis of incorrect assumptions about parallax, and as a consequence made a false distinction between solar, lunar, and stellar refractions. Streete, realizing that the threefold Tychonic table could not stand with a much reduced solar parallax, adopted Tycho's single table of stellar refractions as valid for Sun, Moon, and planets as well as fixed stars; this reduced to zero the refractions above 20° of altitude – where according to F.W. Bessel's table the mean refraction is still 2′ 37″. The Tychonic empirical approach died slowly: Johannes Hevelius (1611–87), in his posthumously published *Prodromus astronomiae* of 1690, would still be maintaining that refractions cease above 45° of altitude. It was Cassini I who, beginning about 1656, brought the treatment of astronomical refraction back into the theoretical path blazed by Kepler; employing, however, the sine law of René Descartes or Willebrord Snell in place of Kepler's less adequate law of refraction.

As we have seen in Chapter 7, Cassini from the early 1650s undertook to revise the elements of solar theory, using observations made with the renovated gnomon of the church of San Petronio in Bologna. One of his early results was a confirma-tion, from approximate measurement of solar diameters, that the eccentricity of the Sun's orbit as determined from the lengths of the seasons cannot be taken whole but must be bisected, as Kepler claimed. This result led Riccioli, who had rejected the elliptical orbit in his *Almagestum novum* of 1651, to adopt it for the tables (differing little from Boulliau's) of his *Astronomia reformata* of 1665.

With respect to refractions, however, Riccioli and Cassini parted ways. Riccioli held to an essentially Tychonic position, with refractions ceasing above 25° of altitude. Cassini, on the contrary, concluded that refractions must continue to the zenith. This was required not only by the sine law of refraction, but also in order to conciliate the altitude of the celestial pole as taken from observations of the circumpolar stars with the altitude of the celestial equator and obliquity of the ecliptic as taken from observations of the solstitial altitudes of the Sun. Assuming that the atmosphere was uniform and of constant height, and using empirical values for the refractions at the apparent altitudes of 0° and 10°, Cassini was able by means of the sine law to calculate a height and an index of refraction for the atmosphere. (He was aware that the atmosphere might not be altogether uniform, but thought knowledge of its variation in density would remain inaccessible to humans.) With these results he then computed the refractions for each single degree of zenith distance.

A further question remained: did refractions vary from winter to summer, or with different states of the weather? As we have seen in Chapter 7, Cassini found that if the horizontal solar parallax were taken to be 12″ or less, then the refractions in summer and winter could not sensibly differ; but if the solar parallax were assumed to be as large as Kepler made it, namely about 1′, then the winter refractions would be greater, the additional correction for parallax in the Sun's meridian altitude being balanced by the additional correction for refraction. In a letter that the Marquis Malvasia published in an appendix to his *Ephemerides* (1662), Cassini explained both options but gave tables in accordance with the second, charmed no doubt by a certain 'harmony' which it presented. For with Cassini's corrected obliquity (23° 29′) and height of the equator, the solar eccentricity deduced from observations of equinoxes proved to be 0.017, implying an equation of centre of 1° 56′ 53″

November hath xxx. Days.

Lunations.

	D.	H.	M.
○ Full Moon.	5	3	56 Mo.
☽ Laſt Quarter.	12	1	30 Mo.
● New Moon.	19	3	51 Mo.
☽ Firſt Quarter.	27	4	59 Mo.

10.15. The entries for November in Thomas Streete's *A Compleat Ephemeris for . . . 1682*. This work is typical of the highest class of popular duodecimo almanacs of the seventeenth century. Unlike the larger and lengthier ephemerides (of which Wing's is a good example), an annual almanac such as this contains Dominical letters, saints' days, weather

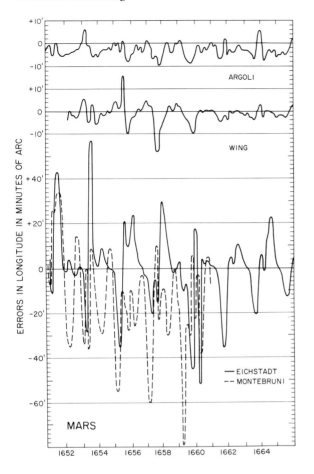

Graph 1

Accuracy of ephemerides during the seventeenth century
The graphs of this and the following two pages, prepared by Owen Gingerich and Barbara Welther, plot the
errors in a number of planetary ephemerides published
during the seventeenth century. The accuracy of the
ephemerides depended both on the accuracy of the tables
or the theory from which they were derived, and on the
accuracy with which the derivation was carried out.

Of the ephemerides whose error-plots are given in
Graphs 1 and 2, those of Andreas Argoli (1570–1657)
ran from 1621 to the end of the century, and were based
on his own *Secundorum mobilium tabulae* (Padua, 1634),
apparently adapted from Kepler's *Rudolphine Tables*.
Those of Lorenz Eichstadt (1596–1660) were a continu-
ation of the ephemerides begun by Kepler; but Eichstadt's
calculations from Kepler's tables appear to have been
very inaccurate. Francisco Montebruni (*fl.* mid-seventeenth century) based his ephemerides on the tables of
Philippe van Lansberge, published in 1632. Vincent
Wing used his own tables for calculating ephemerides,
shifting in the late 1650s from those of the *Harmonicon
coeleste* to the more accurate tables of the *Astronomia
instaurata*.

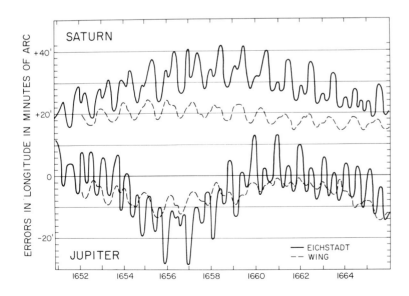

Graph 2

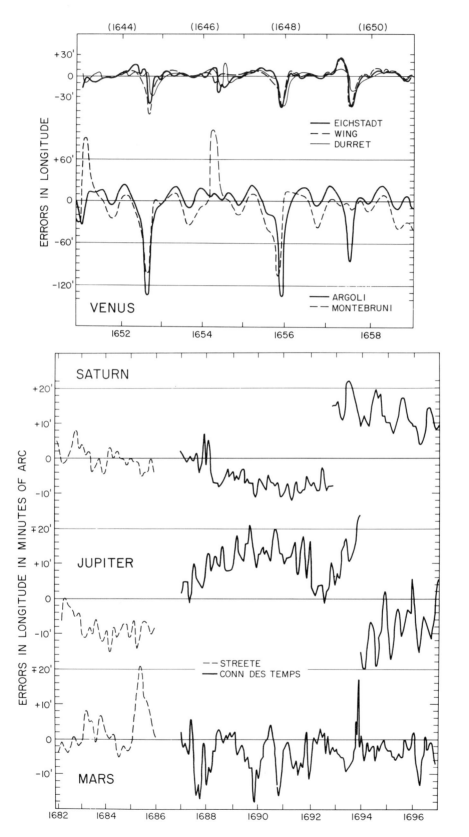

Graph 3

Graph 4

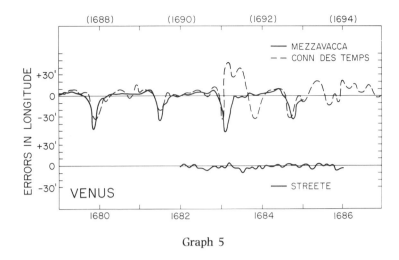

Graph 5

In Graph 3, which shows errors in ephemerides of Venus, it is noteworthy that Durret's positions match those of Eichstadt and Wing very closely; they have apparently been determined from almost the same elements and theory. The errors in the Venusian ephemerides of Argoli and Montebruni are three to four times larger, and so have been plotted separately.

In Graph 4 Streete's almanacs of 1682–85 are compared, for the superior planets, with the *Connaissance des temps*, begun in 1679 by the Paris Academy of Sciences. The errors in both are considerable, even for Mars which, unlike Jupiter and Saturn, is not subject to sizeable long-term perturbations.

Graph 5 compares the Venusian ephemerides of Streete, the *Connaissance des temps*, and Flaminio Mezzavacca (d. 1704), who appears to have copied at least some of his positions from the ephemerides of Argoli. Streete's superior accuracy is evident; it is due in part to the superior solar theory he inherited from Horrocks, and in part to his employing Kepler's third law to determine the mean solar distance of Venus – a practice that derives from Horrocks.

– smaller, and more nearly correct, than any earlier value (Newcomb's value for the eccentricity in 1660 was 0.016 85, corresponding to an equation of centre of 1° 55′ 16″). But if the mean horizontal solar parallax were just 59″, then the new value of the solar eccentricity would be a mean proportional between the radius of the Earth and the semidiameter of the Sun's orbit, and moreover would quite agree with the mean distance of the Moon from the Earth.

The observations made by Jean Richer at Cayenne in 1672 (see Chapter 7) confirmed Cassini's value for the solar eccentricity. On the other hand, from Richer's determinations of the obliquity of the ecliptic, Cassini saw that he must reduce the solar parallax to a fifth or less of 1′ (he settled on a value of 9.5″). He therefore now adopted his earlier table of aestival refractions as the sole table; and this was published in 1684 in *Les élémens de l'astronomie vérifiez*. To 60° of zenith distance the table never

differs from Bessel's table of mean refractions by as much as 3″; at 78° of zenith distance it differs only by 7″. Cassini's table of refractions would be used in the *Connaissance des temps* down to 1761.

Cassini was well aware that revision of the solar elements should be followed by revision of the orbital elements of the other planets. He laboured at the construction of planetary tables during his years at Bologna, but other tasks, particularly his employment as engineer and arbitrator by the Pope, prevented his completing them. After he came to Paris, much of his time and energy were absorbed by the project of revising the map of France, and by the telescopic observations at which he was so expert. He continued, we know, to work at the improvement of the orbital elements of Jupiter, Venus, Mercury. That he did not publish tables of the primary planets during his lifetime is probably attributable not only to various rival claims on his attention, but to the questions about the correct theory of the planets that had led him to propose the *cassinoid* orbit, and to deeper difficulties that were emerging as a result of the application of the micrometer, telescopic sights, and the pendulum clock to astronomical observation, and which by 1693 had led Cassini to the conclusion that orbital elements might at best have to be statistical averages.

The revolution in instrumentation had coincided with the foundation of the Paris Academy. From the beginning the academicians, while repeatedly avowing Kepler's tables to be the best available, were concerned to bring about a revision, from the ground up, of astronomical constants and tables. Huygens's pendulum clock made possible the determination of right ascensions without involvement of refractions; and by means of it Huygens had already satisfied himself that the Earth turns uniformly on its axis, contrary to Tycho's and Kepler's suppositions about the equation of time. Auzout and Picard were using the micrometer to measure the diameters of the luminaries, and so finding the Keplerian theory of the Moon in serious error. Picard first tried placing telescopic sights on graduated arcs in late 1667, but his regular series of observations with such instruments began in November 1668, after he had learned how to cope with errors of collimation. In the summer of 1668 he discovered that the brighter stars were visible by telescope in broad daylight, so that their right ascensions could be obtained quickly and directly from their meridian transits compared in time with the meridian transit of the Sun. Relying on such instruments and procedures, Picard in 1669 set forth his general programme for the advancement of astronomy. It included the setting up of a large mural quadrant for the measurement of meridian altitudes and for timing meridian transits, the construction of a table of refractions that would take account of temperature changes, and a trip to Uraniborg to check Tycho's value for the latitude of the place, and its difference in longitude from Paris, in order that the best use could be made of the Tychonic observations.

From his voyage to Uraniborg in 1671–72 Picard brought back – besides the desired measurements, a copy of Tycho's observations, and Ole Römer as an assistant – a disturbing fact about the Pole Star: it moved in ways not predictable by the

precession. First Picard, and later Cassini I and Cassini II, made close studies of the extra-precessional movements of the Pole Star, Vega, and other stars. They were thus enabled to refute claims by Hooke and Flamsteed to have discovered annual parallax in these stars; but the observed movements remained unexplained. Only with the work of Bradley, between 1728 and 1748, would these anomalies be accounted for in terms of aberration and nutation (see Volume 3); and until then predictive astronomy would have to remain in a kind of limbo, the background against which planetary motions were measured being itself subject to unpredictable motions.

Picard's efforts for the improvement of solar parameters and the correction of the *Rudolphine Tables* came to an end with his death on 12 October 1682, at the age of 62. It was only in the following year that the mural quadrant, so long desired, was finally in place on the east wall of the Observatory. Here, for years to come, Philippe de La Hire (1640–1718) would be measuring meridian altitudes and timing meridian transits, and his observations, which Delambre later found quite comparable in exactitude with those of the middle of the eighteenth century, formed the basis of his tables, a monument to unmitigated empiricism. The solar and syzygiacal lunar tables were published in Latin in 1689, and these and the remaining tables in 1702, with reissues in 1722 and 1727; French editions appeared in 1735 and 1755, German editions in 1725 and 1745. La Hire set out to accomplish what, a century earlier, Peter Ramus had demanded of Tycho, to construct an astronomy without hypotheses. La Hire assumed, of course, the validity of either the Tychonic or Copernican arrangement of Earth, Sun, and planets, but in other respects his tables were based solely on his observations and innumerable linear interpolations. He did not take the orbits to be elliptical; indeed, his observations showed them not to be so shaped, exactly. He did not follow Kepler's areal rule, did not even assume that the orbits or motions were symmetrical about the line of apsides. Nor did he take Kepler's third law for granted, although Cassini I had been confirming it for the satellites of Jupiter and Saturn as well as for the primary planets. Rather than use Cassini's table of refractions, he constructed his own from observations; it differs from Bessel's table of mean refractions by $+9''$ at

30° ZD, $+13''$ at 45°, $+25''$ at 80°; the second differences develop very irregularly; and it does not make allowance for temperature differences, since La Hire believed he had shown such effects to be nil. La Hire's planetary tables are *sui generis*, and as such difficult to judge. The maximum equation of centre for the Sun he found to be 1° 55′ 42″, just 3″ shy of the correct value for 1700; the maximum equation of centre for Mars he put at 10° 40′ 40″, only 14″ short of the correct value for 1700, but in other cases the errors are much larger. Vitiating the tables throughout are the distortions due to aberration, nutation, perturbations, and a poor table of refractions. In Mars and the Sun, La Hire's tables probably predicted more closely than any published tables of the time; in other cases the verdict would be doubtful.

Across the Channel, during the middle and late 1660s, John Flamsteed (1646–1719), was making himself, through private study, into an astronomer. He read Riccioli's *Almagestum novum*. He corresponded with Vincent Wing, challenging (probably on the basis of Horrocks's *Venus in sole visa*) his exaggerated values for the Sun's parallax and equation of centre. He found cause to doubt Streete's assertion of the quiescence of the planetary aphelia and nodes, and determined that Streete's *Caroline Tables*, though in many respects better than other tables, were yet in need of revision. In 1671, through Sir Jonas Moore, Flamsteed obtained a micrometer from Towneley, and also gained access to the papers of Gascoigne, Crabtree, and Horrocks, including Horrocks's theory of the Moon, which by micrometric measurements, particularly of the Moon's diameter, he found to be much more accurate than the lunar theories of Boulliau, Wing, or Streete. From this time forth Flamsteed may be regarded as the principal continuator of the work and thought of Horrocks and Gascoigne – though Towneley's improvements on the micrometer, and skilled observations, also deserve mention.

Flamsteed's first solar tables were printed in 1672 in the volume of Horrocks's *Opera posthuma*; they are based on the maximum equation of centre accepted by Horrocks and Streete, namely 1° 59′. Very soon, however, Flamsteed concluded that this equation would need to be reduced. Independently, it appears, of Cassini's work, but probably influenced by Kepler's *Astronomiae pars optica* (from his

Gresham lectures of 1684 we know that Flamsteed had studied this work with care), he realized that refractions must go to the zenith. Moreover, from distances between Mars and stars in Aquarius, taken with the micrometer in September 1672, he concluded the parallax of Mars at this time to be but 25″, and the horizontal solar parallax therefore not greater than 10″. It followed that all measurements of the Sun's meridian altitudes required new corrections. A little later Flamsteed obtained a copy of Cassini's table of refractions as given in the *Ephemerides* of the Marquis Malvasia; since Cassini here takes the horizontal solar parallax to be 1′, Flamsteed concluded that the Cassinian refractions would need to be revised, but resolved to use them provisionally, until "with better contrived Instruments I may have opportunity of stating them better in easiness". Using Cassini's refractions to correct some of Tycho's observations, he found very nearly the same value for the obliquity of the ecliptic as Cassini had found, 23° 29′; it will be one of his unwavering convictions, supported by many recalculations of ancient and medieval observations, that the obliquity has always remained constant at this value. From a series of 31 observations of the Sun's meridian altitudes made by Cassini between 1655 and 1659, and corrected for refraction and parallax, Flamsteed then constructed new solar tables in which the maximum equation of centre was 1° 54′ 12″ (the correct value in 1660 was 1° 55′ 51″).

Flamsteed was installed at the Greenwich Observatory in July 1676. For the new observatory he had early formed the project of obtaining a large graduated mural arc for determining meridian altitudes and to use with a pendulum clock for timing meridian transits. Not until 1689, however, was a mural arc successfully installed, and in the interim Flamsteed was confined to determining distances between celestial objects by means of a large, movable sextant – not the best instrument for measuring altitudes. Nevertheless, he conceived a way of deriving the elements of solar theory from his observations. This consisted in measuring the distances between the Sun and Venus in daytime and between Venus and a star at night, when the bodies in each case were at equal altitudes. The zenith distances were also needed in order to determine the contraction in the apparent distance between the bodies owing to refraction, but small errors in this measurement would not significantly affect the final result. Having accumulated a supply of such observations, Flamsteed constructed his new solar theory in the spring of 1679, obtaining a maximum equation of centre of 1° 55′ 0″ (the correct value at this time was 1° 55′ 48″), with the apogee at 96° 50′ 0″ (the correct place was 97° 25′ 25″). The tables for this theory were printed in Flamsteed's *The Doctrine of the Sphere*, which was published in 1681 as an appendix to Moore's *A New System of the Mathematicks*, along with new tables fitted to the Horrocksian lunar theory. It is noteworthy that at this time Flamsteed has commenced to calculate equations of centre in conformity with Kepler's areal rule; his actual calculational procedure, however, is not known.

After installation of the meridional mural arc at Greenwich in 1689, Flamsteed again redetermined his solar numbers. The method he now introduced became a standard one; it is independent of an exact knowledge of refractions and only marginally dependent on a knowledge of the obliquity of the ecliptic. It consisted in determining with the pendulum clock the difference between the transit times of the Sun and a given star near the solstitial colure at the beginning of spring, and again in the autumn, when the Sun was as nearly as possible at the same altitude. Finding the difference in right ascension between the Sun's position on 7 March and 15 September 1690 to be 184° 57′ 0″, and by his tables of 1679 computing this same difference to be 184° 58′ 46″, Flamsteed concluded the maximum equation of centre in the older tables to be too small by half the difference, or nearly 1′; and determinations of the Sun's relative right ascension near the time of the solstice led him to advance the apogee of the earlier tables by 35′. As a result of a recalculation, with reduction of right ascensions to longitudes, Flamsteed finally arrived at a maximum equation of centre of 1° 56′ 20″ (too large by 35″ in 1700), and an apogee about 8′ shy of its true position. The solar tables derived from these values were published by Whiston in his *Praelectiones astronomicae* (1707), were used by Newton, and again published by Halley in his *Tabulae astronomicae* (1749), although Halley repeatedly complained that Flamsteed had given too little attention to observations of the Sun's position, to obtaining accurate solar elements, and to detecting the perturbations to which the Earth's orbit is subject.

As for the remaining primary planets, Flamsteed observed them frequently, with a view to "restoring their motions". In 1674, noting that Jupiter was "some 13 or 14 minutes forwarder in the heavens, than Kepler's numbers represent", he was hoping to eliminate the discrepancy by means of Horrocks's correction. In 1682 he was suspecting secular inequalities in the motions of Saturn and Jupiter, "that is, that Saturn moves slower, Jupiter swifter, in our age than formerly". By late 1709, with the help of Abraham Sharp, he had derived from his observations some 1000 places of the planets, "that is about twice as many as we have from all the astronomers that have been before". Various attempts were made, by Flamsteed and by an acquaintance named Bossley, to rectify the elements of the superior planets on the basis of these positions; but Jupiter and Saturn proved recalcitrant. The quarrel with Newton and Halley over the printing of the star catalogue was a major preoccupation that interfered with the work, but it is unlikely that even under more favourable circumstances Flamsteed could have emerged successfully from his bout with the problem of 'the great inquality'. In his last years he was seeking to discover from earlier observations as well as his own a fifty-nine-year perturbational cycle, consisting of two revolutions of Saturn and five of Jupiter, that would save the phenomena. The solution he had need of indeed involves the near-commensurability of the orbital periods of Saturn and Jupiter, and the numbers 2 and 5, but it would not be discerned before it emerged in Laplace's analysis about 1785. Flamsteed published no planetary tables. The only published tables to be based on the Flamsteedian corpus of planetary observations would be those of Halley.

Of the particular observations and calculational procedures used in the derivation of Halley's tables, the author has left us no record. The equations of centre appear to be computed in rather strict conformity with Kepler's areal rule. To save the appearances in Jupiter and Saturn, Halley introduced secular equations into the mean motions of the two planets, giving an acceleration of 57.3′ per thousand years to Jupiter, and a deceleration of 139′ per thousand years to Saturn. The tables were deliv-

Table 10.2 *Orbital elements of the planets adopted by Halley and Cassini compared with Newcomb's values for 1700*

	Eccentricity	Maximum equation of centre		Aphelion	Mean distance
Mercury					
Newcomb (for 1700)	0.205 57	23°40′24″		252°47′24″	0.387 10
Halley (H − N)	+0.000 32	+	2′12″	− 3′ 6″	0
Cassini (C − N)	+0.003 2	+	22′34″	− 12′46″	+0.000 51
Venus					
Newcomb (for 1700)	0.006 917	47′33″		307°20′45″	0.723 33
Halley (H − N)	+0.000 065	+	27″	− 48′19″	0
Cassini (C − N)	+0.000 23	+	1′33″	− 54′25″	+0.000 04
Earth					
Newcomb (for 1700)	0.016 834	1°55′45″		277°47′ 3″	1.000 00
Halley (H − N)	+0.000 086	+	35″	− 8′54″	0
Cassini (C − N)	+0.000 066	+	6″	− 11′ 8″	0
Mars					
Newcomb (for 1700)	0.093 127	10°40′56″		150°32′19″	1.523 69
Halley (H − N)	−0.000 13	−	54″	+ 2′11″	0
Cassini (C − N)	−0.000 23	−	1′37″	+ 4′ 1″	+0.000 04
Jupiter					
Newcomb (for 1700)	0.048 007	5°30′ 9″		189°30′ 3″	5.202 7
Halley (H − N)	+0.000 21	+	1′27″	+ 4′57″	−0.001 7
Cassini (C − N)	+0.000 16	+	1′ 8″	− 3′21″	+0.000 2
Saturn					
Newcomb (for 1700)	0.056 582	6°29′11″		267°10′54″	9.546
Halley (H − N)	+0.000 421	+	2′53″	+ 1°23′46″	−0.006
Cassini (C − N)	+0.000 317	+	2′29″	+ 57′45″	+0.001

ered to the printer in 1717, and printed in 1719, but Halley delayed publication on being appointed Astronomer Royal, in order to take advantage of the opportunity to compare the Moon's places with his tables through a full 'Saros cycle' of eighteen years. The tables, with the added comparisons, at length reached the public in 1749.

The *Tables astronomiques* of Jacques Cassini (Cassini II), a refinement of the unpublished tables of Cassini I, appeared in 1740. In computing equations of centre, Cassini II appears to have followed his "Méthode de déterminer la première équation des planètes suivant l'hypothèse de Kepler". It is equivalent to the method of Cavalieri, in which in a first stage one puts in place of Kepler's equation, namely $M − E = e \sin E$, the approximation $\sin(M − E') = e \sin E'$, then corrects E' by adding the difference between the sine and the arc, given very

nearly by $\frac{1}{6}e^3 \sin^3 E'$. The errors in the tabulated equations of centre, as in Halley's tables, are very small.

Table 10.2 compares Halley's and Cassini's values for eccentricity, maximum equation of centre, aphelion, and mean solar distance with the values for 1700 derived from Newcomb.

Although the solar tables of Halley and Cassini were in 1758 superseded by Lacaille's tables, which included the effects of aberration, nutation, and some perturbations, their planetary tables continued in use till the 1790s. With those of Lalande (printed in the second edition of his *Astronomie*, 1771), they were the chief such tables in use during this period, and were, for instance, those consulted by Lagrange when deriving the constants for his "Théorie des variations séculaires des éléments des planètes" (1781). It was by means of a comparison of Halley's and Lalande's tables of Mercury that Delambre made his first 'coup' in astronomy, predicting the transit of Mercury of 3 May 1786 more accurately than anyone else, and catching a glimpse of it through the clouds after other astronomers, trusting to Lalande's tables, had already abandoned their telescopes. But the day of such purely Keplerian tables was drawing to a close. A week later, on 10 May 1786, Delambre offered his services to Laplace in constructing tables of Jupiter and Saturn that would incorporate the 'great inequality' that Laplace had at last reduced to computation. Here begins a new era, in which it is understood that all planetary tables must incorporate such perturbations as are implied by universal gravitation and come within the range of observational detection, and must be based on observations corrected for aberration and nutation.

Lunar theory and the equation of time

The predictive astronomy of the Moon requires, even more decidedly than that of the planets, a systematic development of the perturbations; but such a development was not to be accomplished during the period of our survey, and the chief improvement in lunar theory from Kepler's time to the middle of the eighteenth century would be due to Horrocks.

We note, however, that Newton's assertion in the scholium to Prop. 35 of Bk III of the *Principia* (second and third editions) that "our Horrocks was the first to state that the Moon revolves in an ellipse about the Earth placed in its lower focus" makes the mistake of transferring to Horrocks what belongs to Curz and Kepler.

True enough, Kepler, while computing the lunar equations of centre and *radii vectores* from ellipse and areal rule, made rather little of this application to the Moon of the hypothesis he had found so successful with the primary planets; and the reason was no doubt that the lunar problem did not thereby appear to be brought much closer to a satisfactory solution. Besides the first inequality, or "anomalia soluta" as Kepler called it, there was the second or menstrual inequality, known since Hipparchus; the third inequality or variation, discovered by Tycho; the inequality of the nodes and inclination, also discovered by Tycho; and the annual equation discovered independently by Tycho and himself. Kepler struggled for years both to

simplify the calculation of these several inequalities, and to account for them causally. Kepler's hypotheses for the second and third inequalities – involving the 'circle of illumination' of the Sun in the one case, and 360 as the archetypical number of days in the year in the other – appear to have been accepted by no one after him. And Kepler himself acknowledged the failure of his theories to account with exactitude for lunar and solar eclipses: "You have below, in the doctrine of eclipses", he writes in the preface to the *Rudolphine Tables*, "also from the observations of the present time, clear evidence that the motions of the Sun, Moon, and *primum mobile* are not precisely equable, but receive small physical intensions and remissions *extra ordinem.*"

This and a number of other Keplerian pronouncements in the same vein, implying as they did that the science of celestial things must remain forever imperfect, would seem to have become for Horrocks a spur to renewed effort. The lunar theory that he developed between 1638 and his death early in 1641 was to prove substantially better than its rivals. In 1672, when it at last entered the public domain with the publication of the *Opera posthuma*, its principal rivals were the lunar theories of Boulliau, Wing, and Streete, which we shall now briefly describe, at least with regard to their predictions of longitude and radius vector.

In the lunar theory presented in Boulliau's *Astronomia Philolaïca*, the Moon is first of all assigned an elliptical orbit, in which equations of centre are computed exactly as for the primary planets. If the lunar motions appear more complicated than those of the primary planets, it is, so Boulliau tells us, because the entire 'lunar system' – that is, the cone composed of equant circles with the ellipse cutting across it – is being conveyed by the Earth about the Sun in the annual motion. The cone and ellipse thereby come to be displaced, twice each month, from their "natural seats", to which they return when the Moon is in syzygy, at which time the lower focus of the ellipse occupies the centre of the Earth. Between times, this lower focus moves about the circumference of a small circle, carrying with it the ellipse and cone, and reaching the greatest distance from the centre of the Earth when the Moon is in quadrature. For this epicyclic motion, which is in fact identical with Tycho's hypothesis for the second inequality, Boulliau introduces the term 'evection'; meaning thereby to stress that the lunar system, while being thus conveyed about, is in itself undistorted, its parts remaining always in the same relation to one another, just as if they constituted a body. The name has persisted, although the theory is wrong. For Boulliau the inequalities of the Moon other than the first or elliptic inequality are simply and solely optical inequalities, arising from the changes in distance and orientation of the lunar system with respect to the observer's eye. That these changes occur is, in Boulliau's opinion, a strong argument for the Earth's annual motion; he cannot sufficiently marvel at Tycho's having supposed the primary planets, each subject only to a zodiacal and an annual inequality, to be conveyed about the Earth by the Sun without undergoing an 'evection', while the Moon, encumbered with at least three inequalities besides the elliptic, rotates about a stationary Earth.

Boulliau's account of the first three inequalities is represented geometrically in Figure 10.16. Here C is the centre of the Earth, and I the focus of the Moon's ellipse, which moves according to the order of the signs in the circle CIA, starting from C, with a speed given by $\angle CBI = 2(L-L') = 2D$, where L is the mean longitude of the Moon and L' the mean longitude of the Sun. To account for the third inequality or Tychonic 'variation', Boulliau supposes that as the focus of the ellipse moves from C to I, the line of apsides does not remain parallel to itself – does not move simply from the position SCT to the position KIE – but rather inclines a little, so that the apogee falls between T and E. The entire ellipse thus oscillates slightly about the point I, so that the true apogee V is in advance of the mean apogee E when the Moon is between syzygy and quadrature, and behind the mean apogee when the Moon is between quadrature and syzygy. (In Figure 10.16, the ellipse is not shown in its inclined position.) Boulliau calls this oscillation the *reflectio*, a term which was retained by Wing and Streete.

If we put $IG = e$, $BC = \epsilon$, M = the Moon's mean anomaly, R = coefficient of the variation, and $D = L - L'$, then the main terms giving the Moon's true distance from mean apogee according to Boulliau's theory may be expressed as follows:

$$v_B = M - (2e + \epsilon)\sin M + (\tfrac{5}{4}e^2 + \tfrac{3}{2}e\epsilon + \tfrac{1}{2}\epsilon^2)\sin 2M -$$
$$\epsilon \sin (2D - M) + (R - \tfrac{1}{2}e\epsilon)\sin 2D +$$
$$(\epsilon^2 + \tfrac{3}{2}e\epsilon)\sin 2(D - M).$$

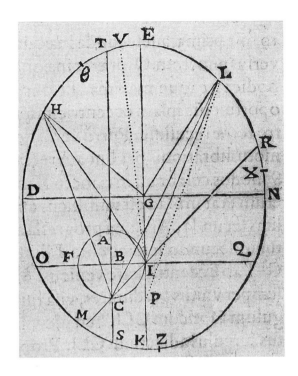

10.16. Diagram of Boulliau's lunar theory, from his *Astronomia Philolaïca* (1645).

Using Boulliau's values for the constants, namely $e = 0.043\,62$, $\epsilon = 0.021\,81$, $R = 40.5'$, we obtain

$$v_B = 374.9' \sin M + 13.9' \sin 2M - 75.0' \sin (2D - M) + 38.9'\sin 2D + 6.5' \sin 2(D - M).$$

The corresponding terms according to present-day lunar theory are

$$v = 377.3' \sin M + 12.8' \sin 2M - 76.4' \sin (2D - M) + 39.5' \sin 2D + 3.5' \sin 2(D - M).$$

Of the several hundred sinusoidal wobbles of the present-day theory here unheeded, some eight have coefficients larger than $1'$ of arc, far the largest of these being the annual equation, $11.2' \sin M'$, where M' is the mean anomaly of the Sun reckoned from apogee.

With Boulliau as with Tycho and Kepler, such account as is taken of the annual equation is incorporated in the equation of time, that is, the number of minutes and seconds that must be added to or subtracted from any given interval of time stated in apparent solar days to obtain the same interval measured in mean solar days. It had been the suggestion of Longomontanus to Tycho, prompted no doubt by the desire to avoid adding one more circle to an already well-laden lunar theory, that the annual inequality be taken into account numerically by simply neglecting that part of the equation of time that arises from the apparent inequality of the Sun's motion along the ecliptic. The Sun's equation of centre turned into time yielded 8.2 minutes at maximum, in which interval the Moon traverses 4.5′ of arc in mean motion, so that the annual inequality would in effect be taken to be 4.5′ sin M'. Yet how such a proposal was to be justified – implying as it does that the *primum mobile* turns faster when the Sun is closer to the Earth – was unclear; and the fact that Tycho in his *Progymnasmata* appeared to wish to use the modified equation of time only when computing the place of the Moon, while retaining the standard equation for other bodies, was an inconsistency that provoked much further discussion.

For Kepler, the obvious conclusion was that the Earth turns more rapidly on its axis when closer to the Sun; the cause being, he speculated, that the solar motive virtue not only pushes the Earth forward in its orbit but also lends added vigour to the vegetative soul in the Earth's viscera that produces its diurnal rotation. The measure of the increment, Kepler supposed, would be given by its producing in the course of the year the $5\frac{1}{4}$ supernumerary days or rotations; if these $5\frac{1}{4}$ days were distributed over the entire area of the eccentric circle, then to the triangular area representing Kepler's maximum 'physical equation' would correspond 21 minutes 40 seconds, in which time the Moon traverses some 11.9′ of arc. so that in effect he would be taking the annual equation to be 11.9′ sin M'. Use of the same equation of time in computing the Sun's position introduces an error equal to about 0.9′ sin M'; for the purpose of eclipse calculations the net value of the annual inequality thus becomes 11.0′ sin M'.

To Boulliau, Kepler's physical speculations were mere figments of imagination, but he nevertheless accepted the notion of an increase in the Earth's rate of diurnal rotation as it approaches perihelion; as justification he offered merely the analogy of a sphere which, descending an inclined plane, rotates more rapidly in the measure that it descends more rapidly. To take account of the increment, he proposed neglecting only that part of the standard or traditional equation of time that originates in the Earth's actual change in orbital speed;

the net effect for the Moon is equivalent to an annual equation of 2.25′ sin M'.

That his lunar theory failed to yield exact predictions, Boulliau freely confessed. He records a considerable number of observations of the Moon by Gassendi, Wendelin, and himself, along with the discrepancies between observation and his theory, some of them as large as 10′. Some of the evidence suggested to him that the *reflectio* varied according to an annual pattern, but he was more inclined to suppose that there was a fourth, as yet unknown inequality, remaining to be discovered. For a time he hoped that a study of the Moon's libration – the small apparent oscillation of the Moon on its axis with respect to the Earth-bound observer's line of sight – would somehow prove a pointer to this fourth inequality in longitude, but the hope proved illusory. In concluding, Boulliau adjured future astronomers to study and seek to resolve this most intricate and baffling of problems.

A major obstacle, for Boulliau and his contemporaries, to the construction of a lunar theory that would yield the Moon's longitude down to a minute or two of arc, lay in the multitude of small inequalities ($+3.4′$ sin $(2D-M-M')-3.2′$sin $(2D+M)-2.8′$ sin $(2D-M')+2.5′$ sin $(M-M')$ $-1.8′$ sin $(M+M')$, to name the larger of these smaller terms) which they had no way of beginning to discern or separate from the large inequalities. Yet inadequate as Boulliau's theory was as regards the longitudes, it was still worse with respect to the distances of the Moon from the Earth. Up to the first-order terms, the radius vector or distance of the Moon from the Earth in Boulliau's theory is given by

$$r = 1 + 0.043\,62 \cos M + 0.021\,81 \sin M \tan (D-M) + 0.021\,81 \sin (2D-M) \tan (D-M),$$

whereas by present-day theory,

$$r = 1 + 0.054\,50 \cos M + 0.010\,02 \cos (2D-M) - 0.008\,25 \cos D \ldots.$$

The two formulae lead to significantly different predictions as to the variation in the apparent diameter of the Moon. The simple fact is that the lunar orbit is not 'evected' in the egregious manner imagined by Tycho and accepted by Boulliau, Micrometric measurements of the Moon's apparent diameter initiated by Gascoigne and carried on later by Towneley, Flamsteed, Picard, La Hire, and others, would show the untenability of the Tychonic mechanism.

The lunar theory that Wing presents in his *Astronomia Britannica* differs from Boulliau's only in minor details. Wing calculates equations of centre in the lunar ellipse by the same procedure – much superior to Boulliau's as we have seen – that he uses for the primary planets. His values for the eccentricity of the ellipse and constant of evection are $e = 0.043\,15$, $\epsilon = 0.021\,58$, both a little smaller than Boulliau's. With regard to the equation of time, he returns to the proposal of Tycho and Longomontanus, claiming (1) that while untenable in a geostatic system, it is entirely consonant with the heliocentric system, in which the diurnal rotation belongs to the Earth and can be subject to intension and remission with the Earth's approach to and recession from the Sun; (2) that the magnitude of the effect given by Tycho was firmly and accurately based on Tycho's eclipse observations. Wing like Tycho, then, in effect espouses an annual equation for the Moon of $4.5' \sin M'$.

Streete in the lunar theory of his *Astronomia Carolina* introduces several novelties. He abandons the elliptical orbit, putting the circular eccentric *ALPA* of Figure 10.17 in its place. Here *A* is the apogee, *P* the perigee, and *C* the centre of the eccentric; *I* is the place of the Earth when the Moon is in the syzygies; and the circle about *I* as centre is an 'equant circle' with radius *DI* such that *AI:IP :: IP:DI*, or if $CI = e$, $DI = (1-e)^2/(1+e)$. $\angle ACD$ measures the mean anomaly, and $\angle AID$ the true anomaly, the Moon being at *L* on *ID* extended. This entire system is then subjected to an evection in the Tychonic manner, the diameter of the circle of evection being obtained by dividing *CI* in extreme and mean ratio and taking the largest segment: $2BT = 2\epsilon = \frac{1}{2}e(\sqrt{5} - 1)/2$. The variation or *reflectio* is $\angle IQT$, obtained by laying off from *I* the line *QI* equal to *AP*, so that it intersects at *Q* the line *QT* through the centre of the Earth and parallel to the line of apsides *AP*. Taking account of equation of centre, evection, and *reflectio*, and using Streete's value for *e*, namely 0.0707, we find

$$\angle QTL = M - 372.4' \sin M + 6.1' \sin 2M - 75.1' \sin (2D - M) + 37.6' \sin 2D - 8.6' \sin 2(2D - M) + 7.0' \sin 2(D - M) + \ldots.$$

As for the equation of time, Streete constructs it of the same two parts used by all astronomers down

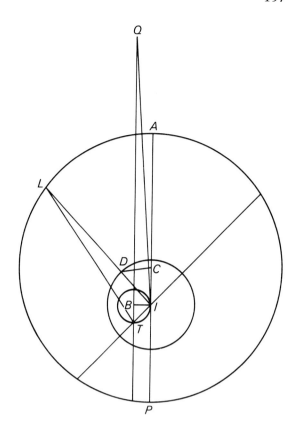

10.17. Streete's lunar theory.

to Tycho, the part originating in the Sun's anomaly being simply the Sun's equation of centre turned into time. But in this second part the signs of addition and subtraction are everywhere reversed from what they would be in the traditional table; for Streete, like Kepler, Boulliau, and Wing, supposes that the Earth's rate of axial rotation increases as the Earth comes closer to the Sun. His procedure for calculating the equation of time amounts to giving the Moon an annual equation of about $8.7' \sin M'$. His values for the radius vector are as unsatisfactory as those of Boulliau and Wing.

We turn now to Horrocks's lunar theory, the elements of which are preserved in an undescribed figure on the verso of a page in his unpublished *Philosophicall Exercises*, and in a set of undemonstrated precepts enunciated in two letters, one from Horrocks to Crabtree dated 20 December 1638, the other from Crabtree to Gascoigne and written in June or July 1642; extracts from both letters are printed in the *Opera posthuma*. The theory invokes a Keplerian elliptical orbit in which the eccentricity varies while the apogee librates back and forth round its mean position. Figure 10.18, which is taken from Section 14 of the second part of the *Philosophicall Exercises*, shows how the two oscillations are correlated. *A* is the centre of the Earth. The point *C* moves counterclockwise about *D* in the small circle *BCE* at a speed regulated by $\angle BDC = 2(L' - a)$, where L' is the instantaneous longitude of the Sun (which is in a direction from *A* parallel to the line through *E* and *C*), and a that of the lunar apogee; the motion starts from *B* when $L' - a = 0$. The momentary eccentricity is *AC*, and the momentary displacement of the apogee from its mean position is $\angle CAB$.

Essentially the same diagram appears on p. 5 of van Lansberge's *Theoricae motuum coelestium*, in the section dealing with lunar theory; Horrocks had studied this work, and so the idea of putting the eccentricity and apogee into oscillation probably first came to him from the pages of van Lansberge's book. But in his hands the very inaccurate Lansbergian theory (Horrocks found its predictions to disagree with observation at times by as much as a degree) underwent a radical change. Not only does the orbit become elliptical with equations of centre computed according to the areal rule (or the Horrocksian approximation thereto), but the motion in the small circle is measured by $2(a - L') = 2(D - M)$ rather than by $2D$, as in van Lansberge's theory and in the theories of all those who followed Tycho in their account of the evection. The semi-annual inequality in the apogee implied by Horrocks's theory is shown in Figure 10.19, taken from the *Opera posthuma*. It was by comparing Kepler's theory with observation of eclipses that Horrocks arrived at his theory, first discovering the oscillation of the apsidal line, then the variation in the eccentricity. Thus he computes equations of centre in agreement with the formula

$$2E \sin(L - A + \delta) - \tfrac{5}{4} E^2 \sin 2(L - A + \delta) + \ldots,$$

where *E* is the variable eccentricity and δ the equation to be applied to the mean apogee *A*, and

$$E \cos \delta = e + \tfrac{1}{2}\epsilon \cos 2(D - M),$$
$$E \sin \delta = \tfrac{1}{2}\epsilon \sin 2(D - M).$$

Eliminating *E* and δ from the first two terms of the equation of centre, we obtain

$$2e \sin M - \tfrac{5}{4} e^2 \sin 2M + \epsilon \sin (2D - M) + \tfrac{5}{4}e\epsilon \sin 2D - \tfrac{5}{16}\epsilon^2 \sin 2(2D - M).$$

With Horrocks's final values of e and ϵ as reported by Crabtree, namely $e = 0.05524$ and $\epsilon = 0.02324$, this expression becomes

$$371.7' \sin M - 13.1' \sin 2M + 79.9' \sin$$
$$(2D - M) + 5.5' \sin 2D - 0.6' \sin 2(2D - M).$$

To this quantity Horrocks adds $36.5' \sin 2D$, giving a total variation term of $42.0' \sin 2D$; and he adopts Kepler's 'physical equation of time', which as we have seen makes for an annual equation in the Moon equal to $11.9' \sin M'$. The parameters admit of improvement, but the theory has the decided advantage over the theories of Boulliau, Wing, and Streete that it is free from first-order terms with arguments of the wrong form. The improvement in the value of the radius vector is especially noteworthy. From Horrocks's theory we have

$$r = 1.00159 + 0.05524 \cos M - 0.00153 \cos$$
$$2M + 0.01162 \cos (2D - M),$$

whereas according to present-day theory

$$r = 1.001485 + 0.05450 \cos M$$
$$- 0.00148 \cos 2M + 0.01002 \cos (2D - M)$$
$$- 0.00825 \cos 2D \ldots.$$

Horrocks neglects the term with argument $2D$, corresponding to the variation; as Flamsteed puts it, he treats the variation as a 'physical' equation only, affecting the longitude of the Moon but not the shape of its orbit. But even with this fault, the theory was much superior to its competitors. Flamsteed noted that on 23 February 1672 the Moon's observed longitude differed from Horrocks's theory by only 2', while Boulliau's and Streete's tables were in error by nearly 15'. And of greater importance, whereas the theories of Boulliau and Streete called for an increase in the Moon's diameter between 6 November 1671 and 23 February 1672, Horrocks's theory agreed with the observations in calling for a decrease.

Accompanying the first published account of Horrocks's lunar theory, in the *Opera posthuma* of 1672, were the numerical tables that Flamsteed had fitted to the elements of the theory as transmitted in Crabtree's letter of 1642. On one point of interpretation Flamsteed went astray, thereby deeply changing the structure of Horrocks's theory: in place of the varying eccentricities intended by Horrocks, he used their projections onto the mean line of apsides, so diminishing their values by the factor $\cos \delta$. The mistake was later

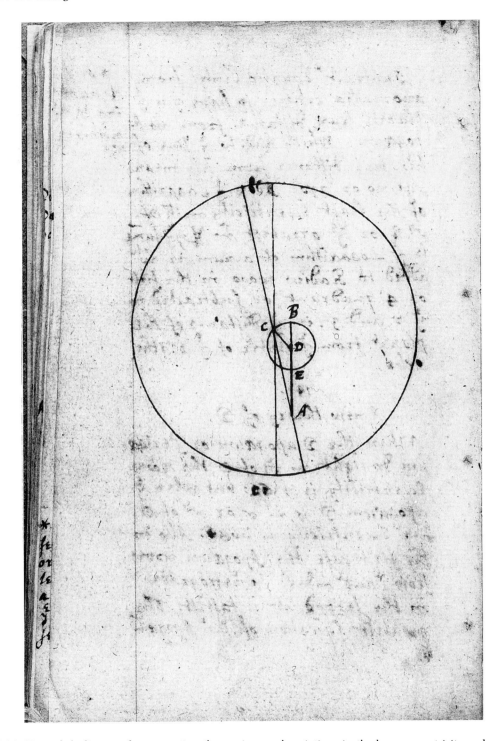

10.18. Horrocks's diagram for computing the semi-annual variations in the lunar eccentricity and apse.

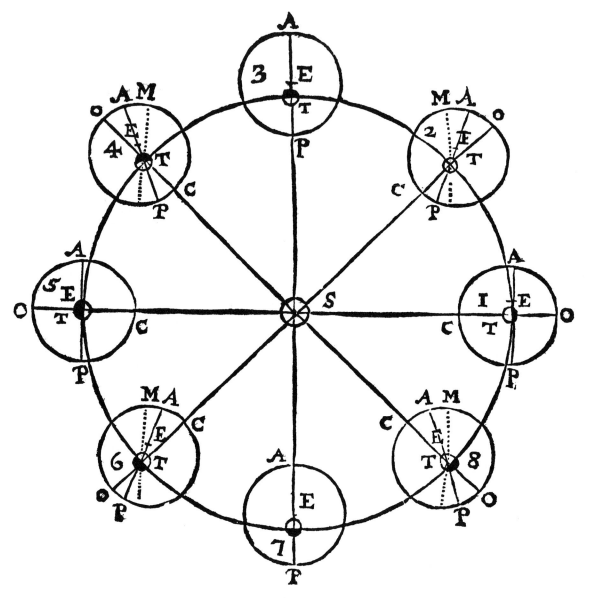

10.19. Diagram to illustrate the Horrocksian theory of the evection (from Crabtree's letter to Gascoigne of June or July 1642).

perceived and corrected by Halley. In another respect Flamsteed improved on Horrocks's theory, by insisting on a return to the astronomical equation of time of Ptolemy and the astronomers before Tycho, and introducing a separate annual equation for the Moon of value $11.8'$ sin M'. This alteration was based, of course, on the assumption of the uniform axial rotation of the Earth; a few years later, by means of the pendulum clock, Flamsteed

would verify this assumption with care, as had Huygens before him. A somewhat revised set of tables of the Horrocksian lunar theory appeared with Flamsteed's *The Doctrine of the Sphere*; and here, as mentioned in the preceding section, Flamsteed computes equations of centre in strict conformity with the areal rule. These tables were republished in 1707 in Whiston's *Praelectiones astronomicae*.

The Horrocksian scheme for the Moon was adopted by Streete in 1674 in his *The Description and Use of the Planetary Systems, Together with Easie Tables*; this work is a description of and set of precepts for using a mechanical device for finding the positions of the Moon and planets. To Flamsteed's charge of plagiarism Streete replied, in a postscript written in June 1675, that his greatest equation of the apogee, greatest and least eccentricity, and mean motions differed from those of Horrocks, and that he computed equations of centre "according to the Hypothesis of the Learned Doctor Seth Ward, ... and not that of Horrox. I am very well assured that my explanation and correction of the Theory of Horrox, is not altogether needless or impertinent."

The perspicuous advantage of the Horrocksian theory over the available alternatives is evident enough when we make the comparison with the theory initially constructed by Cassini I, and published by Cassini II in his *Élémens d'astronomie* and *Tables astronomiques* of 1740. The Cassinis retain the Tychonic mechanism for the evection, and therefore, in order to obtain a fit between the theory and the observed diameters of the Moon, are forced to introduce further *ad hoc* oscillations of the Earth away from the focus of the ellipse, in such a way as not to affect the longitudinal equation required for the evection. The resulting theory is perhaps as complicated as any kinematical scheme used in pre-Newtonian astronomy. We have not attempted to assess its accuracy.

The Horrocksian lunar theory was presupposed by Newton in Prop. 66 of Bk I of his *Principia* (1687), and in 1694 he attempted to improve on it by adding on small sinusoidal oscillations in longitude, as determined by a comparison with observations supplied by Flamsteed. This attempt was not particularly successful: the emended theory was not demonstrably superior to the original in its prediction of lunar longitudes. Newton's rules for calculating the place of the Moon were incorporated in the tables of Leadbetter's *Uranoscopia* (1735); in the tables that Flamsteed constructed about 1702 and which, having been given by Halley to P.-C. Le Monnier, were published by the latter in his *Institutions astronomiques* of 1746; and in Halley's *Tabulae astronomicae* (1749). The errors, as mentioned earlier, were found by Halley to be sometimes as high as 7' or 8': too large to permit

the determination to within 1° of the longitude at sea.

In the longer run there was to be yet one more role for the Horrocksian theory to play: its oscillating ellipse, Euler tells us, was suggestive in leading him to the invention of the method of variation of orbital elements, so fruitful in the thought of Lagrange and Laplace, and in the celestial mechanics of the nineteenth century.

Cometary theory

If during our period Newton's theory of gravitation had not yet led to significant improvement in the predictive astronomy of planets and the Moon, the opposite is true for comets. It was through Newton's labours that cometary motions were first brought securely within the realm of lawfulness, so that precise prediction became a possibility. It was in following Newton's theory that Halley was able to make the first successful prediction of the return of a comet.

From Antiquity until the late sixteenth century, the generally accepted view of comets in Europe had been the Aristotelian one, according to which they were meteorological phenomena, confined to the Earth's upper atmosphere. A few ancients such as Apollonius of Myndus (a contemporary of Aristotle) and Seneca the Younger (Roman philosopher, statesman, orator, and tragedian who died *c.* AD65) thought comets were celestial bodies similar to planets; Seneca urged that there was nothing against supposing that they followed regular, if as yet unknown, orbits. But these views received little attention before the appearance of the comet in 1577.

The earliest known attempt to determine the distance of a comet from the Earth was made by Georg Peurbach (1423–61), mathematician and astronomer of the University of Vienna. Accepting the Aristotelian view, Peurbach supposed the comet of 1456 ("Halley's") to be in the upper region of the air, beneath the sphere of fire and hence within the sphere of the Moon; the latter, so he believed, was 33 000 German miles distant from the Earth. The sphere of fire, he thought, could hardly be as much as 27 000 German miles thick; the comet, therefore, must be at least 1000 miles distant. This claim he was able to corroborate by observing the comet during a period of three hours; its displacement with respect to the stars, and hence its diurnal parallax, proved to be negligible.

Peurbach's pupil, Johannes Regiomontanus (1436–76), wrote a treatise entitled *Problemata XVI* on the determination of cometary parallax. Like Peurbach, he assumed comets to be sublunary. The chief flaw of the treatise (which was published posthumously in 1531) is that it fails to take into account the proper motion of comets. If as seems likely Regiomontanus observed the comet of 1472, he must have discovered that its rapid proper motion made most of the methods described in his *Problemata* inapplicable; and this fact can have been what deterred him from publishing the treatise. It was published in 1531 and again in 1544, and was referred to by later astronomers including Tycho. These astronomers would take account of the proper motion of a comet by noting its positions in relation to stars on successive days when it was at the same altitude and hence subject to the same diurnal parallax; the motion with respect to the stars in the interim was assumed to be uniform.

Observation of the spectacular comets of 1531, 1532, and 1533 led to the ascertainment of a hitherto unnoticed fact: the tails of comets are always directed away from the Sun. This discovery was announced by Girolamo Fracastoro in his *Homocentricorum* (1538) and by Peter Apian in his *Astronomicum Caesareum* (1540). Girolamo Cardano wrote of it in his *De subtilitate* (1550) and *De rerum varietate* (1557), explaining the direction of the tail as a kind of optical effect produced by the Sun's rays shining through the crystalline globe of the comet's body – a view often echoed later. Because he found the proper motion of the comet of 1532 to be slower than that of the Moon, Cardano concluded that comets are supralunary.

The constant relation between the direction of the cometary tail and the Sun could be taken as supporting the notion that comets were celestial bodies. In earlier times astrologers had interpreted the direction of the tail as indicating the region of the Earth that was to be affected by the comet, but the new discovery made this interpretation implausible.

General agreement that comets were supra- rather than sub-lunary objects first came about through studies of the diurnal parallaxes of the comet of 1577. Michael Mästlin (1550–1631), professor of astronomy at Tübingen, observed this comet on all favourable occasions between 12 November 1577 and 8 January 1578. For each determination of position, he ascertained two alignments of the comet with pairs of stars; his sighting instrument was simply a thread held up against the sky. The method was potentially as accurate as naked-eye astronomy could be, except insofar as limited by the sometimes gross inaccuracy of the available star catalogues. On making due allowance for proper motion, Mästlin was able to conclude that the diurnal parallax of the comet was indetectably small, so that the comet must be above the Moon.

Mästlin reported these results in his *Observatio et demonstratio cometae aetherae*, which appeared in 1578. Here, in addition, Mästlin proposed an orbit for the comet; he is thus the first astronomer to devise an astronomical theory to account for apparent cometary positions. He found that he could deduce the observed positions very nearly if he assumed the comet to be moving in a circular arc just outside of Venus's orbit, the centre of the arc being identical with the centre of Venus's circle. To account for the anomaly that remained, he introduced a small circle on the deferent, along the diameter of which the comet was supposed to librate. Although he assigned the comet a circumsolar orbit, Mästlin nevertheless regarded it as a merely ephemeral object that ceased to exist when it ceased to be seen, and as therefore incapable of periodic returns.

Tycho Brahe's study of the comet of 1577 was first reported briefly in a German tract of 1578, then more fully in his *De mundi aetherei recentioribus phaenomenis* of 1588. His observations showed the parallax of the comet to be negligible, and they further showed the apparent path of the comet to be a great circle on the celestial sphere. Tycho would later put forward the view that the apparent paths of all comets are great circles, at least in the middle portions of their appearances. This empirical result, he urged, argued for the immobility of the Earth.

Several years after Mästlin, and with the latter's work before him, Tycho devised a theory of motion for the comet which resembled Mästlin's theory in putting the comet in a retrograde circular path just outside the circumsolar orbit of Venus. The two theories differed greatly, however, in the sidereal periods they assigned to the comet. As for the anomaly, Tycho did not give a detailed account of it, holding that comets did not merit such treatment:

For it is probable that comets just as they do not have bodies as perfect and perfectly made for perpetual duration as do the other stars . . ., so also they do not observe so absolute and constant a course of equality in their revolutions – it is as though they mimic to a certain extent the uniform regularity of the planets but do not follow it altogether.

That the theory of Mästlin, a Copernican, and the theory of Tycho, who denied motion to the Earth, should agree so well in the path they assigned to the comet, must be regarded as accidental. Almost any line or circle can be fitted to the observations of position, especially if longitude alone is taken into account; and quite different paths were assigned to the comet of 1577 by different investigators.

The two most influential cometary theorists in the first half of the seventeenth century were Kepler and Descartes, and both followed Mästlin and Tycho in taking comets to be ephemeral bodies. It was Kepler who introduced the idea that the paths of comets were rectilinear; he argued for this idea in his several writings on comets, and it became the dominant one during the middle years of the century. Thus the astronomy of comets was severed completely from the astronomy of planets, just after the two studies had for the first time been brought into a close relation with one another.

Kepler's strong attachment to the hypothesis of rectilinear trajectories for comets had roots in his cosmological notions. The cosmos, he held, was finite and spherical, with the fixed stars placed in the celestial sphere, and the planets moving in nearly coplanar orbits about the central Sun, their motions determined by a providentially designed mechanism. What to make of such merely ephemeral bodies as comets? Kepler thought they might be formed out of the impure parts of the aether, coagulated into globules by an animal spirit, and then set into motion when the Sun shone upon them. Perhaps their function was to remove waste products from the cosmos. Re-entrant orbits were consonant with permanence; what more fitting for a transient body than a rectilinear trajectory?

Kepler at first favoured Cardano's theory that cometary tails are an optical effect produced by the passage of light through the comet's head, but he later rejected this theory as optically implausible, and proposed that:

. . . the train or beard is an effluvium from the head, expelled by the rays of the Sun into the opposed zone, and in its continued effusion the head is finally exhausted and consumed, so that the tail represents the death of the head.

In his *De cometis libelli tres* of 1619, Kepler undertook to fit straight-line paths to observations of the comets of 1607 and 1618. It was part of his endeavour to show that the motions of comets could be accounted for only if the Earth were granted an annual motion, but he succeeded only in illustrating the difficulties that could arise if the Earth were taken to be immobile and one or another path was proposed for the comet. For locating cometary trajectories, Kepler used a trial-and-error procedure; he was willing to allow that the comet underwent acceleration or deceleration, providing only that its path be rectilinear. Such theories as he proposed for particular comets were too rough to yield precise agreement between theoretical prediction and observation.

As empirical evidence for rectilinear trajectories, Kepler could cite Tycho's generalization that comets start their apparent path in one part of the sky, move swiftly past the Earth along an apparent great-circle arc, and disappear into an opposite region. Indeed, the visible portions of a comet's path before or after perihelion may be close to rectilinear; there are exceptions, but they had not yet been noted or attended to. Both the empirical and dynamical constraints on cometary theorizing were as yet slight, and any hypothesis put forward was unavoidably conjectural.

Gassendi, accepting Tycho's claim that the apparent paths of comets were great-circle arcs, advocated rectilinear trajectories for them in 1658 in his *Syntagma philosophiae*, Bk V, "De cometis & novis sideribus". But, accepting as he did the atomistic cosmology of Epicurus, he added the notion that comets are permanent bodies that move on uniformly forever in an infinite universe. Descartes in his *Principia philosophiae* (1644) proposed slightly curved, sinuous trajectories. A comet, he urged, was a dead star whose vortex had collapsed, and which was then captured by one vortex after another. If it fell far into a vortex, it could become a planet; but if it remained close to the periphery, it continued as a comet, and its path was nearly rectilinear but slightly concave toward the central star. Difficulties with this proposal were later raised by Huygens, Newton, and others: some comets come closer to the Sun than Descartes would allow, and some move contrary to the direction of the solar vortex.

The comet of 1664 provided a stimulus and test for renewed theorizing, some of it on the assumption that comets were permanent bodies that could return periodically. Thus Cassini I, in his *Hypothesis motus cometae novissimi*, sent from Italy to Paris and then transmitted to the Royal Society in London in February 1665, proposed a vast circular orbit for the comet, centred round the star Sirius. By means of Kepler's procedure, but assuming the Earth to be immobile, he first located a possible rectilinear path for the comet, in which it would be moving uniformly. This path he assumed to be almost coincident with the comet's true, circular path, which he likened to that of a planet:

The arrival of the comet at perigee and its highest apparent velocity occurred before midday on 29 December 1664. . . . Thus the same happened with it as usually happens with superior planets, which are swiftest about the time when they are at perigee in their epicycles and in retrograde motion, by which they move contrary to the Sun. Therefore on preceding and succeeding days the increments of distance from the Earth, and the velocities of diurnal motion, are equal. . . . Hence, in order that the uniformity of the celestial motions may be evident, since the comet's motion appears retrograde for the same reason as the motions of the other planets do – for the comet may be near the Earth in the lower part of some epicycle and so be conspicuous, while in the upper part of the epicycle where its motion might appear direct it cannot be perceived because of its incredible distance from the Earth – the centre of this epicycle must lie on a straight line between the Earth and Canis Major. Moreover, since this centre must be placed near some notable body in the universe (just as the Earth [is circled by] the Moon, the Sun by the inner planets, Jupiter by the Medicean stars, and Saturn by his own satellite), Sirius, the most brilliant of all the fixed stars, presents itself. And so from the first I set myself the task of considering whether the proper motion of the comet seemed to be arranged about that star.

To account for the comet's departure from the great-circle arc passing through the initial observations, and for its motion in latitude, Cassini assumed the plane of its orbit to be inclined at 30° to the plane passing through the Earth and the arc of the motion of Sirius. He did not, however, derive quantitative consequences from this assumption, but rather sought to obtain predictive accuracy by the device of giving a motion to the nodes of the plane of the cometary orbit on the ecliptic. The curvature of the trajectory, he allowed, was very slight; as Newton would later remark, Cassini's hypothesized path for the visible appearances of the comet was essentially a straight line.

Cassini developed similar theories for the comet of 1665 and the post-perihelion appearances of the comet of 1680. The principle that comets appear to move very nearly on the arcs of great circles, and Cassini's success in using Keplerian procedures for locating a rectilinear path, prevented him from regarding the pre-perihelion comet of November 1680 and the post-perihelion comet of December 1680 and after as the same comet, for the two apparent paths came nowhere near to lying on the same great circle. As will appear in Chapter 13 below, the question whether these comets were the same or not would play a considerable role in the advance of Newton's mind toward the moment in which he will undertake to write his *Principia*.

Auzout in Paris used great-circle plotting as well as the Keplerian procedure to produce an ephemeris of the comet of 1665 while it was still appearing. Like Cassini (although doubtingly) he imagined that comets might return periodically, but he thought it unlikely that they moved in circles about stars like Sirius, and also considered it unlikely that cometary motions could be accurately predicted without taking into account the annual motion of the Earth. The Royal Society in London received a copy of the letter in which he expressed these thoughts, and a partial English translation of it, later to be read by Newton, was printed in 1665 in the *Philosophical Transactions*.

Hevelius in his *Prodromus cometicus* (1665) and *Cometographia* (1668) revealed himself a follower of Kepler in cometary theory, but he allowed physical causes to bend the trajectories from their intrinsic rectilinearity into parabolic and hyperbolic paths. From the observations of many earlier comets he attempted to show that their apparent motions entailed the annual motion of the Earth. Like Kepler he allowed the proper motions of comets to be accelerated during approach to the Sun and decelerated during motion away from the Sun. He explained the changing speed by supposing the disk-shaped head of the comet to behave rather like a magnetic compass, with one of the faces turning so as always to be perpendicular to the solar rays; thus aethereal resistance would be least when the comet was passing the Sun and slipping through the aether edge-forward.

The reflections of Wren and Hooke regarding the comet of 1664 will be dealt with in Chapter 12 below: they would lead to Hooke's formulation of a theory of planetary motion, and this in turn would impinge on Newton's thought in an important way. Be it noted here that Wren and Wallis dealt with the mathematical problem of determining the path of a comet from four observations, assuming the Earth's annual motion and uniform rectilinear motion of the comet. Wren used geometrical constructions, and applied his method to the comet of 1664; Wallis devised both a geometrical and a trigonometrical method of computation, but apparently did not undertake to apply his method to actual observations. The Wren–Wallis procedures would play a role, albeit an abortive one, in Newton's study of comets.

Concluding summary

In the seventeenth century, after Kepler, planetary astronomy gradually became Keplerian in important respects: adoption of the true Sun in place of the Mean Sun as the chief point of reference, adoption of elliptical orbits, adoption of one or another rule of motion closely approximating to Kepler's areal rule, adoption at length of Kepler's third law as the basis for determining mean solar distances.

Gassendi's observation of the transit of Mercury of 1631 showed Kepler's tables to be the best available at the time, but van Lansberge's and Longomontanus's tables went on being used into the 1660s, recognition of their inaccuracy increasing the while. Boulliau's tables, published in 1645, were based on elliptical orbits, but with a different rule of motion from Kepler's; they were copied by Riccioli in Italy and by Shakerley and John Newton in England, and had the effect of popularizing the elliptical orbit. Inconsistencies in Boulliau's derivation of his rule of motion, however, were pointed

out by Neile and Ward in the 1650s; Boulliau had to propose a modified rule whose artificiality and *ad hoc* character seem obvious today. Wing, Mercator, and others proposed other substitute rules of motion; in 1670 Mercator could point out that any such rule, to be successful, must yield results very close to those of Kepler's areal rule. In the mid-1660s, the astronomers of the newly formed Academy of Sciences in Paris, with a new concern for improving the accuracy of astronomical constants, felt justified in taking Kepler's tables as their starting-point, and ignoring all other tables. In the late 1660s in England, the young Flamsteed concluded that Streete's tables, which were close to Kepler's but were the first to adopt mean solar distances based on Kepler's third law, were the most accurate of those available.

Improvements on the solar and planetary parameters found in Kepler's *Rudolphine Tables* required above all a reduction in solar parallax and the adoption of an improved table of refractions. Horrocks in the late 1630s, developing a line of inquiry that he found in Kepler's writings, took the important step of reducing solar parallax to a twelfth of Tycho's value, and went on to work out many of the consequences which in part found their way into the tables of Streete. Further progress in the same direction came from Flamsteed, and from Cassini I and his fellow academicians. Moreover, Cassini, following another theoretical path opened up by Kepler, produced the table of refractions that was needed. The improvements in parameters thus made possible were embodied in the tables of Cassini and Halley, published at last in the 1740s.

Instrumental innovations – the micrometer, telescopic sights, and the pendulum clock – seemed to promise a yet higher degree of precision for planetary theories. But the way to making use of the new observational precision remained blocked by puzzling motions to which the stars of the stellar background were subject. These motions were not accounted for until Bradley announced his discovery of aberration in 1729, and in subsequent years explained the 18.6-year period of the motion of nutation, dependent on the rotation of the lunar nodes.

Another puzzle consisted of the rather large, unexplained anomalies in the mean motions of Jupiter and Saturn. These anomalies would remain an obstacle to the refinement of planetary theory until 1785 when Laplace showed them to be consequences of the perturbations of the two planets, produced by their mutual attractions in accordance with Newton's inverse-square law of universal gravitation.

Lunar theory, by the time Tycho had done with it, was more complicated than anyone would have liked; and it was still so inaccurate as to cause Kepler despair. Despite the efforts of a series of astronomers from Kepler to Streete and Wing, the situation remained unchanged up to the publication in 1672 of Flamsteed's version of the lunar theory of Horrocks. Horrocks's theory was both simpler in conception and more accurate than anything that had appeared earlier. Newton would account qualitatively for its basic pattern in the first edition of his *Principia*; and in the 1690s he would use it as a basic pattern on which to superimpose further perturbational refinements. However, as we shall see, he failed to achieve a theory decisively more accurate. Further progress would have to await systematic derivation of perturbations in the middle of the eighteenth century.

Whereas in the case of the planets and the Moon, the chief anomalies had come to be rather accurately identified before 1687, the problem of cometary motion before this date was different: lack of firm ground for hypothesizing one shape of trajectory rather than another left too much to assumption. The rectilinear trajectories proposed by Kepler and Gassendi, the sinuous path imagined by Descartes, and the slightly curved paths recommended by Cassini and Hevelius, remained, all of them, fundamentally conjectural. Tycho's assumption that the apparent paths of comets approximate to great circles tended to govern thinking about comets, and to reinforce the hypothesis of rectilinear or quasi-rectilinear paths. The hypothesis implied that comets are not attracted to the Sun with the same accelerative force as other bodies at the same distance. Little headway could be made until this hypothesis was abandoned, and Newton's theory of universal gravitation was brought to bear on the problem. As we shall see in Chapter 13 below, Newton's ultimate success, after many false starts, in finding a way to apply this theory to cometary observations must be accounted one of his major achievements.

Further reading

Jane L. Jervis, *Cometary Theory in Fifteenth-century Europe* (Studia Copernicana, vol. 26) (Dordrecht, 1985)

J.A. Ruffner, *The Background and Early Development of Newton's Theory of Comets* (Ann Arbor, 1973)

V.E. Thoren, Kepler's second law in England, *British Journal for the History of Science*, vol. 7 (1974), 243–56

D.T. Whiteside, Newton's early thoughts on planetary motion: a fresh look, *British Journal for the History of Science*, vol. 2 (1964), 117–37

Curtis Wilson, From Kepler's laws, so-called, to universal gravitation: empirical factors, *Archive for History of Exact Sciences*, vol. 6 (1970), 89–170

Curtis Wilson, Horrocks, harmonies, and the exactitude of Kepler's third law, *Science and History: Studies in Honor of Edward Rosen* (Studia Copernicana, vol. 16) (Wrocław, 1978), 235–59

11

The Cartesian vortex theory
ERIC J. AITON

When the solid spheres were abandoned by Tycho Brahe (in whose system the orbits of Mars and the Sun intersect), the alternatives open to astronomers were the acceptance of either void space or a fluid aether. Johannes Kepler, in his *Astronomia nova* (1609), postulated a quasi-fluid heavens in the form of a magnetic vortex (see Chapter 5). For he supposed that the Sun propagates an immaterial species (arising from its magnetic force and similar to rays of light), which turns with the Sun: "after the manner of an impetuous vortex, which extends over the whole width of the world, and at the same time bears along the planets with it in a circle with a stronger or weaker thrust according as, by the law of its emanation, it is denser or rarer." While the rotating species, together with an independent magnetic mechanism introduced to vary the distances of the planets from the Sun, served to explain the orbital motions, they could not explain the rotation of the Sun itself, nor the diurnal motion of the Earth, for which purpose Kepler had to postulate an internal animistic force. Moreover, experiments with real magnets could not produce the effects required; in particular, magnets could not be made to rotate bodies about them. As a consequence, Kepler's magnetic hypotheses were almost universally rejected by his followers, thus leaving the planetary motions without any satisfactory physical explanation. It was at this opportune time that the Cartesian vortex theory appeared on the scene and rapidly secured an entrenched position.

Following his meeting with Isaac Beeckman in 1618 and the famous dream of 10 November 1619, René Descartes (1596–1650), who had been educated at the Jesuit College of La Flèche, spent nine years developing a mathematical physics independent of Aristotle's philosophy. Then in the winter of 1628–29, he provided a metaphysical foundation, in which the real distinction between mind and body justified the elimination of substantial forms from physics and their relocation in the realm of mental images or appearances. During the period 1629–33, Descartes composed a treatise on the system of the world, introducing the celestial vortices and the explanation of terrestrial gravity as an effect of the subtle or celestial matter. But the condemnation of Galileo Galilei led him to withhold this work, *Le Monde, ou traité de la lumière*, which appeared only in 1664. From 1640 he was less secretive about his subtle matter, probably as a result of having found in the concept of relativity of motion a means of reconciling the Copernican theory of the Earth's motion (clearly implied in his system) with Scripture. Descartes gave the vortex theory to the public in his comprehensive and definitive treatise on physics and cosmology, *Principia philosophiae* (1644), which appeared again in 1647 in an approved French translation by the Abbé Picot; our discussion of Descartes's physical account of the universe will be based primarily on this work.

Descartes's theory of scientific explanation

From his metaphysics Descartes deduced a number of general principles or first causes of all actual or possible physical phenomena. For example, God's immutability implied the conservation of the total quantity of motion in the universe. Although Descartes's general principles were deduced *a priori*, he nevertheless agreed that conclusive demonstrations in physics were impossible. When he wrote of deducing the effects from the principles, this was not therefore to be understood in the sense of the necessary implication of mathematics. There were, he explained, many different ways of deducing a particular effect from the principles; in other words, there were different hypothetical explanations, compatible with the principles without being necessary consequences of them. Experiment and observation played a dual role; first, to provide a test between competing possible explanations, and secondly, to determine which of the infinity of possible effects God had chosen to bring into being, for it was those real effects that we should try to explain in accordance with the principles.

In a letter of 28 October 1640 to the influential Minim, Marin Mersenne (1588–1648), Descartes epitomized the complementary *a priori* and hypothetical parts of physics with the remark: "I do believe that one can explain one and the same particular effect in many different ways which are possible, but I believe that one cannot explain the possibility of things in general except in one way only, which is the true one." Descartes could have expected his idea to be freely quoted, for Mersenne promoted communications between scientists both by his voluminous correspondence and by the arrangement of meetings in his cell at the Convent de l'Annoncide.

As an example of Descartes's method of hypothetical explanation in accordance with *a priori* general principles, we may consider the theory of refraction described in the *Dioptrique* published in 1637 as an appendix to the *Discours de la méthode*. Light is conceived as an impulse (or tendency to motion) transmitted instantaneously through the medium of subtle matter. The law of refraction is explained by a comparison of the instantaneously transmitted impulse with the motion of a tennis ball (travelling with finite speed) meeting an obstacle (in the form of a frail canvas which it breaks through) at the interface of the two media. In other words, Descartes used a mechanical model, to which he could apply the rules of motion, as a representation of the light rays.

Descartes's laws of nature and theory of motion

Descartes held the essence of body (supposed divisible to infinity) to be extension. From a purely geometrical point of view, motion and rest were simply modes or states, but in the context of dynamics, Descartes envisaged motion as possessing an independent reality of its own, so that it could pass from one body to another. For he conceived all efficient physical causation to be located in the impulsion of moving bodies. The motion of a body was in effect a force (quantified as the product

of 'size' and 'speed') that enabled the body to act on others by impact. Bodies either in motion or at rest also possessed an inertial force which enabled them to resist a change of state. This inertial force was seen by Descartes as the cause of cohesion. For it was by virtue of its resistance to change of state that a part of a body could not separate from its neighbours without the intervention of an external impulsion.

The principles of dynamics, in the form of three laws of nature and seven rules of impact, were deduced *a priori* by Descartes, starting from the axiom of God's immutability. For Descartes, God was the first cause of motion and in consequence of his immutability always conserved the same quantity of motion in the universe. Descartes's first and second laws of nature, together equivalent (at least formally) to Newton's first law of motion, also depended directly on God's immutability, but were evidently inspired (at least in part) by literary sources and experience. The first law of nature states that "each thing perseveres always in the same state, as much as it can [*quantum in se est*], and only changes it by colliding with others". The inclusion of the phrase 'quantum in se est', used by Lucretius in his poem *De rerum natura* to mean 'naturally' or 'without external force', indicates one source of inspiration. The second law of nature states that "each part of matter in itself never tends to move along curved lines but along straight lines". The reason was that, at each instant, God conserved the motion precisely as it was at that instant. It should be noted that the law related only to a tendency to motion, for in Descartes's universe, real motion was always along closed curves. An immediate inference drawn by Descartes was the existence of centrifugal force; so that, whenever a body was constrained to move in a circle, it tended continually to recede from the centre.

The third law of nature contains the basic principle governing the transfer of motion from one body to another. This law states

. . . that if a body which moves and strikes another has less force to continue its motion in a straight line than the other to resist it, it loses its determination without losing any of its motion, and that if it has more force, it moves this other body with itself and loses as much of its motion as it gives to the other.

The precise meaning of the term 'determination',

used in this law, is revealed by consideration of Descartes's mechanical model for the explanation of the law of refraction in optics. First, he explained that motion was different from determination and their quantities had to be examined separately. Second, the motion that was reduced in a certain ratio when the ball broke through the canvas was measured by the total speed, which shows that motion was seen by Descartes as a scalar quantity. Third, only the part of the determination which made the ball tend perpendicularly towards the canvas was changed, while that part parallel to the canvas was conserved. Evidently, for Descartes 'determination' was a vector quantity (in modern terms 'momentum'), which could be resolved into components.

Of the seven rules of impact, only the fourth is of importance in relation to the planetary motions. This concerns a moving body B striking a larger stationary body C. According to Descartes, the body B loses all its determination without losing any of its speed (so that it rebounds), while C remains at rest. The explanation is that, the greater the speed of B, the greater the resistance of C and, as C is larger, its force of resistance cannot be surmounted. Although this explanation is consistent with Descartes's idea of the identical nature of impressed force and inertial force (impact involving a contest between them), he felt the need to offer another explanatory principle in his letter of 17 February 1645 to Claude Clerselier (1614–84), a faithful disciple who became the first editor of his correspondence; namely, "when two bodies meet which have in themselves incompatible modes, some change must occur in these modes to make them compatible, but this change is always the least that it can be". The modes were the motion and the determination. Applying the principle to the fourth rule, Descartes argued that, if B were to push C before it, it would have to transfer more than half its motion and more than half its determination, whereas if B were reflected, losing all its determination but retaining its motion, while C remained at rest, the total change in the two modes would be less. However, as Descartes explained, this rule would only apply if the body C were not surrounded by air or any fluid body.

Having established the general principles of motion which describe abstractly the manner in which God conserves the quantity of motion in the universe, Descartes turned his attention to the actual world of planets immersed in a fluid aether. Common experience with ordinary fluids such as water brought to mind two important properties that needed explanation. In the limiting case of an ideal fluid, such as he supposed the aether to be, these properties would become, first, that a body floating in a fluid could be moved by the smallest force and, second, that a body moving through the fluid would not be impeded by it.

Descartes defined a fluid as a body divided into many small parts which move independently in several different ways. Then he explained that a fluid does not impede the motion of a solid body through it, because "bodies which are already in process of moving do not prevent the occupation by other bodies of the places they are themselves disposed to leave". In the case of the aether, the parts moved very quickly out of the path of the moving body, but in the case of ordinary fluids such as water, the parts did not move so quickly and consequently some resistance was offered. A solid body floating in a liquid was regarded by Descartes to be held in equilibrium rather than to be in a state of rest. Consequently, the body had no resistance to motion and the least external force could move it by breaking the equilibrium. In particular, a current of the fluid itself, however small, would always enable the fluid to carry the body with it. For suppose that the body, floating in equilibrium, is pushed in a certain direction by an external force, however small. Then this force, added to the force of the fluid pushing the body in the same direction, is greater than that of the fluid pushing the body in the opposite direction, and by the third law of nature, the determination of the latter portion of fluid is lost, so that it is reflected from the body. As this fluid no longer prevents the body from moving, the body begins to move. Descartes claimed, correctly, that this explanation was not inconsistent with the fourth rule, according to which a body could not move a larger stationary body, because in fact no body in a state of rest was involved.

Descartes's primary concept of conservation of motion and his opposition of states of motion and rest, as also his rules of collision, implied motion to be absolute. In his explanations of the planetary motions, apart from that of the Earth around the Sun, he tacitly adopts this concept of motion. Yet in describing the nature of motion itself, he dismisses the definition according to which motion is "the action by which a body passes from one place to another" as a common notion or belief based only on appearances. In reality, he declares, motion is "the transport of a part of matter or of a body from the neighbourhood of those which touch it immediately, and which we may consider as at rest, into the neighbourhood of some others". In particular, a body carried by a fluid had less motion than it would have if it failed to follow the current, for "it separates much less from the parts which surround it when it follows the course of the liquid than when it does not follow it". This idea of a limited relativity of motion enabled Descartes to reconcile the motion of the Earth with the official doctrine of the Church.

While recognizing that the systems of Tycho and Copernicus explained the appearances equally well (whereas the system of Ptolemy failed to explain some appearances, such as the phases of Venus, for example), he believed that of Copernicus to have greater clarity and simplicity. For this reason, he preferred to incorporate the Copernican system in his own hypothesis, which he claimed to be "the simplest and most convenient of all, both for comprehending the appearances and for seeking their natural causes". According to his own hypothesis, Descartes explained, the Earth was at rest relative to the surrounding celestial matter, but was nevertheless carried around the Sun by this matter in the same way that "a vessel which is not carried by the wind nor by oars, and also is not retained by an anchor, remains at rest in the middle of the sea, although perchance the flow and ebb of this great mass of water carries it imperceptibly with itself".

Despite his explicit description of the true nature of motion as essentially relative, there are indications that Descartes did not in fact take the relativity of motion seriously. These may be found in a letter to an unknown correspondent written shortly after the publication of the *Principia philosophiae*, in which he claimed that he had satisfied the ecclesiastical objections by truly denying the motion of the Earth while accepting the Copernican system. For he remarked that all the passages of Scripture that seemed to preclude the motion of the Earth were concerned not with the system of the world but only with the mode of expression. The implication seems to be that Descartes supposed the Earth really to move and that

he regarded the explanation based on the relativity of motion, like the Biblical texts themselves, as a matter of words rather than philosophical truth.

Descartes's vortex theory

Aristotle did not doubt that the early philosophers who attempted to explain the formation of the universe had in mind the analogy of the whirlpool and whirlwind, "reasoning from what happens in liquids and in the air, where larger and heavier things always move towards the middle of the vortex". Descartes's hypothetical explanation of the planetary motions was clearly inspired by the analogy of a water vortex. For he remarks in Pt III, Chap. 30 of the *Principia philosophiae* that

. . . in the winding of rivers, where the water folds back on itself, turning in circles, if some straws or other very light bodies float amidst the water, we can see that it carries them and moves them in circles with it; and even among these straws one can notice that there are often some which also turn about their own centre; and that those which are closer to the centre of the vortex which contains them make their revolution more quickly than those which are more distant; and lastly that, although those vortices of water conspire always to rotate in rings, they almost never describe entirely perfect circles, and extend themselves sometimes more in length and sometimes more in width, so that all the parts of the circumference which they describe are not equally distant from the centre. Thus we can easily imagine that all the same things apply to the planets, and it needs only this to explain all their phenomena.

Having thus briefly indicated his explanation of the planetary motions, Descartes turned his attention to the whole fabric of the visible world, seeking the causes of the phenomena of stars, planets and comets in general. These he found in a hypothetical explanation of the formation of the universe which has similarities to the account of Leucippus preserved by Diogenes Laertius. In deference to the theologians, Descartes claimed that his hypothesis was false (for he admitted that God created the universe perfect in the beginning) but he justified his fable as contributing to our knowledge in the same way that a study of the development of plants aids our understanding of those created perfect by God in the Garden of Eden.

Even if his hypothesis concerning the initial state of the universe were to be taken seriously (and it is easy to believe that this was his real intention), the hypothesis, Descartes remarked, might still be false, but this would not harm the consequences taken from it. For whatever may have been the disposition of matter and motion in the beginning (even a state of chaos), the operation of the laws of nature would have transformed the universe into a succession of possible states until the present one had been reached. Thus if we start with a possible state (and the one chosen by Descartes seemed to him to be clearly possible), application of the laws of nature should enable us to derive a chain of possible states (some of which have never perhaps been actual) until we reach the final state God has chosen to bring into existence.

Choosing what he considered to be the most simple, intelligible and reasonable hypothesis, Descartes supposed that, in the beginning, all matter was divided into equal parts, intermediate in size between those which now compose the universe. Two rotations (equal in force) were then imparted to them, a rotation of each about its own centre and rotations of many together about centres disposed in the same way as are at present the centres of the fixed stars; that is, scattered indefinitely throughout the universe (Figure 11.1). As a consequence of the operation of the laws of nature, the primitive matter in time attained three distinct forms or elements, characterized as luminous, transparent and opaque. The second element consisted of tiny globules in rapid motion. The spaces between them were filled by the first element, whose minute parts were of indeterminate shape and size but of immense impetuosity. The aether or celestial matter consisted mainly of the second element, the Sun and stars entirely of the first, and the Earth and planets of the larger and slower moving parts of the third element.

Two vortices could not touch at the poles; for if they were in the same sense, they would combine, and if they were in different senses, they would hinder each other's motion. Consequently, the poles of each vortex were, in general, near the equators of its neighbours. The first element entered at the poles, moving towards the centre and from there outwards towards the equator, passing into other vortices through their poles. The circulating globules of the second element could not similarly pass from one vortex to another, because in passing from the equator of one to a pole of another, they would not be able to preserve their motion. When the matter of the first element arrived at the centre, it pushed outwards the more slowly moving globules, which already tended to recede from the axis of rotation of the vortex on account of their centrifugal force. Consequently, a permanent reservoir of matter of the first element was formed at the centre of the vortex, turning on its axis and constituting the central body; in the case of the solar vortex, the body of the Sun. The globules closer to the centre were smaller and moved faster, up to a certain distance, beyond which the more distant moved faster. In the solar vortex, those at the distance of Saturn were the slowest. Below Saturn, the rotation of the Sun augmented the speed of the globules. These were therefore smaller, for if they were the same size, they would have more centrifugal force and hence would rise.

Owing to their irregular shape, parts of the first element easily hooked together, transferring a large part of their motion to the smallest parts. These hooked parts were formed near the poles, where less motion was needed for the circulation. They formed small columns whose cross-sections were curvilinear triangles with concave sides, so that they could thread their way through the spaces between globules in contact. These channelled parts formed spirals in opposite senses coming from the north and south poles and were used by Descartes to explain the Earth's magnetism. On entering the Sun, these channelled parts were thrown up by the agitated first element. Acquiring the character of the third element by agglutination, they floated on the surface, forming sunspots for a while, until they were dissipated again by the agitation of the matter of the Sun.

A star could be completely covered with spots. As it would then no longer push the globules outwards, the vortex collapsed and the star, passing into another vortex, became a planet or a comet. It was pushed towards the centre of the vortex until it acquired the same agitation (that is, force or quantity of motion) as the circulating aether. If this happened before it reached the slowest layer (the sphere of Saturn) it ascended again, becoming a comet and passing out of the vortex into another. The part of the comet's path in the solar vortex (the observable part) was thus a single curve (Figure 11.2). If the star descended below Saturn before

11.1. Descartes's conception of a typical region of the universe, as expounded in Part III of *Principia philosophiae* (1644).
Several stars are portrayed, each surrounded by its own vortex.

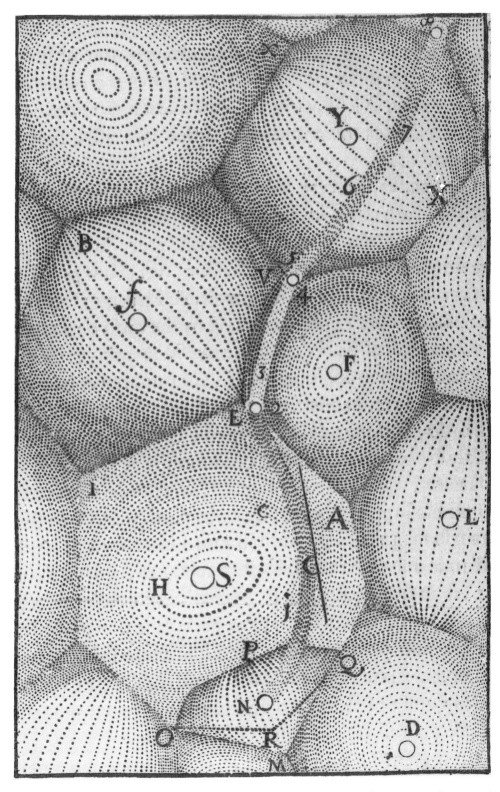

11.2. Diagram to illustrate the path of a typical comet from one vortex to the next, from Part III of Descartes's *Principia philosophiae*.

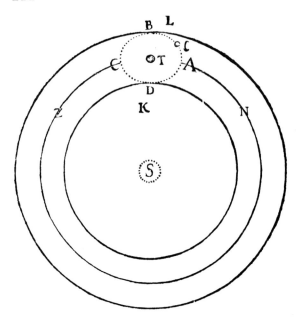

11.3. Diagram to illustrate the formation of the terrestrial vortex, from Part III of Descartes's *Principia philosophiae*.

acquiring the same agitation as the aether, it became a planet, circulating in a stable orbit. If it descended further, it would be surrounded by smaller globules, which on account of their greater speed, would increase its force and make it rise again. If it rose, it would be surrounded by slower globules, which would diminish its force (presumably very slightly, since the motion of a body was not impeded by an ideal fluid) and make it fall again.

The equilibrium position depended on the density, so that the planets floated at different distances from the Sun, the least dense being nearest. There was no question of a gravity of the planet towards the Sun balancing the centrifugal force. For the planet had exactly the same centrifugal force as the aether. Like the aether, the planet circulated, not because it was heavy but because its tendency to recede from the Sun was impeded by the external constraints of the neighbouring vortices.

To explain why planets nearer the Sun move faster, Descartes had proposed that the first element in the Sun augmented the speed of the aether. However, the linear speed of the sunspots was slower than that of any planet. To overcome this difficulty, Descartes postulated a solar atmosphere extending as far as Mercury. As an imperfect fluid, this atmosphere could retard the spots without affecting the layers of globules close to the Sun.

The Earth's diurnal motion

Descartes's theory of the Earth's diurnal motion lacks consistency, for at different points in the *Principia philosophiae* he described two different explanations. First he attributed this motion to the circulation of the terrestrial vortex but later explained that the Earth had been able to retain the rotation received at the time of its creation:

. . . without any noticeable diminution, because the larger a body is, the longer also can it retain the agitation which has been impressed on it, and the duration of five or six thousand years that the world has existed, compared with the size of a planet, is not so much as a minute compared with the smallness of a top.

In the case of a child's top, Descartes attributed the diminution in speed to the effect of the surrounding air, while the planet could be slowed down, for example, by the effect of the channelled parts causing magnetism as they entered the body of the Earth. The circulation of the aether itself could not be a cause of retardation because this circulation was supposed by Descartes to be faster than the diurnal motion.

The satellites

There remained the problem of the satellites possessed by at least three planets of the solar system, namely Jupiter, Saturn and the Earth. Galileo had established the existence of the four moons of Jupiter, while the puzzling appearance of Saturn's rings in the telescopes of the time seemed to indicate that Saturn had two moons (see Chapter 6). In order to explain the motions of the satellites Descartes supposed that inside the solar vortex were smaller vortices containing Jupiter, Saturn and the Earth in their respective centres. The rapid circulation of the satellites of Jupiter he attributed to the rotation of this planet (like the Sun and the Earth) on its axis, whereas the satellites of Saturn, he suggested, either circulated very slowly or not at all, because Saturn (like the Moon) always turns the same face towards the centre of the vortex that contains it; that is, in the case of Saturn, the solar vortex.

Concerning the origin of the satellite systems, Descartes suggested two possibilities. Either the systems were formed before entering the solar vortex or the satellites were captured inside the solar vortex. He seems to have preferred the former in the cases of Jupiter and Saturn but the latter in the case of the Earth and the Moon.

The terrestrial vortex

Descartes's explanation of the formation of the Earth–Moon system in Pt III, Chap. 149 of the *Principia philosophiae* introduced another inconsistency. Although he had previously supposed that the planets moved with the same speed as the circulating aether, he now proposed that the Moon, having the same density as the Earth, had to circulate at the same distance from the Sun but, owing to its smaller size, faster than the Earth.

Let *S* (Figure 11.3) be the Sun and *NTZ* the circle along which the Earth and the Moon were revolving in the solar vortex. Then, in whatever part of this circle the Moon may have been in the beginning, Descartes explains:

... it soon had to come to A, *close to the Earth* T, *... and finding at the point* A *that the Earth with the air and the part of the heavens which surrounds it offered some resistance, it had to deviate towards* B; *I say towards* B *rather than towards* D, *because in this way, the course which it has taken deviated less from the straight line. And while the Moon thus moved from* A *to* B, *it disposed the matter of the heavens contained in the circle* ABCD *to turn with the air and the Earth about the centre* T, *and there form a small vortex, which with the Moon and the Earth has ever since continued its course about the Sun along the circle* TZN.

An attempt to explain the tides led Descartes to predict an interesting property of the Earth–Moon system, whose empirical verification was quite beyond the capability of observational techniques at that time. For he declared that the Earth was slightly displaced from the centre of its vortex towards a point diametrically opposite the Moon and moved in a small epicycle. When the Moon was at *B* (Figure 11.4), for example, the aether would not flow so freely between *B* and *T* (because of the presence of the Moon) as between *T* and *D*, unless the Earth moved slightly towards *D* so as to equalize the pressure on the two sides. As a consequence of the constrictions on both sides, the oceans were depressed, causing low water to occur directly beneath the Moon and at the point diametrically

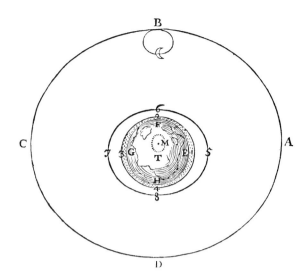

11.4. The Earth–Moon system, from Part IV of Descartes's *Principia philosophiae*.

opposite. Although the interval between the meridian passage of the Moon and high water varies considerably from place to place, Descartes's theory explained the semi-diurnal period depending on the lunar day.

Another tidal problem led Descartes to consider the cause of the lunar variation, an inequality discovered by Tycho in 1598. For a simple application of this inequality in the Moon's motion sufficed to explain the monthly inequality in the tides. Descartes first demonstrated that the Moon moves faster when full or new, that is, at *B* or *D* (Figure 11.3), than in the quadratures, that is at *A* or *C*. For the aether in the terrestrial vortex *ABCD* is composed of globules similar to those in the solar vortex at *N* or *Z*, and hence larger and less agitated than those in the solar vortex below *D* but smaller and more agitated than those above *B*. This, he inferred, had the effect that they mingled more easily with those at *N* and *Z* than with those below *D* or above *B*, so that

... the circle ABCD *is not exactly round but longer than broad in the form of an ellipse, and as the celestial matter which it contains goes more slowly between* A *and* C *than between* B *and* D, *the Moon which it carries with it must also go more slowly there, and make its excursions greater.*

The use of the term 'ellipse' to describe the oval quite clearly does not imply a knowledge of Kepler's first law. For the Earth is supposed to occupy the centre, or at least to move in a small epicycle about this point, and there is no suggestion that the Earth occupies a focus of a geometrical ellipse. Moreover, Descartes recognized that the variation was an inequality peculiar to the Moon.

Terrestrial gravity
Terrestrial gravity or weight, explained in Pt IV, Chaps. 20–27 of the *Principia philosophiae*, appeared to Descartes to be an effect of the celestial matter analogous to that to be seen in whirlpools, where solid bodies floating in the water were pushed towards the centre. Thus he supposed that the celestial matter, circulating rapidly around the Earth, pushed towards it "all the bodies that we call heavy". This matter, he added, employed its excess of speed over that of the Earth "as much to turn faster than the Earth in the same sense, as to make divers other motions on all sides". Owing to its greater speed, the celestial matter had more centrifugal force than earthly bodies released in the air, so that it displaced these earthly bodies downwards by rising to take their place. As the falling body gained speed, the new impressions of the celestial matter decreased, so that after a few feet, a limiting speed was attained. Since the celestial matter contained in the pores of earthly bodies rendered them lighter, bodies of the same volume could have different weights, while the internal motions of the particles of liquids rendered them lighter still.

It might seem that the motion of the celestial matter in the same sense as the rotation of the Earth should have the effect of pushing heavy bodies towards the axis of rotation and the horizontal components of the "divers other motions" the effect of pushing them towards the centre, so that the combined effect should appear as a tendency of bodies to fall to a point somewhere between the centre and the foot of the perpendicular to the Earth's axis of rotation. Descartes explains, however, that bodies fall towards the centre of the Earth because

. . . *although the parts of the celestial matter move in several different ways simultaneously, they nevertheless accord to compensate and oppose each other, so that they extend their action equally on all sides where they can extend it; and because of this alone, that the mass of the Earth by its hardness is repugnant to their motion,*

they tend to recede equally on all sides from its neighbourhood, along straight lines taken from the centre.

Thus, according to Descartes, the reaction of the Earth on the globules converted the centrifugal force engendered by the circulation into a radial force. As Christiaan Huygens (1629–95) pointed out later, this explanation was inconsistent with the idea that the celestial matter could freely penetrate the pores of terrestrial bodies.

On the question of whether weight varied with distance from the Earth, Descartes had remarked to Mersenne in 1638 that this could be settled by experiment, though he doubted the existence of a tower high enough for a difference to be measurable. He evidently believed that bodies were heavy only in the vicinity of the Earth, for 'planets' such as the Moon and Venus, being bodies similar to the Earth, should be heavy and fall towards the Earth, "if it were not that their great distance had entirely removed the inclination". The form of expression used here by Descartes is explained by his determination, at this time, to avoid discussing the dangerous subject of the celestial matter and vortices.

Descartes's explanation of the planetary motions follows the pattern of his scientific method as outlined in the letter to Mersenne quoted near the beginning of this chapter. First, he set out to explain the possibility of things in general in the true way, which he considered to be unique; that is, by deduction from *a priori* principles such as God's immutability and the laws of nature. This meant that causes had always to be sought in the effects of impulsion of matter in motion. Having established the *a priori* principles, he then set out to formulate hypothetical explanations (in accordance with these principles) for particular phenomena. Since he believed that many explanations were possible for the same particular effect, it sufficed, in each case, to offer one that worked. This may perhaps account for Descartes's apparent disregard of the inconsistencies between the hypotheses introduced to explain different phenomena.

Descartes's hypotheses were qualitative and it seems that his intention was only to give a general indication of the causes of phenomena. Thus in a letter of 1648 or 1649 to an unknown correspondent, he remarked that in the *Principia philosophiae* he had not described all the motions of each planet, but he had assumed all those that the observers had found and he had attempted to explain their causes. His description, however, refers only vaguely to "divers apogees or aphelions, and perihelions or perigees". Descartes nowhere refers to any of Kepler's laws and there is indeed no evidence that he even knew them in any precise sense.

Vortices from Descartes to Newton

The first work on Cartesian physics written by a disciple was the *Fundamenta physices* (Amsterdam, 1646) of Henricus Regius (1598–1679), which appeared in two further editions under the title *Philosophia naturalis* (1654 and 1661) and also in a French translation *Philosophie naturelle* (Utrecht, 1686). Although Regius followed the Cartesian explanations of phenomena, he regarded these simply as mechanical hypotheses without dependence on *a priori* general principles. The Cartesian metaphysics was relegated to the end of the book, where it appeared in a section on the mental functions of the human mind. In the preface to the French translation of his *Principia philosophiae*, Descartes felt himself obliged to point out the fundamental errors of the *Fundamenta physices* and to disown Regius as a disciple. Emphasizing again that in his view physics must be founded on metaphysics, Descartes asked of his readers that they should not attribute any opinion to him unless they could find it in his own writings, and that even then they should not accept it as true unless they could see clearly that it was deduced from true principles. The separation of physics and metaphysics effected by Regius set a fashion, however, that was followed by other disciples of Descartes, such as Jacques Rohault (1620–72), whose *Traité de physique* (1671), in numerous editions, became the standard work on Cartesian physics.

When Clerselier published the first volume of the *Lettres de Descartes* in 1657, he noted in the preface that the Cartesian doctrines were taught in the universities of Holland and hoped that they would have even greater success in France. It was not, however, in the universities but in the informal assemblies of philosophers which took place regularly in Paris that Cartesianism was first received with enthusiasm in France. H.L. Habert de Montmor continued the meetings which had been conducted by Mersenne. In a letter of 1660, Huygens refers to his participation in these Tuesday meetings, whose members (about twenty or thirty in number) included Clerselier, Géraud de Cordemoy and Rohault. In the period immediately before his death, Pierre Gassendi (1592–1655) had attended Montmor's meetings, where he wrongly attributed the idea of the vortex to Epicurus and claimed that Descartes had derived his inspiration from this source. An opponent of Descartes, Gassendi himself revived the doctrine of atomism but he regarded the cause of the motion of the atoms to be internal and of the nature of mind. It was possible to regard Gassendi's atomism and the corpuscular physics of Descartes as essentially the same; this was the view taken, for example, by Robert Boyle.

The most popular of the informal assemblies seems to have been the 'Wednesdays' of Rohault, vividly described by Clerselier in 1659 in the preface to the second volume of the *Lettres de Descartes*, where he remarked that ladies often took the first rank. These were the *femmes savantes* satirized by Molière. Descartes's *Le Monde* and *Traité de l'homme*, early works on the mechanical view of physical nature and human physiology respectively, which he had suppressed following the condemnation of Galileo in 1633, were published in Paris in 1664 at the same time as the *Opera philosophica Renati Descartes* in Amsterdam, while Clerselier brought out his third volume of the *Lettres de Descartes* in 1666.

Rohault, who was Clerselier's son-in-law, described his own scientific method in *Traité de physique* as follows:

. . . to discover what can be the nature of a subject, it is simply necessary to seek to find in it something that can serve to render account of all the effects of which experience makes clear to us that it is capable . . . and if that which we have supposed or established for explaining the particular nature of a subject does not satisfy all that appears to us, or even is found evidently contrary to a single experience, we must estimate our conjecture or thought absolutely false.

Rohault abandoned the *a priori* or metaphysical part of Descartes's scientific method. The difference of approach may be illustrated by an example. Whereas Descartes had deduced his first law of nature (in effect the principle of inertial motion) *a priori* from the axiom of God's immutability, Rohault preferred to find an *a posteriori* proof in the fact that observations of the heavens over thousands of years had failed to reveal any cessation of motion. While Rohault concerned himself with the explanation of particular phenomena, so that the Cartesian system is replaced by a collection of mechanical models, he did follow Descartes (at least in practice) in supposing that there was only one way of explaining the possibility of things in general; that is, by the impulsion of matter in motion. For the details of the explanations of the planetary motions, he followed Descartes closely.

For a few months in 1665 Pierre Sylvain Régis (1632–1707) received instruction from Rohault before setting up a similar course of lectures on Cartesianism in Toulouse, then in Montpellier and finally in 1680 as Rohault's successor in Paris. After some delay caused by ecclesiastical opposition, his *Système de philosophie* was published in Paris in 1690. This work was a complete system of natural and moral philosophy and appeared to the reviewer in the *Journal des sçavans* to show the influence of Gassendi as well as that of Descartes. Régis tended to an empiricist theory of knowledge which supported Rohault's view of physics as hypothetical. According to Régis, however, physics should be systematic, so that the several hypotheses advanced to explain particular phenomena should be consistent with each other. As we have seen, this ideal was not always attained by Descartes himself, while Claude Perrault (whom Régis may have had in mind) claimed the right to consider particular problems in isolation. In seeking the cause of terrestrial gravity, for example, he believed himself to be justified in confining his attention to "the globe that we inhabit".

Nicolas Malebranche (1638–1715), the most original Cartesian after Descartes, took a further step towards positivism in eliminating the idea of efficient causation from physics. If, as Descartes supposed, motion was a state or mode of a body, then it could not also be in Malebranche's view an independent entity capable of passing from one body to another. Consequently, impact could not be the true cause of changes in motion. Malebranche found a solution in the doctrine of Occasionalism, which had been introduced by Louis de la Forge and Cordemoy following a reading of Descartes's *Traité de l'homme*. According to Malebranche's interpretation, the impact served as the occasion for God (the true cause) to bring about the change. The 'moving force' of a body was simply "the will of God which conserves it successively in different points". Nevertheless the occasional causes were to be sought in impact, for Malebranche held that "it is by the motion of visible or invisible bodies that all things happen". Another innovation of Malebranche was the replacement of Descartes's globules by small elastic vortices, which enabled him to give plausible accounts of the phenomena of heat and light. The elasticity of the small vortices was derived from the centrifugal force of their rapid internal circulations.

The planetary motions

In the period between Descartes and Newton the Cartesians contributed nothing new to the explanation of the motions of the planets. Like Descartes himself, they neglected Kepler's laws but described the lunar variation and its application to the explanation of the monthly inequality in the tides. For example, in the first five editions of his *Recherche de la vérité* (1674–1700), Malebranche states that the periodic times of the planets are not "entirely in the proportion of their distances". Only in the sixth edition (1712) did he refer to Kepler's third law; a marginal note shows his source to have been Huygens's *Cosmotheoros* (1698).

On one occasion a reference was made to Kepler's elliptical orbit in a predominantly Cartesian assembly. For at a meeting of the Abbé Bourdelot's Academy in 1677, M. de Castelet expressed concern that false propositions were accepted uncritically on the authority of Descartes, and in particular what he supposed to be Descartes's explanation of the elliptical orbit. Mistaking Descartes's explanation of the lunar variation (which served to explain the monthly inequality in the tides) for an attempted explanation of the elliptical orbit, he claimed that Descartes's theory was disproved by the occurrence of the Moon's perigee sometimes in the syzygies and sometimes in the quadratures. It is interesting to note that the small inequality in the tides arising from the Moon's elliptical orbit had been observed by Joseph Childrey in 1652. Although de Castelet's lecture, published later in the year in the form of a letter to the Abbé Bourdelot, received a favourable review in the *Journal des sçavans*, it nevertheless failed to draw the attention of the Cartesians to the need for an explanation of the elliptical orbit. A reply was published in two letters by a Cartesian, Claude Gadroys, but even with the help of a letter from Gian Domenico Cassini (Cassini I) setting out the lunar theory and explicitly referring to Kepler's ellipse, he only succeeded in obscuring the issue, claiming that the distance of the Moon could have no effect on the tides.

Terrestrial gravity

The presentation and criticism of papers on the cause of terrestrial gravity was the main business of the Wednesday meetings of the Academy of Sciences between 7 August and 20 November 1669. These papers reflected the differing viewpoints of the members, who had been chosen (as B. le B. de Fontenelle (1657–1757) remarked in his official history) so that one system of philosophy should not dominate to the exclusion of others. Gilles Personne de Roberval (1602–75) outlined the positivist position, which simply accepted the fact that bodies were heavy, leaving aside any consideration of the cause. If, however, he were asked to speculate beyond the evidence of the senses, he thought the opinion that gravity was a natural attraction to be the most probable. Bernard Frenicle de Bessy and Edme Mariotte defended the Platonic view of gravity as a natural inclination of parts of the same whole to join together. Claude Perrault proposed an explanation in which the aether circulated naturally in two independent vortices whose axes were perpendicular and whose layers increased in speed with distance from the Earth. According to Perrault, heavy bodies were deflected downwards towards the more slowly moving layers by virtue of their resistance to motion. Quite clearly, this hypothesis was based on Aristotelian principles rather than those of Descartes. In particular, the centrifugal force of the aether, which is an essential element in Descartes's theory, is completely lacking. Only Huygens defended the Cartesian theory. This he did in a highly original manner, his principal innovation being the introduction of a quantitative element. Following an experimental investigation of a water vortex in which he sought to establish that the circulation of the aether could, in principle, explain the effect of gravity, Huygens introduced his new hypothesis with the intention of removing the principal defects of Descartes's theory. Although designed initially as an explanation of terrestrial gravity, Huygens's

hypothesis was later also applied to the problem of the motions of the planets in the solar vortex, both by G.W. Leibniz and himself, when they had accepted Newton's demonstrations of the existence of a *vis centripeta* or gravity of the planets towards the Sun.

An experiment to illustrate his theory of gravity had been described by Descartes himself in his letter of 16 October 1639 to Mersenne. Having filled a round vessel with small lead shot and a few pieces of wood or stone, he rotated the vessel rapidly, when he observed that the lead shot (representing the aether) pushed the larger pieces of wood or stone towards the centre. The weakness of the experiment was that it failed to represent the greater speed of the aether compared with terrestrial bodies, which was an essential feature of the Cartesian explanation. In its final form, Huygens's experiment demonstrated that the greater speed of the aether could in principle explain the fall of heavy bodies towards the Earth. Having filled a cylindrical vessel with water, he placed in it a small globe of the same density as the water, constrained by strings attached to the vessel in such a way that it was free to move only radially. He then rotated the vessel and brought it suddenly to rest, when he observed that the water continued to circulate but the globe, prevented from circulating by the strings, moved towards the centre.

A satisfactory hypothesis would have to explain first, for what reason heavy bodies (like the globe in the experiment) were prevented from circulating with the aether and, second, why bodies fell towards the centre of the Earth and not towards the axis of rotation. Describing his hypothesis, which he published in 1690 in his *Discours sur la cause de la pesanteur*, Huygens declared:

I suppose that in the spherical space which includes the Earth and the bodies which surround it, up to a very great extent, there is a fluid matter which consists of very small parts, and which is diversely agitated in all senses, with much rapidity. Which matter not being able to leave this space, which is surrounded by other bodies, I say that its motion must become in part circular about the centre; not such however that it turns all in the same sense, but so that the greater part of its different motions are made in spherical surfaces about the centre of the said space, which by this also becomes the centre of the Earth.

Heavy bodies, he explained, were prevented from moving horizontally by the rapidity with which the impulsions in different directions followed one another, while the smallness and mobility of the parts of the aether enabled it to sustain all the circulations in different directions (on spherical surfaces) like boiling water.

The weight of a body, Huygens supposed, was equal to the centrifugal force of a quantity of aether, equal in volume to its solid parts. Applying his theorems on centrifugal force, he deduced that the speed of the aether needed to produce the effect of gravity was seventeen times the speed of a point on the equator due to the Earth's diurnal motion. This extreme speed of the aether enabled Huygens to explain the constant acceleration of falling bodies. For the speed of fall (at least in the cases within our experience) being negligible compared with that of the aether, these bodies were subject to a constant force. Another example of the explanatory power of Huygens's theory was provided by Jean Richer's observations in 1672 of the variation in the length of the seconds-pendulum oscillating at Cayenne on the equator and at Paris. According to Huygens's theory, the diurnal motion of the Earth took away $\frac{1}{289}$ part of the weight of a body at the equator and this was consistent with Richer's observations. Another prediction of his theories of gravity and centrifugal force was that, assuming the Earth to be a perfect sphere, the plumb-line at Paris should be inclined at 5′ 54″ to the vertical. From the absence of any observable deviation, he deduced the Earth to be an oblate spheroid.

While the Cartesian theory of gravity in the quantitative form developed by Huygens had enabled him to make predictions leading to new discoveries, the Cartesian theory of the planetary motions had not been improved in any way by the followers of Descartes and was too lacking in explanatory detail to be of any use to practical astronomers, whose primary aim was the precise prediction of planetary positions. Astronomers in fact tended to reject physical theories and confine themselves to the traditional role of 'saving the appearances'. Those who included physical hypotheses in their expositions generally favoured a mixture of ideas derived from Kepler and Roberval, though the influence of the Cartesian vortex theory is also sometimes evident. For example, the Englishman Vincent Wing, who in his *Harmonicon coeleste* (1651) had stated that "the motion of the Planets is essentiall, and given them by God . . . in the beginning", in his last work, *Astronomia Britannica* (1669) added some axioms which amount to a mechanical hypothesis, in which the 'celestial matter' is carried around 'in the manner of a vortex' by virtue of the motion of the Sun about its axis. Both Giovanni Alphonso Borelli and Robert Hooke, whose works were known to Isaac Newton, were influenced mainly by Roberval, though Hooke took from Descartes the idea that a comet "passes from one part of the heaven to another, and so passes through the spheres of the activity of multitudes of central bodies".

The key to the popular appeal of the Cartesian vortex theory may perhaps be found in the promise that it offered of an intelligible physics of the planetary motions based on the mechanics of the artisan. Fontenelle, in his *Entretiens sur la pluralité des mondes* (1686), the first literary masterpiece having for its aim the popularization of science, compared nature to an opera. He imagines the philosophers present at a performance where they see Phaeton lifted up by the winds without discovering the wires to which he is attached or knowing anything of what is done behind the scenes. In turn the philosophers give their opinions. One attributes his rise to a 'hidden virtue' (a form of explanation ridiculed by Molière in the *virtus dormitiva* of opium in his *La Malade imaginaire*), another to his love for the top of the theatre, where alone he is happy. Yet another declares that he is made of certain numbers that make him rise. But finally Descartes and some of the moderns explain that "Phaeton rises because he is drawn up by wires, and that a greater weight than he descends". Thus, Fontenelle concludes: "we no longer believe that a body can move, unless it is drawn, or rather pushed by another body; we no longer believe that it can rise or descend, except by the effect of a counter-weight or a spring." Quite clearly, Fontenelle (who published his work anonymously) intended his readers to have no doubt that the practical common-sense approach of Descartes was eminently reasonable and far preferable to the fanciful explanations of the traditional philosophies.

Further reading

E.J. Aiton, *The Vortex Theory of Planetary Motions* (London and New York, 1972)

R. Descartes, *Principles of Philosophy*, transl. by V.R. Miller and R.P. Miller (Dordrecht, 1983)

R. Descartes, *Oeuvres de Descartes*, ed. by C. Adam and P. Tannery (Paris, 1897–1910)

R. Dugas, *La Mécanique au XVIIe siècle* (Neuchatel, 1954)

E. Gilson, *Études sur le rôle de la pensée mediévale dans la formation du système Cartésien* (Paris, 1951)

P. Mouy, *Le Développement de la physique Cartésienne* (Paris, 1934)

J. Pelseneer, Gilbert, Bacon, Galilée, Képler, Harvey et Descartes: leurs relations, *Isis*, vol. 17 (1932), 171–208

J.F. Scott, *The Scientific Work of René Descartes* (London, 1952; reprinted 1976)

Robert S. Westman, Huygens and the problem of Cartesianism, in *Studies on Christiaan Huygens*, ed. by H.J.M. Bos *et al.* (Lisse, 1980), 83–103

12

Magnetical philosophy and astronomy from Wilkins to Hooke

J.A. BENNETT

Among English natural philosophers of the mid-seventeenth century there was a strong non-mechanical current of thought that informed much of their speculation in cosmology. John Wilkins (1614–72) identified this distinctive natural philosophical tradition, when he wrote in 1654 (in *Vindiciae academiarum*) that among the novel theoretical systems that had "strenuous Assertours" in the University of Oxford, were "the Atomicall and Magneticall in Philosophy, the Copernican in Astronomy &c". If the 'Atomicall' represented the growing influence of mechanism, 'Magneticall' derived from a slightly older novelty – "the Magnetic Philosophy", as Kepler had put it, "of the Englishman William Gilbert".

Earlier in this volume (see Chapter 4) Pumfrey has traced the direct use of the conclusions of Gilbert (1544–1603) regarding "the great magnet of the Earth" – her magnetic properties of polarity, directionality and motion – in cosmology, and has shown that by the mid-seventeenth century this kind of cosmological speculation was in rapid decline. There remained a residual influence whose hold was most tenacious in England, where Gilbert's (and Kepler's) ideas had always been more prevalent than elsewhere. The conceptual link with Gilbert became more subtle. The idea that unseen influences operated between heavenly bodies – forces of attraction or repulsion – drew support from analogies with magnetism, the great exemplar of invisible influence at a distance, but they were not required to be magnetic in nature. Since the very term 'magnetic' was sometimes given this wider meaning, it is better to say that cosmic influences were not necessarily (though they still might be) identical with the influence of the loadstone. Cosmic attractions or repulsions, however, derived from the magnetical philosophy, and their theoretical use survived a growing commitment to mechanism (the belief that all natural

phenomena arose from mechanical causes).

The magnetical philosophy would always, of course, be non-mechanical, even though the animism of Gilbert had not remained an important component. 'Magnetic' action (whose meaning, as we have said, need not be confined by the loadstone) could be regarded as a natural cause, requiring neither further explanatory reduction nor postulated non-material action, and the metaphysical choice was not between animism and materialism, but between the kinds of reduction to be demanded of a proper materialist explanation. This choice depended on how the concepts of matter and its action were defined, and the survival of the magnetic model in England became more significant as natural philosophy on the Continent grew more fully committed to the rigid definitions of mechanism. Magnetism and mechanism, as defining archetypes of material action, were theoretically incompatible, but human nature proved sufficiently accommodating to allow their fruitful co-existence, even in the theorizing of individuals.

The survival of aspects of the magnetical philosophy in England, co-existent with a growing interest in mechanism, is partly explained by the nature of the English community of natural philosophers, characterized as it was by informal collaborative groups, who met regularly, who discussed, observed and experimented together, and who formed part of a long-standing tradition. The most important early centre for such collaboration was Gresham College in London, where the tradition stretched back to the time of Gilbert and forward to the foundation of the Royal Society. This kind of social organization permitted the formation of shared methods and concepts, practised within a working rather than a written tradition, and maintained through meetings, collaborations and group activity. It was possible for England to preserve distinctive elements in natural philosophy, and to

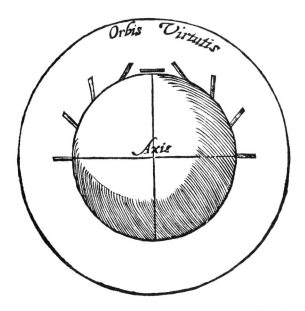

12.1. The *orbis virtutis* that surrounds a spherical loadstone (which serves as a model of the Earth), from William Gilbert's *De magnete* (1600).

witness the prolonged survival of ideas that had become unpopular elsewhere. In 1657 Christopher Wren (1632–1723) delivered a lecture at Gresham College, in which Gilbert was presented as "the sole Inventor of Magneticks" and as a central influence in Kepler's magnetic cosmology and elliptical astronomy, whose perfection was the goal of contemporary astronomy. Wren contrasted Gilbert's positive contributions to cosmology with the negative thrust of Galileo's polemic, and he awarded the title 'Father of the new Philosophy' to Gilbert rather than – specifically – to René Descartes (1596–1650).

Wilkins himself, some time before he identified (in the passage quoted above) the "Atomicall and Magneticall" interests at Oxford, provides an example of the new and more subtle link between Gilbert and contemporary ideas on gravity. His *Discourse Concerning a New Planet* and the second edition of his *Discovery of a World in the Moone*, both of which were published in 1640, draw substantially on Gilbert and on Kepler. The subject of gravity arises in the *Discovery*, because Wilkins discusses the possibility of a journey to the Moon, feasible because gravity is not an inherent property of matter, but is described rather, with strong Gilbertian overtones, as "a respective mutuall de-

sire of union, whereby condensed bodies, when they come within the sphere of their owne vigor, doe naturally apply themselves, one to another by attraction or coition". This attraction, by analogy with the *orbis virtutis* Gilbert had assigned to a terella (a spherical loadstone, serving as a model for the Earth, see Figure 12.1), is effectively contained within a finite spherical boundary, from which a lunarnaut will escape after some twenty miles.

Yet Wilkins does have the idea of a distance-related law. He cites a report in Francis Bacon's *Sylva sylvarum* (1627) that heavy bodies are moved more easily in deep mines than on the surface, and Bacon's suggestion that a similar diminution occurs above the Earth's surface. Wilkins goes on:

... you must not conceive, as if the orbe of magneticall vigor, were bounded in an exact superficies, or as if it did equally hold out just to such a determinate line, and no farther. But ... it is probable, that this magneticall vigor dos remit of its degrees proportionally to its distance from the earth, which is the cause of it.

Further, while the loadstone is the principal explanatory analogue used in discussing gravity, Wilkins is clear that gravity and magnetism are not the same:

This great globe of earth and water, hath been proved by many observations, to participate of Magneticall properties. And as the Loadstone dos cast forth its owne vigor round about its body, in a magneticall compasse: So likewise dos our earth. The difference is this, that it is another kind of affection which causes the union betwixt the Iron and Loadstone, from that which makes bodies move unto the earth.

One of Wilkins's pupils, Walter Charleton (1620–1707), illustrates the uneasy coexistence of 'magnetic' ideas on gravity with the new mechanical philosophy. In the same year as the *Vindiciae academiarum* was published (1654), there appeared Charleton's mechanistic tract *Physiologia Epicuro–Gassendo–Charltoniana*, where gravity is treated at times as an inherent attractive power, at other times as mediated mechanically by material agents. To read that gravity is "a certain Magnetick Attraction of the Earth", or that "the Terrestrial Globe is naturally endowed with a certain Attractive or Magnetique Virtue", and to note Charleton's frequent references to Gilbert, "the Father of Magnetique Philosophy", tends to locate

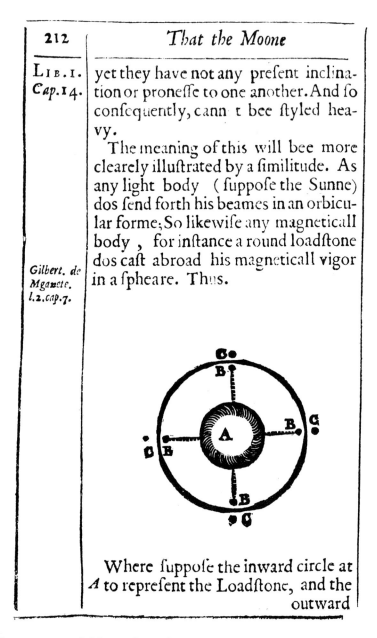

Lib.1.
Cap.14.

yet they have not any prefent inclination or pronefle to one another. And fo
confequently, cann t bee ftyled heavy.

   The meaning of this will bee more
clearely illuftrated by a fimilitude. As
any light body (fuppofe the Sunne)
dos fend forth his beames in an orbicu-
lar forme; So likewife any magneticall
body , for inftance a round loadftone
dos caft abroad his magneticall vigor
in a fpheare. Thus.

Gilbert. de
Mgaxete.
l.2.cap.7.

Where fuppofe the inward circle at
A to reprefent the Loadftone, and the
                                                       outward

12.2. A spherical loadstone, surrounded by a sphere of "magneticall vigor", as illustrated by John Wilkins (1640). Wilkins argued that the Earth's gravity followed this model: that it decreased with distance, and that outside the "sphere of vigor" bodies would be weightless.

Charleton in the 'Magneticall' rather than the 'Atomicall' camp, even though he specifically says that the connection between gravity and the loadstone is only analogical. But Charleton also postulates mechanical explanatory accounts of both the loadstone and gravity, as might be expected of someone presenting himself so positively

as a mechanist in the tradition of Epicurus and Pierre Gassendi.

   This ability to retain attractive or 'magnetic' theoretical components, while moving to a mechanical position under pressure for metaphysical rigour, is noticeable in other English natural philosophers and would prove an especially useful compromise in cosmology. A more thoroughgoing mechanism, where Gilbert was not an important component in a rich conceptual tradition, would not admit the concept of attractive virtue to any explanatory role, however mitigated as a short-hand or temporary expedient.

Two of Wilkins's other protégés, Wren and Robert Hooke (1635–1702), would become critically involved in discussions of a cosmic role for a 'magnetic' gravitation. Wren's interest in Keplerian astronomy dates from his association with the circle who met under Wilkins's patronage in Oxford in the 1650s, and at a technical level, probably derived from his contact with the Professor of Astronomy, Seth Ward. By 1657 Wren thought that "the Perfection of . . . Dioptricks, and the Elliptical Astronomy, seem most worthy our Enquiry". His subsequent lectures at Gresham College included *Praelectiones Greshamenses in astronomiam Kepleri*.

Certainly by 1658 Wren had a first-hand knowledge of Kepler's work, since he published a statement of the correct 'area' version of Kepler's second law. (Ward had advocated the use of an 'equant' approximation.) Wren's publication derived from his interest in a French mathematical challenge to Englishmen, which took the form of a problem Wren considered to be linked with Keplerian astronomy. In an appendix to his solution Wren posed a counter-challenge: 'Kepler's problem', which derived from the area law and to which Wren published a geometrical solution in 1659. From statements made at around the same period, we know that Wren was also interested in Kepler's 'celestial physics', and that he saw Kepler's account of planetary dynamics as derived from Gilbert's magnetical philosophy. We have seen that he contrasted the contributions of Gilbert and Galileo to the progress of the new astronomy: Galileo had adopted a negative approach, by removing the difficulties and anomalies of Copernicanism, but Gilbert had advanced new proofs and opened up its development in new directions.

We also know that Wren was becoming increasingly familiar with the writings of Descartes, and his interest and approval are evidence of a growing awareness of and commitment to mechanism. He suggested to Robert Boyle, for example, observations of the Torricellian tube (a mercury column of the form later adopted as a barometer), as a test of the Cartesian explanation of the Moon's link with the tides – a mechanical pressure transmitted by a material medium. He also undertook an empirical, and later theoretical, study of the laws of impact, held by the mechanists to lie at the root of all natural phenomena.

In Cartesian mechanics the principle of rectilinear inertia meant that a body moving in a curve, as the planets are observed to do, must exhibit a centrifugal tendency, and the thoroughgoing mechanism of Descartes's cosmology led to an account of orbital motion in terms of the competing centrifugal tendencies of bodies moving in a vortex filled with matter. Constrained centrifugal tendencies or forces, and their dynamic interaction, thus played a key role in the explanation of planetary motion.

From a different theoretical tradition, however, Wren was able to suggest a central attractive force as the constraining principle of orbital motion and to take the vital step of analysing planetary motion in terms of two components – as he later recalled to Edmond Halley, "a composition of a Descent towards the sun, & an imprest motion". There is evidence that one clue towards this analysis for Wren may have been Galileo's relation of the 'Platonic' cosmogony, whereby God allowed the planets to 'fall' towards the Sun from a unique point in space and, at the appropriate moments, transformed their rectilinear paths into circular ones. If so, Descartes's analysis of circular motion demonstrated the continuing need for a constraining force, which the Gilbertian cosmological tradition would have identified as a central attractive force.

Wren and Hooke were friends and they collaborated in and discussed together many ventures in mathematical science and natural philosophy, not least their ambitions for reforming planetary dynamics. We can locate fairly precisely the circumstances of their collaboration in cosmology and Hooke's more extensive and explicit accounts are a valuable commentary on the meagre record of Wren's ideas. However, it is clear that Wren's involvement, as he claimed himself, with the explanatory programme combining rectilinear inertia and attractive force was prior to that of Hooke.

From 1662 Hooke was spasmodically involved in investigating variations of gravity with distance from the Earth. The work derived from Bacon's report in *Sylva sylvarum* regarding moving weights in mines, and more specifically from some relevant experiments by Henry Power, reported to the Royal Society. Hooke, as the Society's Curator of Experiments, was instructed to pursue Bacon's inference from the mine reports and to apply Power's method – involving weighing weights suspended down mine shafts – above the Earth. In 1662 and 1664 Hooke weighed bodies suspended from high in Westminster Abbey and the old cathedral of St Paul's – experiments which derived from the Society's instructions (rather than from his own interests), which had no implications beyond a localized study of the Earth's gravity, and which failed to isolate any such variation.

The restricted reference of these early experiments is in striking contrast to the greatly expanded context, with cosmological implications, that Hooke gave the programme of investigating gravity's variation with distance at the very first meeting of the Royal Society (21 March 1665/6) following their recess during the Plague. The contrast locates Hooke's entry into the cosmological debate and raises the question of what had happened to alter his horizons so radically since his previous report of August 1664. Central to this change was a collaboration with Wren over explaining the motion of a comet that appeared in December 1664.

Both Wren and Hooke were present at the Royal Society when the comet's appearance was first reported, and Hooke sent an account to Robert Boyle. It was a spectacular appearance, which aroused widespread interest, and a number of observations and reports were received by the Society. At first the Society asked Hooke to compile a 'history', but in January Wren's interest was such that he was given the responsibility of contributing on the Society's behalf to the growing international discussion. Wren, in Oxford, was in contact with John Wallis (1616–1703), his fellow Savilian Professor (Wren held the chair of astronomy, Wallis of geometry), and Wallis was promoting the ideas of Jeremiah Horrocks, who in the late 1630s had accounted for the motion of the 1577 comet in terms of a rectilinear motion (as postulated by Kepler, see Chapter 10) modified by the magnetic action of the Sun – a combination of an attractive force and a circular motion derived from the Sun's rotation. Early in January 1664/5 Wren proposed to Wallis a geometrical problem aimed at locating a comet's path in space from four observations, on the assumption that its motion is uniform and rectilinear.

On 1 February Wren produced before the Royal Society a 'theory' of the comet's motion – essentially his own solution to the problem he had proposed to Wallis and an application of his construction to the contemporary comet (Figure 12.3). (Hooke later published this construction in *Cometa*.) It is clear from subsequent activities at the Royal Society that this 'theory' was a first attempt and that a fuller account was expected from Wren. In fact Wren began to lose interest as he planned a visit to France – his only journey abroad – and in March the Society turned once again to Hooke, who had been lecturing at Gresham College on the comet since late February.

However, at the very end of March there was a new and unexpected development – the appearance of a second comet. This immediately reawakened Wren's interest: was this the same, or another, but physically related, comet? Both he and Hooke were now caught up in the problem. Wren was occupied with observations in the early weeks of April, and his correspondence with Hooke and with Robert Moray on the subject shows that he was actively working on a theory of comets and that he was not at all committed to a strict rectilinear path. This activity continued at least until May, during which time Wren was in contact with Hooke, but although the Society continued to encourage both men, only Hooke's ideas appeared in print in any fullness.

Hooke's ideas in fact appeared in his *Cometa*, whose publication was stimulated by a comet of 1677, but which contained a large section on the comets of 1664–65, prepared for lectures at Gresham College early in 1665 and for a paper read at a Royal Society meeting in August 1666. The various notions espoused within this section are not at all consistent with each other and have been drawn together from different sources. A careful study of internal and external evidence shows that at an early stage, around March 1664/5, Hooke thought that comets moved in circular paths, but his subsequent conclusion was that their motion was basically rectilinear but modified by gravitational influences. It was late in April that Hooke had become directly concerned with Wren's rectilinear construction of the path of the 1664 comet,

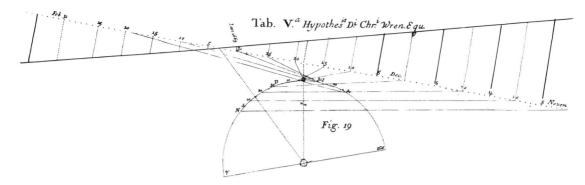

Tab. V.ᵃ Hypothesⁱˢ Dⁱ Chrⁱ Wren.Equ.

Fig. 19

12.3. Wren's 'theory' of cometary motion, applied to the comet of 1664–65. The semicircle represents the orbit of the Earth, the continuous straight line the path of the comet, and the dotted straight line its projection onto the plane of the ecliptic.

and after presenting Wren's account and claiming a qualified success for it, Hooke says:

Now though according to my former Delineation the Comet seemed to take a circuit, as if it would within three years return to its former position, yet I am not wholly convinced that it moves in a circle or Ellipse, but I rather incline to the incomparable Kepler's opinion, that its natural motion tends towards a straight line, though in some other suppositions I differ from him.

For Hooke, or for Wren, a comet's uniform rectilinear motion made physical sense as an unencumbered inertial motion, and any deviations, indicated by a less-than-perfect fit with observations, were naturally explained by gravitational influences of bodies within whose 'magnetic' range the comet might pass. The comet posed for Hooke, in a particularly dramatic manner, the fundamental dynamical components – a rectilinear inertia and a central attractive force – of planetary orbital motion, which he came to see as a special case of his cometary theory:

. . . particularly by tracing the way of this Comet of 1664. it is very evident that either the observations are false, or its appearances cannot be solved by that supposition, without supposing the way of it a little incurvated by the attractive power of the Sun, through whose system it was passing, though it were not wholly stayed and circumflected into a Circle.

The comet also brought Hooke into direct, relevant discussion with Wren, who had already been engaged with the problem. There are other related aspects of their shared interests around this time; in particular, as we shall see, the motion of the conical pendulum. It is also worth noting that during the Society's recess caused by the Plague of London, Hooke had performed experiments on the variation of gravity in mines, in collaboration with Wilkins.

We should notice how freely Hooke uses notions of attractive force in *Cometa* – a freedom hardly compatible with his avowed mechanism. He often invokes central attractive forces, reinforced by magnetic analogies:

. . . I suppose the gravitating power of the Sun in the centre of this part of the Heaven in which we are, hath an attractive power upon all the bodies of the Planets, and of the Earth that move about it, and that each of those again have a respect answerable, whereby they may be said to attract the Sun in the same manner as the Load-stone hath to Iron, and the Iron hath to the Load-stone.

Hooke frequently asserted and demonstrated his commitment to what he had called in the *Micrographia* (1665) "the real, the mechanical, the experimental Philosophy". He therefore needed to be able to 'launder' his attractive forces from time to time by appealing to a mechanical explanation. Hooke specifically distinguishes his mechanism from previous accounts of gravity on Earth (Aristotle's innate qualities, magnetic forces, Henry More's "Hylarchick Spirit", the effect of a rotating aether) and from previous explanations of planetary motion (intelligences, solid orbs, magnetism and vortices). He postulates a mechanical vibrative motion of the parts of the Earth, communicated to other bodies – close to the surface and beyond – through a subtle aether, and these bodies respond if their own vibrative motions are in harmony with those of the Earth. Thus in place of the circulating aether of Descartes, Hooke proposes a stationary but vibrating aether, the vibration falling off with distance. But the function of this explanation is largely symbolic: it exists to preserve Hooke's credentials as a mechanist. At the level of working explanations, Hooke is happy to use attractive forces.

The broader implications of Hooke's experiments on variation in gravity, as presented to the Royal Society in March 1665/6, can now be understood:

Gravity, tho' it seems to be one of the most universal active principles in the world, and consequently ought to be the most considerable, yet has it had the ill fate, to have been always, till of late, esteemed otherwise, even to slighting and neglect. But the inquisitiveness of this latter age hath begun to find sufficient arguments to entertain other thoughts of it.

He then surveys the recent ideas of Gilbert, Bacon and Kepler, and asks whether gravity "be magnetical, electrical, or of some other nature distant from either". He proposes to compare the variation of gravity with distance with the results of a parallel series of experiments on magnets:

. . . so if this analogy between the decrease of the attraction of the one, and of the gravity of the other, be found real, we may perhaps by the help of the load-stone, as it were, epitomise all the experiments of gravity, and determine, to what distance the gravitating power of the earth acts; and explicate perhaps divers other phaenomena of nature by ways not yet thought of.

Hooke's magnetical experiments now began, but new urgency was given to his cosmological programme by a paper from Wallis, read to the Royal Society in May. Wallis proposed to complete Galileo's derivation of tidal phenomena from the motion of the Earth (see Chapter 6), by adding a small epicyclic motion about the centre of gravity of the Earth and Moon. His idea was that the primary orbital trajectory was not traced out by the centre of the Earth, but by her common centre of gravity with the Moon, so that the lunar motion was reflected in a small motion of the Earth.

An important question that arose from Wallis's hypothesis was how the Earth and Moon, having no physical connection, could move as one body.

Wallis maintained that he remained agnostic on the cause of such a connection – "whether by any Magnetick, or what other Tye, I will not determine; nor need I, as to this purpose" – but he naturally claimed support from the magnetical analogy:

. . . it is harder to shew How they have, than That they have it. That the Load-stone and Iron have somewhat equivalent to a Tye; though we see it not, yet by the effects we know. . . . How the Earth and Moon are connected; I will not now undertake to shew (nor is it necessary to my purpose;) but, That there is somewhat, that doth connect them, (as much as what connects the Load-stone, and the Iron, which it draws,) is past doubt to those, who allow them to be carryed about the Sun, as one Aggregate or Body, whose parts keep a respective position to one another.

It was following the reading of Wallis's paper at the Royal Society that Hooke "mentioned . . . that the motion of the celestial bodies might be represented by pendulums", and was asked to demonstrate this at the following meeting.

It was in this way that Hooke came to read to the Royal Society his famous paper "concerning the inflection of a direct motion into a curve by a supervening attractive principle". He was now specifically concerned with planetary dynamics:

I have often wondered, why the planets should move about the sun according to Copernicus's supposition, being not included in any solid orbs . . . nor tied to it, as their center, by any visible strings; and neither depart from it beyond such a degree, nor yet move in a straight line, as all bodies, that have but one single impulse, ought to do.

For Hooke the answer lay in the continuous attractive influence of the central body, a supposition that would bring the study of planetary motion within the compass of ordinary mechanics, and that would end in a rigorous, predictive account of the solar system:

For if such a principle be supposed, all the phaenomena of the planets seem possible to be explained by the common principle of mechanic motions; and possibly the prosecuting this speculation may give us a true hypothesis of their motion, and from some few observations, their motions may be so far brought to a certainty, that we may be able to calculate them to the greatest exactness and certainty, that can be desired.

It is significant that Hooke mentions also that "By this hypothesis, the phaenomena of the comets as well of the planets may be solved."

This was the first public statement of his analysis of circular motion – "compounded of an endeavour by a direct motion by the tangent, and of another endeavour tending to the center". The celestial bodies "might be represented by pendulums", because a conical pendulum demonstrated the combination of a tangential impulse and a central restoring force, and at the meeting Hooke used such a pendulum to generate circular and elliptical 'orbital' motions. The link with Wallis's paper on the common centre of gravity of Earth and Moon was explained by Hooke attaching a smaller pendulum to the first one, the larger bob to represent the Earth, the smaller the Moon. The primary trajectory – circular or elliptical – was then performed by neither bob, but ". . . a certain point, which seemed to be the center of gravity of these two bodies, howsoever posited (considered as one) seemed to be regularly moved in such a circle or ellipses, the two balls having other peculiar motions in small epicycles about the said point".

That Hooke immediately proffered the conical pendulum model, on hearing Wallis's paper, indicates that the analogy was probably already familiar to him, and elsewhere he dated his interest in the conical pendulum to 1665. Its use as a model of compounded planetary motion was firmly attributed to Wren by Thomas Sprat, who said in his *History of the Royal Society* (1667) that Wren had discovered that a pendulum

. . . would continue to move either in Circular, or Elliptical Motions . . . and that by a complication of several Pendulums depending one upon another, there might be represented motions like the Planetary Helical Motions, or more intricate: And yet that these Pendulums would discover without confusion (as the Planets do) three or four several Motions, acting upon one Body with differing Periods.

Corroboration of Wren's priority over Hooke, with respect to the conical pendulum, comes from Henry Oldenburg, who, after the publication of Christiaan Huygens's *Horologium oscillatorium* in 1673, told Huygens that Hooke had demonstrated the properties of the conical pendulum to the Royal Society some years earlier and that Wren had discussed it before Hooke ("devant lui", so possibly "in front of him"), as several people were prepared to testify. Wren had also discussed with Huygens in 1661 (or possibly 1663) the application of the conical pendulum to a clock.

If the importance of Hooke's friendship and collaboration with Wren is enhanced, it is worth remembering also Wren's links with Wallis early in 1664, when they were discussing the rectilinear motion of comets. We noted that Wallis was publicizing the ideas of Horrocks on cometary motion, and Wren was associated with him in the Royal Society's current interest in Horrocks's astronomical manuscripts. In the very letter where Horrocks first suggests that comets are emitted from the Sun in straight lines – a letter referred to by Wallis – he also uses the conical pendulum as a model of elliptical planetary motion. Wren's understanding of the dynamics of the pendulum's motion was by then very different from that of Horrocks, but Horrocks may well have been influential in turning it into an explicit planetary model, which Wren developed by substituting for a single pendulum "a complication of several Pendulums depending one upon another".

Questions of priority are unimportant and probably irresolvable; we have access to only a small selection of the relevant exchanges. It is important, however, that even the incomplete record demonstrates that the magnetical tradition in cosmology was a common resource among those principally concerned – Wilkins, Wallis, Wren and Hooke.

We have seen that Hooke's *Cometa* was imbued with the magnetical analogy. So too was his most famous statement of his programme for a 'System of the World' – and the first one to be published – made in another of the *Lectiones Cutlerianae*; namely, the *Attempt to Prove the Motion of the Earth*, which first appeared in 1674. Here Hooke organized his proposal into three 'suppositions'. According to the first, "all Coelestial Bodies whatsoever, have an attraction or gravitating power towards their own Centres", and this attraction not only holds their parts together, but extends to other heavenly bodies "that are within the sphere of their activity". This concept of a 'sphere of activity' can be traced through Wilkins's 'sphere of vigor' to Gilbert's *orbis virtutis*. The second supposition is that of rectilinear inertia, and the third states that the attraction decreases with distance, but according to a function which, Hooke says, "I have not yet experimentally verified". The experiments, of course, were those begun back in 1662. Hooke reveals another step in the development of his programme, when he says: "He that understands the nature of the Circular Pendulum and Circular Motion, will easily understand the whole ground of this Principle." But as before it was a prospect for the future – an unfulfilled promise of "a System of the World ... answering in all things to the common Rules of Mechanical Motions".

Hooke did take one further step, again very possibly in association with Wren. Huygens, of course, stood outside the magnetical tradition, so that his discussion of circular motion in *Horologium oscillatorium* was in terms of constrained centrifugal force. The expression he derived for this force (velocity squared divided by radius) could, however, be applied to the centripetal attractive force required by Hooke and Wren. The derivation of an inverse-square force law then became a simple substitution, invoking what we call Kepler's third law and the simplifying assumption that planetary orbits are circular. Both Wren and Hooke, as good Keplerians, knew this assumption was inadmissible, and as Hooke's diary shows, they discussed the problem from time to time in the 1670s and the early 1680s.

Isaac Newton recalled how in about 1677 he and Wren had "discoursd of this Problem of Determining the Hevenly motions upon philosophicall principles", a discussion which had left Newton with the impression that Wren was working on the assumption of an inverse-square force law. Newton's exchange with Hooke, in a series of letters written between 1679 and 1680, is more famous and better documented. Like Huygens, Newton had stood outside the magnetical tradition. Unlike Wren and Hooke, he had not become involved in natural philosophy through joining an established working tradition and discussing and collaborating with its members. Newton's social isolation meant that his principal contact with a natural philosophical alternative to the Aristotelian curriculum of Cambridge, came from reading the books of the mechanical philosophy, and his early thoughts about orbital motion were informed by Cartesian vortices and the dynamics of constrained centrifugal force. Hooke was able to place before him – and he did so emphatically in his letters – a different dynamical picture, that of a centripetal, inverse-square force continuously attracting a body away from its rectilinear inertial motion. It was a formula that Newton would turn to powerful effect in the *Principia*, but which sprang from the residual influence of an unfamiliar tradition, the magnetical philosophy.

Further reading

J.A. Bennett, Hooke and Wren and the system of the world, *British Journal for the History of Science*, vol. 8 (1975), 32–61

J.A. Bennett, Cosmology and the magnetical philosophy, 1640–1680, *Journal for the History of Astronomy*, vol. 12 (1981), 165–77

R. Hooke, *Lectiones Cutlerianae* (London, 1679; reprinted by R.T. Gunther in *Early Science in Oxford*, vol. 8 (Oxford, 1931))

R. Hooke, *The Posthumous Works*, ed. by R. Waller (London, 1705; reprinted New York, 1969)

F.R. Johnson, *Astronomical Thought in Renaissance England* (Baltimore, 1937)

A. Koyré, *Newtonian Studies* (London, 1965)

P.D. Lawrence and A.G. Molland, David Gregory's inaugural lecture at Oxford, *Notes and Records of the Royal Society of London*, vol. 25 (1970), 143–78

S. Ward, *Vindiciae Academiarum* (Oxford, 1654; reprinted by Allen G. Debus, *Science and Education in the Seventeenth Century: The Webster–Ward Debate* (London, 1970))

R.S. Westfall, Hooke and the law of universal gravitation, *British Journal for the History of Science*, vol. 3 (1966–67), 245–61

R.S. Westfall, *Force in Newton's Physics* (London, 1971)

D.T. Whiteside, Newton's early thoughts on planetary motion: a fresh look, *British Journal for the History of Science*, vol. 2 (1964), 117–37

J. Wilkins, *The Mathematical and Philosophical Works of the Rt Rev. John Wilkins* (2nd edn, London, 1802; reprinted London, 1970). This includes his *A Discourse Concerning a New Planet* and *The Discovery of a New World*

PART IV

The Newtonian achievement in astronomy

13

The Newtonian achievement in astronomy
CURTIS WILSON

The principal foundation of the predictive astronomy of the planets and satellites in the solar system, as developed during the past three centuries, is the law of universal gravitation; its discoverer was Isaac Newton. Born on Christmas Day 1642 (OS) in Woolsthorpe, Lincolnshire, Newton received his earliest education in his own village, then at age twelve entered the grammar school in Grantham, where he studied Latin, Greek, and probably some mathematics – how much is unknown. On 5 June 1661, at age eighteen, he was admitted to Trinity College, Cambridge. Here in self-directed study he encountered the new mathematics of the day and the new natural philosophy; soon he was making contributions of his own to both. In 1669 he succeeded Isaac Barrow in the Lucasian chair of Mathematics; he would resign that position only in 1701, five years after he had left Cambridge to become Warden of the Mint. His *Principia mathematica philosophiae naturalis*, in which the law of universal gravitation was first announced to the world, appeared in 1687. But when, and how, did he discover this law?

Until recently, it was the received account that Newton first hit on this idea in 1666, and that it came to him as a simple generalization, based on the comparison of the fall of a stone or apple at the surface of the Earth with the 'fall' of the orbiting Moon away from a rectilinear path and toward the Earth's centre. Today we know that such an account is deeply misleading. There is no evidence that Newton before 1679 seriously entertained the idea that the heavenly bodies mutually attract one another over the vast distances of interplanetary space, and no evidence that before late 1684 the fundamental argument for universal gravitation, as it would be presented in the *Principia*, had occurred to him. In his thoughts about planetary motion, Newton before 1679 assumed aethereal vortices to carry the planets about, and accepted the Cartesian account of a centrifugal force developed in circular motion. The possibility that the Earth's gravity might extend to the Moon he did indeed consider; but this possibility was connected in his mind with a hypothetical aethereal mechanism to counteract the centrifugal tendency of the orbiting bodies, and not with universal gravitation.

In brief, Newton before 1679 was so far from having entertained the idea of universal gravitation that he had not yet considered the orbital paths produced by different laws of centripetal attraction as a mathematical problem. This problem was posed to him by Robert Hooke (1635–1703) in 1679. Still, in the five years following, there is no sign that Newton regarded universal gravitation as the obvious generalization it has since been taken to be. Only in late 1684 did the fundamental argument for universal gravitation begin to take shape in Newton's mind, enabling him to proceed with confidence in the writing of the *Principia*. In the first section of this chapter I shall seek to indicate the course of Newton's thinking insofar as it can be reconstructed up to this point.

If the discovery of a powerful argument for universal gravitation came as a relief to Newton, providing as it did a relatively secure basis for further deduction, it imposed at the same time an enormous burden on the mathematical astronomer. As Newton put the matter in an early revision of the tract *De motu* that was to grow into the *Principia*:

By reason of [the] deviation of the Sun from the centre of gravity [of the solar system] the centripetal force [on a planet] does not always tend to that immobile centre [that is, the Sun considered as immobile], and hence the planets neither move exactly in ellipse[s] nor revolve twice in the same orbit. So that there are as many orbits to a planet as it has revolutions, as in the motion of the

Moon, and the orbit of any one planet depends on the combined motion of all the planets, not to mention the action of all these on each other. But to consider simultaneously all these causes of motion and to define these motions by exact laws allowing of convenient calculation exceeds, unless I am mistaken, the force of the entire human intellect.

The second section of this chapter will outline Newton's efforts to cope mathematically with the astronomical consequences of universal gravitation: to derive the modification of Kepler's third law required because the planet attracts the Sun as well as the Sun the planet; to prove that spheres attract as though their masses are concentrated at their centres; to account for the chief perturbations in the motions of the Moon; to explain the precession of the equinoxes on the assumption that the Earth is a spheroid flattened at the poles; and to fit parabolic trajectories to the observations of comets.

Here we must both admire Newton's ingenuity and penetration, and regret limitations in the mathematical techniques at his disposal. G.B. Airy in 1834 characterized section 11 of Bk 1 of the *Principia*, in which Newton undertakes to deal with the perturbational problem, as "the most valuable chapter that has ever been written in physical science", on the grounds that Newton's geometrical procedures afford an insight into the relation between dynamical causes and geometrical effects that could not be reached in any other way. But the unavailability of analytical devices that would be introduced only later, such as trigonometric series, meant that many of Newton's derivations were qualitative rather than quantitative, and left as an open question for future mathematical astronomers the sufficiency of universal gravitation to account exactly for all the phenomena. Among Newton's quantitative derivations, those of the precession of the equinoxes and of the motion of the lunar apse (the latter not published in Newton's lifetime) were dependent on "fudge factors"; conclusive derivations of these phenomena were first worked out by J. le R. d'Alembert and A.-C. Clairaut, respectively, in the late 1740s.

I The discovery of universal gravitation

On entering Cambridge University in June 1661, Newton acquired a small notebook (now in Cambridge University Library) in which he proceeded to record systematic notes on his studies, or to copy standard "class-notes". The earliest notes are in Greek and pertain to Aristotelian and neoplatonic readings. There follow notes in Latin on sixteenth- and seventeenth-century textbooks of Aristotelian and scholastic branches of knowledge, all of them standard parts of the curriculum of the day.

At some time in late 1663 or early 1664 Newton began the reading of Walter Charleton's *Physiologia Epicuro-Gassendo-Charltoniana* (1654) (see Chapter 12). This was the first in a series of extra-curricular readings that introduced Newton to the new mechanical and corpuscularist philosophies of René Descartes, Thomas Hobbes, Robert Boyle, and others. Charleton's book is largely derived from the writings of Pierre Gassendi (1592–1655), who had done more than anyone else to reconcile the teachings of Epicurean atomism – long stigmatized as atheistic – with Christian orthodoxy. Like Gassendi, Charleton urged that certainty was unattainable in natural philosophy, and that arguments and opinions could at best be probable. In particular, he attacked the Cartesian claim that extension is the essential attribute of matter, and that since all that exists is either substance or attribute, extension empty of matter is a contradiction in terms. The Gassendist and Charletonian position was that space and time are neither substance nor accident, but rather conditions for the existence of anything. At an early stage Newton adopted the atomic doctrine along with the essentials of the Gassendist ontology of space and time, and these doctrines form a basis for all his further inquiries in natural philosophy. With the vague probabilism he found in Charleton, however, he was presumably less happy.

Under the stimulus of reading Charleton, Newton began a new section of his notebook under the title "Questiones quaedam philosophicae"; many of the topics for inquiry – among them gravity and levity, atomism, and the vacuum – are taken directly from Charleton's book.

In the autumn and winter of 1664 Newton began the study of Descartes's *Principia philosophiae* (1644). Here he found much to stimulate his thought and much to criticize; a few years later he would be undertaking explicitly "to dispose of his [Descartes's] fictions". But there are important respects in which Newton accepted and built on Cartesian principles. From Descartes he took over the principle of inertia, following the Cartesian formulation in his articulation of it. From Descartes he adopted the principle that quantity of motion, measured by quantity of matter and speed conjointly, is conserved; and from Descartes he adopted the measure of force as quantity of motion engendered or destroyed in a body undergoing impact. However, Newton departed from Descartes in realizing that the conserved quantity of motion is a vectorial rather than a scalar quantity.

In his *Principia philosophiae* Descartes had formulated rules of impact, all of them quite new and most of them quite wrong; for instance, he had claimed that in a collision between a moving and a stationary body where the latter was the larger, the stationary body would remain at rest while the moving body simply reversed its direction of motion. Newton, treating quantity of motion as vectorial, proceeded to derive the correct rules for both elastic and inelastic impact, some years before 1669 when these rules were first published in writings by Wren, Wallis, and Huygens.

Descartes's premise that space is everywhere filled with matter led him to suppose that motion can only be that of the whole of the matter in a closed path or vortex. It was by means of a vortex in the aethereal matter surrounding the Sun that Descartes proposed to account for the nearly circular revolutions of the planets about the Sun, all in the same direction and nearly in the same plane. By the 1660s and 1670s the hypothesis of planetary vortices was no longer confined to Cartesians, but had become the standard replacement for the circulating "immaterial virtue" of Kepler's celestial physics, and the chief English writers on astronomy during these years, Thomas Streete in his *Astronomia Carolina* (1661) and Vincent Wing in his *Astronomia Britannica* (1669), both explicitly assumed that the planets are carried about the Sun in an aethereal vortex. Newton, who was reading Streete's book in late 1664 and who read Wing's book shortly after it appeared, seems to have made the same assumption; no evidence exists to show that he entertained an alternative hypothesis before 1679.

According to the account of circular or vortical motion given by Descartes in Part III of his *Principia philosophiae*: "Any body which is moving in a circle constantly tends to move [directly] away from the centre of the circle which it is describing." Thus

when a stone is whirled in a sling, the 'endeavour' or *conatus* of the stone to recede is experienced as a tug of the sling on one's hand. Actually Descartes recognized two 'endeavours' in the whirling stone: (1) an endeavour to move along the tangent to the circular path, and (2) an endeavour to recede from the centre (*conatus recedendi a centro*). He appears to have acknowledged that the endeavour from the centre has its origin in the endeavour along the tangent, but nevertheless treated the two endeavours as distinct, and both of them as real.

Newton accepted from Descartes the reality of the *conatus recedendi a centro*. So in his *Waste Book*, wherein from early 1665 he began to set down his reformulations and corrections of Cartesian dynamics, he wrote: "it appeares that all bodys moved circularly have an endeavour from the centre about which they move . . . ." As far as the evidence goes, Newton before 1679 never conceived the dynamics of circular motion differently, never became aware that the *conatus a centro* is only an apparent force, to be associated solely with a rotating frame of reference.

Yet in important respects Newton's thought on circular motion departed from the Cartesian account. In the first place, for Descartes as for Newton a *conatus* was a resisted force; and the *conatus a centro* developed by the planets and satellites in their orbital motions must be opposed by some other force which keeps them from receding. According to Descartes, the resistance is provided by pressure in the vortex itself. Also according to Descartes, aethereal pressure is responsible for the phenomena of light. From early on Newton questioned the consistency of these two suppositions, and doubted that circulating aether could fufil all the functions that Descartes assigned to it. As he wrote in his "Questiones quaedam philosophicae": "Light cannot be by pression & c for then we should see in the night as well, or better, than in the day. We should see a bright light above us because we are pressed downward." In place of the pressure supplied by the vortex, Newton imagined a flux of aethereal matter descending with great rapidity through the vortex into the central body, passing through the pores of all bodies it meets with and bearing them downward. This hypothesis is given an initial formulation in Newton's "Questiones" under the heading "Of Gravity and Levity", and is articulated once again in "An Hypothesis explaining the Properties of Light" which Newton sent to the Royal Society in December 1675. In the latter Newton was primarily concerned to account for the colours of thin films; these could not be "handsomely explained", he thought, "without having recourse to aethereal pulses". The aethereal medium, he proposed, was not "one uniform matter" but rather compounded of various "aethereal Spirits", for "the Electric and Magnetic effluvia and gravitating principle seem to argue such variety". These aethereal spirits may be condensible, so that: ". . . the whole frame of Nature may be nothing but various Contextures of some certaine aethereall Spirits or vapours condens'd as it were by praecipitation. . . . Thus perhaps may all things be originated from aether." Gravitation towards the Earth may be due to a certain aethereal spirit that is condensed in the Earth's body:

. . . so may the gravitating attraction of the Earth be caused by the continuall condensation of some other such like aethereall Spirit, not of the main body of flegmatic aether, but of something very thinly and subtily diffused through it . . .; the vast body of the Earth, which may be every where to the very centre in perpetuall working, may continually condense so much of this Spirit as to cause it from above to descend with great celerity for a supply. In which descent it may beare down with it the bodys it pervades with force proportionall to the superficies of all their parts it acts upon; nature making a circulation by the slow ascent of as much matter out of the bowells of the Earth in an aereall form which for a time constitutes the Atmosphere, but being continually boyed up by the new Air, Exhalations and Vapours riseing underneath, at length . . . vanishes againe into the aethereall Spaces, and there perhaps in time relents, and is attenuated into its first principle. For nature is a perpetuall circulatory worker . . . And as the Earth, so perhaps may the Sun imbibe this Spirit copiously to conserve his Shineing, and keep the Planets from recedeing further from him. And they that will, may also suppose, that this Spirit affords or carryes with it thither the solary fewell and materiall Principle of Light; And that the vast aethereall Spaces between us, and the stars are for a sufficient repository for this food of the Sunn and Planets. But this of the Constitution of aethereall Natures by the by.

Newton later refers to this hypothesis in a letter to Edmond Halley of 20 June 1686, in which he is contesting Hooke's claim to have supplied him with the inverse-square law in 1679. But Newton, as he protested to Halley, already knew of the inverse-square law and had framed his hypothesis of 1675 to be in accordance with it:

. . . I hinted a cause of gravity towards the Earth, Sun & Planets with the dependance of the celestial motions thereon: in which the proportion of the decrease of gravity from the superficies of the Planet (though for brevities sake not there exprest) can be no other then reciprocally duplicate of the distance from the centre.

In his letter to Halley of 27 July 1686, Newton explained how the hypothesis of 1675 implied the inverse-square proportion: ". . . the diminution of [the aether's] velocity in acting upon the first parts of any body it meets with [is] recompensed by the increase of its density arising from that retardation."

That Newton, at the same time that he postulated a circulation of aether into the central body, with a return outward again in a different form, was assuming a vortex around the central body, is made clear at a later point in the "Hypothesis" of 1675, where he observes that different substances may be "sociable" or "unsociable" with one another: "The like unsociablenes may be in aethereall Natures, perhaps in the aethers in the vortices of the Sun and Planets . . . ." Newton's concern here, we may guess, was with the possibility that the vortex about the Sun would interfere with or destroy the vortices about individual planets such as the Earth. The "sociableness" and "unsociableness" referred to would appear to involve attractions and repulsions over small distances. As stated earlier, in none of Newton's writings dating from before 1679 do we find evidence that he entertained the hypothesis of attractions extending over the distances of interplanetary space. That during this period he assumed vortices about the Sun and planets is confirmed by his notes on the endpapers of Wing's *Astronomia Britannica* (1669), in which he attributes the inequalities in the motion of the Moon to the compression of the circumterrestrial vortex by the circumsolar vortex.

While accepting vortices along with the Cartesian account of the dynamics of circular motion, Newton went beyond Descartes in deriving, apparently by 1665 or 1666, a quantitative evaluation of the *conatus recedendi a centro*, equivalent to the formula for centrifugal force published by

Christiaan Huygens in 1673, or to our formula v^2/r. His earliest derivation makes use of the accompanying diagram in Figure 13.1, in which a ball (at *b*) is imagined to be moving along the sides of a regular polygon inscribed in a circle, and to be reflected at the vertices by a polygon of the same number of sides circumscribed about the circle. Newton shows that the "force" of *b* on the barrier (equal to *b*'s change in quantity of motion due to impact) is to the force of *b*'s motion (its quantity of motion along a side of the inscribed polygon) as the side of the latter polygon to the radius of the circle. In a complete cycle the total change of *b*'s motion will be to its momentum along a side as the periphery of the polygon to the radius of the circle. And so "... if [the] body were reflected by the sides of an equilaterall circumscribed polygon of an infinite number of sides (i.e. by the circle it selfe) the force of all the reflections [is] to the force of the body's motion as all those sides (*id est* the perimeter) to the radius". This result implies that the total change in velocity in one revolution is equal to $2\pi v$, where v is the speed of the revolving body. If we divide by the period $T(= 2\pi r/v)$, we obtain for the instantaneous change in velocity the familiar formula v^2/r. In evaluating the acceleration, Newton more often uses the expression r/T^2, which is proportional to v^2/r.

In a manuscript written probably toward the end of the 1660s and now in Cambridge University Library, Newton made a number of applications of this result. First, he compared the acceleration of gravity with the *conatus recedendi a centro* of a body at the Earth's equator. From the measured period of a simple pendulum of given length, he concluded that gravity would cause a body to fall 196 inches in the first second of descent – a rather accurate value. But the *conatus recedendi* of bodies at the equator owing to the Earth's rotation, Newton found by his formula, would lead to a rise of only $\frac{5}{9}$ of an inch in the first second of motion. He concluded that the catastrophic consequences of the Earth's rotation imagined by anti-Copernicans would not occur.

In the same manuscript Newton also carried out a comparison of the force of gravity at the surface of the Earth with the Moon's endeavour to recede from the Earth, using the *conatus recedendi* of an object at the Earth's equator as a middle term, and employing 3500 miles for the Earth's radius – an inexact value taken from Galileo's *Dialogue*. The

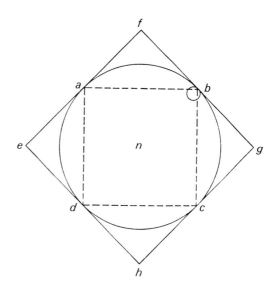

13.1. Newton's derivation of the formula for the centrifugal force generated by a body in circular motion.

conatus recedendi of a stone at the Earth's equator proved to be about 12.5 times greater than the Moon's endeavour from the Earth, while the acceleration of gravity of the stone was about 350 times greater than its *conatus recedendi*. "And so", Newton concluded, "the force of gravity is 4000 and more times greater than the endeavour of the Moon to recede from the centre of the Earth." The multiplier actually implied by Newton's data is 4375 rather than 4000.

There is no explicit indication in the manuscript that Newton had expected the stone's gravity and the Moon's *conatus* to be to one another inversely as the squares of their respective distances from the Earth's centre, that is, as about 60^2 to 1; his use of 4000 rather than 4375, however, suggests that he wanted to see the departure from 3600 as small as possible. Years later, probably after 1712, Newton wrote that he was here testing the inverse-square law of gravitation, and that the result "answered pretty nearly". Henry Pemberton and William Whiston in accounts written still later say that he was somewhat disappointed, and so turned his mind to other studies. Whiston adds that this disappointment "made Sir Isaac suspect that [the power restraining the Moon in her orbit] was partly that of Gravity, and partly that of Cartesius's Vortices...".

We may interpret Newton's claim of a "Moontest" in the late 1660s in a way consistent with what else we know of his thought at this time if we suppose that the inverse-square force was not that of universal gravitation, but such as would be caused by the aethereal flux into the Earth described in his "Questiones" and again in "An Hypothesis explaining the Properties of Light" of 1675. That hypothesis did not exclude but rather incorporated vortices.

In the manuscript we have been citing Newton also applied his formula for evaluating *conatus recedendi* to the circumsolar planets, and here indeed arrived at an inverse-square law. From Streete's *Astronomia Carolina* he had first learned of Kepler's third law, according to which the periods of the planets are as the $\frac{3}{2}$ power of their mean solar distances, or $T^2 \propto r^3$; Streete, following Jeremiah Horrocks, had insisted that this relation is exact. But the *conatus recedendi*, assuming a circular orbit, will be as r/T^2. Hence the endeavours of the planets from the Sun or centre will be as $r/(r^{3/2})^2 = 1/r^2$. Here again the result is consonant with a theory combining an aethereal vortex about the Sun with an aethereal flux into the Sun, in the manner of Newton's "Hypothesis explaining the Properties of Light" of 1675.

What of Kepler's two other, so-called laws – the rules according to which each planet moves (1) in an ellipse with the Sun at one focus, and (2) in such a way that the radius vector from Sun to planet sweeps out equal areas in equal times? (The second of these rules will hereinafter be referred to as "the areal rule".) When did Newton learn of these rules, and did he, as has sometimes been implied in popular accounts, assume them to be empirical laws established with exactitude? Newton met with the assertion of the first law in Streete's *Astronomia Carolina*; we note, however, that Streete insists that the planetary ellipses are at rest with respect to the stars whereas Kepler had allowed that their apses progress, as they in fact do, owing to mutual planetary perturbations. Streete does not mention Kepler's areal rule. Instead, he presents what we may call a "modified equant rule": the equant principle, taken exactly, requires the planets's motion to be angularly uniform about a fixed point, the empty focus of the ellipse; but this principle has to be modified to yield agreement with the Tychonic observations of Mars in the octants of anomaly, that is, around 45° either way from aphelion and perihelion. Another modified equant

rule was proposed by Wing in his *Harmonicon coeleste*, which Newton perused in late 1664 or early 1665 (see Chapter 10).

Newton's notes on the endpapers of Wing's *Astronomia Britannia* of 1669 show him doubting the strict accuracy of Kepler's third law – Wing's values for the solar distances of the planets did not agree exactly with that law – but inclined to believe it. They also show him devising a general equant rule for determining the positions of the Moon and planets, and in the process assuming that the orbits need not be taken to be precisely elliptical. Newton's unwillingness to grant the elliptical orbit the status of an empirical law is confirmed by his later remark to Halley, in the letter of 20 June 1686, that: "Kepler knew the Orb to be not circular but oval and guest it to be Elliptical." It is unlikely that Newton ever read Kepler's *Astronomia nova* or understood anything of the complicated course of reasoning, calculation, and guesswork that led Kepler to the elliptical orbit; but it is not likely, either, that such a reading would have persuaded him, or anyone grounded in the Newtonian laws of motion. Newton's laws of motion being assumed, his claim in his letter to Halley to have been the first to establish the ellipticity of the orbits is correct.

Returning to the Newton of the 1660s and 1670s, we must urge that, even had his "Moontest" confirmed an inverse-square law, he would not have been in a position to commence the writing of his *Principia*; crucially lacking was a recognition of the relation between central forces and the areal rule. At some point in the 1670s the areal rule came into his ken (perhaps from Nicholas Mercator's mention of it in the *Philosophical Transactions* for 25 March 1670); and by 1676 he had devised a series method of computing true anomaly from mean anomaly on the basis of it. But not having yet seen its connection with central forces, he must have regarded it as only one more hypothetical rule for computing planetary positions. Without the premiss that the areal rule was equivalent to the existence of a central force, he could have no way of proceeding to the determination of the laws of central force implied by different orbital figures, and therefore the entire development in the first three sections of the *Principia* was closed to him.

We should note that the central metaphysical doctrine of the *Principia* – the assumption of the absoluteness of space and time, and their independence of the necessarily relative measures we apply to them – was in place for Newton by the late 1660s. An unfinished essay dating from the last years of the decade and beginning "De gravitatione et aequipondio fluidorum", shows Newton objecting vehemently to the relativism of motion whereby Descartes sought to accommodate the immobility of the Earth – known to be a matter of church doctrine since the condemnation of Galileo by the Holy Office in Rome – to a heliocentric theory of planetary motion (see Chapter 11). According to Newton, Descartes is here forgetting his own dynamics, and so contradicting himself:

. . . unless it is conceded that there can be a single physical motion of any body, and that the rest of its changes of relation and position with respect to other bodies are so many external designations, it follows that the Earth (for example) endeavours to recede from the centre of the Sun on account of a motion relative to the fixed stars, and endeavours the less to recede on account of a lesser motion relative to Saturn and the aethereal orb in which it is carried, and still less relative to Jupiter and the swirling matter which occasions its orbit . . .; and indeed relative to its own orb it has no endeavour, because it does not move in it. Since all these endeavours and non-endeavours cannot absolutely agree, it is rather to be said that only the motion which causes the Earth to endeavour to recede from the Sun is to be declared the Earth's natural and absolute motion.

Implied here is the idea that true rotations are distinguishable from merely apparent, relative rotations, because the true rotations give rise to centrifugal forces; it is the idea that Newton will illustrate, in the Scholium to the Definitions of the *Principia*, by means of the experiment of the rotating bucket. Also in the "De gravitatione", Newton asserts absolute space and time to be necessary consequences of God's omnipresence and eternity, just as in the General Scholium written some forty years later and added to the second edition of the *Principia*. Newton adopted the fundamental ontological premises that underlie the *Principia* while he was still in his twenties, and found no reason to add to, subtract from, or modify these tenets later on.

At some time during the 1670s Newton read *Theoricae mediceorum planetarum ex causis physicis deductae* (*The Theory of the Medicean Planets Deduced from Physical Causes*, 1666) by the Italian G.A. Borelli. This work proposed that the elliptical orbits of the planets and satellites could be accounted for by the interaction of a centrifugal force due to the curvature of the orbits, with a constant force towards the central body. The two forces were conceived to be in disequilibrium in such a way that one alternately overbalanced the other, the effect being to modify the circular path that would have resulted from an exact balance between them into an elliptical or quasi-elliptical orbit. The rotation round the centre was sustained by the continued action of a trans-radial impetus which in the case of the circumsolar planets Borelli associated with the rotation of the Sun and its rays.

How Newton viewed this hypothesis we do not know. In his letter to Halley of 20 June 1686, while complaining of Hooke's claim to priority in the matter of the inverse-square law, Newton said: "Borell wrote long before him that by a tendency of the Planets towards the Sun like that of gravity or magnetism the Planets would move in Ellipses . . . ." Also: "he [Hooke] has published Borell's Hypothesis in his own name. . . . Borell did something in it and wrote modestly, he [Hooke] has done nothing and yet written in such a way as if he knew and had sufficiently hinted all but what remained to be determined by the drudgery of calculations and observations, excusing himself from that labour by reason of his other business." Newton's charge that Hooke plagiarized Borelli's treatise is quite wrong.

Hooke's ideas on planetary motion had appeared in works published in 1674 and 1678 (see Chapter 12), and it is fairly certain that Newton read the relevant passages (he denies it when he first writes to Hooke in 1679, then asserts it in his letter to Halley of 20 June 1686). It has sometimes been said that Hooke in these writings enunciates the law of universal gravitation. In his *Cometa* of 1678, however, Hooke makes clear that the attraction he there has in mind, although approaching universality more closely than the attraction proposed in any earlier hypothesis, is still not universal. He supposes that the gravitational power, like magnetism, can be lost, and that partial destruction of this power accounts for the quasi-rectilinear trajectories of comets. A mere generalizing induction, we notice, does not here lead Hooke all the way to a strictly universal gravitation; nor will it lead Newton so far. The analogy of magnetism told against such a generalization, and so did the supposedly rectilinear or quasi-rectilinear trajectories that astronomers since Kepler had assigned to comets.

Of seminal importance, however, will be Hooke's proposal of the idea of obtaining the orbital motions of planets by combining an attraction to a centre with inertial motion along the tangent – an idea first formulated, it would appear, by Christopher Wren (see Chapter 12). In a correspondence that Hooke initiated with Newton in 1679, he so insisted on this way of conceiving orbital motion as to goad Newton into discovering a major consequence of the idea, namely the areal rule; and this consequence was to provide the basis for all Newton's future work in planetary dynamics.

The correspondence with Hooke, 1679–80

Hooke wrote to Newton on 24 November 1679 because, as a secretary of the Royal Society, he had recently been assigned the task of carrying on correspondence with members of the society not in London. Hooke was aware that, with Newton, his relations were unlikely to be easy. Twice before, first when Newton presented to the Royal Society his "New Theory of Light and Colours" in 1672, and again when he submitted to the society his "Hypothesis explaining the Properties of Light" in 1675, Hooke had charged him both with being mistaken and with plagiarizing his, Hooke's, ideas. The quarrel had been in a manner patched up in 1676, but Hooke rightly suspected that the wounded feelings in Newton had not disappeared. In fact, unbeknownst to Hooke, and in the wake of the controversies that the publication of his optical theory had led to, Newton had resolved to avoid all further correspondence concerning his discoveries in natural philosophy; they would appear soon enough, he averred, if they were published after his death.

In his letter Hooke sought to be cordial. He promised that Newton's ideas would be communicated to the Society no further than Newton himself should prescribe, and he urged that difference of opinion in philosophical matters not be the occasion of enmity: " 'tis not with me I am sure". Hooke claimed to be himself open to criticism:

For my own part I shall take it as a great favour if you shall please to communicate by Letter your objections against any hypothesis or opinion of mine. And particularly if you will let me know your thoughts of that of compounding the celestiall motions of the planets of a direct motion by the tangent and an attractive motion towards the centrall body . . . .

After chatting about other proposals and projects of interest to himself and the Royal Society, Hooke toward the end remarked: "Mr Flamsteed by some late perpendicular Observations hath confirmed the parallax of the orb of the Earth."

Newton replied on 28 November. He acknowledged himself "every way by the kindness of your letter tempted to concur with your desires in a Philosophical correspondence". Unfortunately, for the last half year he has been in Lincolnshire, attending to family affairs (his mother died in the summer of 1679), and before that:

I had for some years past been endeavouring to bend my self from Philosophy to other studies . . . which makes me almost wholy unacquainted with what Philosophers at London or abroad have of late been imployed about. And perhaps you will incline the more to beleive me when I tell you that I did not before the receipt of your last letter, so much as heare (that I remember) of your Hypotheses of compounding the celestial motions of the Planets, of a direct motion by the tangent to the curve & of the laws and causes of springyness . . . .

In a letter of 1686, as indicated earlier, Newton will claim that he *had* previously read in Hooke's writings about his hypothesis concerning the celestial motions. But it is evident that he had not yet been struck by it as an hypothesis that might admit of being shown to be true.

After further insisting that he has laid philosophy aside, Newton (to sweeten his answer, as he put it later to Halley) commented on several of the items in Hooke's letter. In particular, he was glad to hear that annual parallax, earlier discovered by Hooke, has been confirmed by John Flamsteed (1646–1719). (In fact, both Hooke and Flamsteed were mistaken; see Volume 3.)

In requital of this advertisement I shall communicate to you a fansy of my own about discovering the Earth's diurnal motion. In order thereto I will consider the Earth's diurnal motion alone without the annual, that

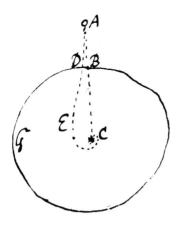

13.2. The spiral trajectory of a falling body as proposed by Newton in his letter to Robert Hooke of 28 November 1679.

having little influence on the experiment I shall here propound. Suppose then BDG [Figure 13.2] represents the Globe of the Earth carried round once a day about its center C from west to east according to the order of the letters BDG; and let A be a heavy body suspended in the Air and moving round with the Earth so as perpetually to hang over the same point thereof B. Then imagin this body A let fall and its gravity will give it a new motion towards the centre of the Earth without diminishing the old one from west to east. Whence the motion of this body from west to east, by reason that before it fell it was more distant from the centre of the Earth then the parts of the Earth at which it arrives in its fall, will be greater than the motion from west to east of the parts of the Earth at which the body arrives in its fall: and therefore it will not descend in the perpendicular AC, but outrunning the parts of the Earth will shoot forward to the east side of the perpendicular describing in its fall a spiral line ADEC, quite contrary to the opinion of the vulgar who think that if the Earth moved, heavy bodies in falling would be outrun by its parts and fall on the west side of the perpendicular. The advance of the body from the perpendicular eastward will in a descent of but 20 or 30 yards be very small and yet I am apt to think it may be enough to determin the matter of fact.

And Newton went on to make a number of detailed suggestions for the carrying out of the experiment.

In the final paragraph of his letter Newton returns to the theme of his weariness of 'philosophy'.

"If I were not so unhappy as to be unacquainted with your Hypotheses", Newton writes, "I should so far comply with your desire as to send you what Objections I could think of against them if I could think of any". And he could with pleasure hear and answer any objections made against his own notions "in a transient discourse for a divertisement. . . . But yet my affection to Philosophy being worn out, . . . I must acknowledge my self avers from spending that time in writing about it which I think I can spend otherwise more to my own content and the good of others. . . ."

Hooke's response is dated 9 December. Newton's "deserting Philosophy . . . seems a little unkind", but the secretary does not despair of him: ". . . you that have soe fully known those Dilights cannot chuse but sumetime have a hankering after them and now and then Desire a taste of them. . . ." And Hooke assured Newton that he valued the great favour and kindness of his letter, and particularly his notion about detecting the diurnal motion of the Earth; the passage of his letter concerning this proposal was read to the Royal Society, and the members generally agreed that, as Newton had suggested, a heavy body let fall from a great height would fall to the east of the vertical and not to the west as most have hitherto imagined.

But as to the curve Line which you seem to suppose it to Desend by (though that was not then at all Discoursed of) Vizt a kind of spirall which after sume few revolutions Leave it in the Centre of the Earth my theory of circular motion makes me suppose it would be very differing and nothing att all akin to a spirall but rather a kind [of] Elleptueid [!].

Hooke lists the conditions presupposed for this purely imaginary motion: the falling body must be in the plane of the Earth's equator; the northern and southern hemisphere must be divided along this plane so as to leave about a yard's space between, and this space must be evacuated; the gravitation to the centre must remain as before; the falling body is to have impressed on it the diurnal motion of the superficial parts of the Earth from which it was let fall. Under these conditions, Hooke conceives that the body's path will "resemble an Ellipse", such as the curve *AFGH* in Figure 13.3. If, however, a resisting medium were present, the curve would be somewhat like the line *AIKLNOP*

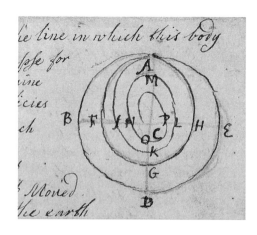

13.3. Hooke's diagram of the trajectory of a body let fall into an evacuated space imagined as separating the northern and southern hemispheres of the Earth.

and after many revolutions would terminate in the centre C.

Hooke has one more correction to make in Newton's account of the falling body, and it shows a sharp intelligence at work. If the body is not in the plane of the equator, but at latitude 51° 32' as in London, the ellipse-like curve will be in a plane inclined to the plane of the equator by 51° 32', so that the fall of the body will not be exactly east of the vertical but rather southeast and indeed more south than east. For, in Figure 13.4, let *NLQS* represent the meridian of London, *RQ* the plane of the equator, *L* London, and *PL* the parallel in which London moves about the axis, *NS*. Then the body let fall at *L* will descend in the plane *LC* at right angles to the plane *NLQSR* of the meridian, and not in the surface of the cone *PLC* with apex at *C* and circular base *PL*: "I could adde many other conciderations which are consonant to my Theory of Circular motions compounded by a Direct motion and an attractive one to a Center. But I feare I have already trespassed to much upon your more Usefull thoughts . . . ."

We note that, if the vertical direction is determined by a plumb bob, the southerly deviation of the falling body will probably be indetectable, since both plum bob and falling body are affected by the Coriolis force involved. But Newton considered Hooke's correction to be correct.

Newton's response, dated only four days later (13 December), plunges abruptly into the questions at issue:

13.4. Hooke's diagram to show why a body dropped from a height in a northern latitude should fall south as well as east of the point from which it is let fall.

I agree with you that the body in our latitude will fall more to the south than east if the height it falls from be any thing great. And also that if its gravity be supposed uniform it will not descend in a spiral to the very centre but circulate with an alternate ascent and descent made by it's vis centrifuga *and gravity alternately over-ballancing one another. Yet I imagin the body will not describe an Ellipsoeid but rather such a figure [Figure 13.5] as is represented by* AFOGHIKL &c.

Here *A* is the initial position of the body, and *AM* its initial velocity. Newton argues that, assuming a uniform gravity, the point of closest approach to *C* in the initial fall will be in the quadrant *BCD*, as at *O*:

The innumerable & infinitely little motions (for I here consider motion according to the method of indivisibles) continually generated by gravity in its passage from A *to* F *incline it to verge from* GN *towards* D, *& the like motions generated in its passage from* F *to* G *incline it to verge from* GN *towards* C. *But these motions are proportional to the time they are generated in, & the time of passing from* A *to* F *(by reason of the longer journey & slower motion) is greater than the time of passing from* F *to* G.

Newton then goes on to say that if the gravity be supposed not uniform but greater nearer the centre, the point *O* of closest approach may fall in the line *CD* or in the angle *DCE* or in the following quadrants, or even nowhere: "For the increase of gravity in the descend may be supposed such that the body shall by an infinite number of spiral revolutions descend continually till it cross the center by motion transcendently swift."

How did Newton arrive at the results just given? In the final paragraph of his letter he gives a hint:

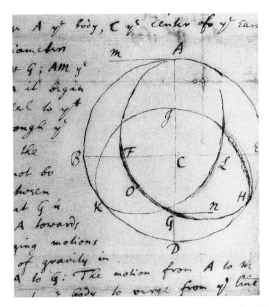

13.5. Newton's diagram of the trajectory of a body let fall into an evacuated space imagined as separating the northern and southern hemispheres of the Earth, under the assumption of a uniform gravity.

Your acute Letter having put me upon considering thus far the species of this curve, I might add something about its description by points quam proximé. *But the thing being of no great moment I rather beg your pardon for having troubled you thus far with this second scribble wherein if you meet with any thing inept or erroneous I hope you will pardon the former and the latter I submit and leave to your correction . . . .*

It would be important to know what Newton meant by the "description of points *quam proximé*", but all we can do is guess, and the following seems a likely interpretation. In Figure 13.6 let the body be initially moving in the line *AB*. Imagine the time to be divided into a large number of tiny intervals, and suppose that gravity acts at the end of each interval, impressing a new element of velocity towards the centre *S*. At *B*, for instance, the body will receive an impulse; the velocity it thus acquires, let us suppose, is such as would carry it from *B* towards *S* through a distance equal to *cC* in the next interval, if no other motion were present. Since, however, it is already moving with the velocity that carried it from *A* to *B* in the first interval, and this velocity will carry it through the equal distance from *B* to *c* in the second interval, the net result is that the body at the end of the second interval will

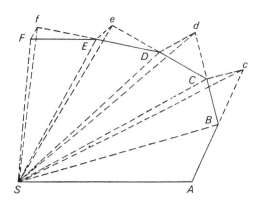

13.6. A diagram by Newton to illustrate the analysis of the motion of a body subject to a centripetal force.

be found on the diagonal of the parallelogram formed with the sides *Bc* and *cC*, namely at *C*. If the gravity is uniform, the velocities successively imparted, *cC*, *dD*, *eE*, etc., are equal; if it varies as some function of the distance from *A*, the velocities imparted will vary according to the assumed function.

The method we are suggesting is graphical and so but crudely approximative. In the diagram that accompanies his letter, and in which he is seeking to portray the case of uniform gravity, Newton makes the central angle between successive farthest departures from the centre nearly 240°, impossibly large; an exact method would have shown him that this angle cannot exceed ($360°/\sqrt{3} =$) 207° 50′.

Newton's recourse to graphing, however, must have quickly led to an important insight. In the diagram we have drawn, it is easy to show that the triangles successively swept out about the centre, namely *SAB*, *SBC*, *SCD*, etc., are equal. As his later statements attest, Newton at this time concluded that a force directed to a fixed point – commonly called a 'central force' – implies Kepler's areal rule; and conversely.

These two propositions will appear as the first two theorems of Newton's *Principia*; and in fact Newton's entire treatment of orbital motion will be based upon them. His proofs of them are not such as would satisfy a present-day mathematician. They hold, in fact, only when the arc *BF* of the preceding diagram is vanishingly small. Nevertheless, they satisfied Newton and his contemporaries.

Hooke's response was dated 6 January, nearly a month later.

Your Calculation of the Curve by a body attracted by an aequall power at all Distances from the centre Such as that of a ball Rouling in an inverted Concave Cone is right and the two auges [= apsides] will not unite by about a third of a Revolution. But my supposition is that the Attraction always is in a duplicate proportion to the Distance from the Centre Reciprocall, and Consequently that the Velocity will be in a subduplicate proportion to the Attraction and Consequently as Kepler Supposes Reciprocall to the Distance. And that with Such an attraction the auges will unite in the same part of the Circle and that the neerest point of accesse to the centre will be opposite to the furthest Distant. Which I conceive doth very Intelligibly and truly make out all the Appearances of the Heavens.

We interrupt our quotation in order to inquire: on what basis did Hooke conclude that the attractions of the heavenly bodies vary inversely as the squares of the distances from their centres?

As noted in Chapter 12, Hooke was probably aware that, if the planetary orbits were assumed to be concentric circles, Kepler's third law would imply the inverse-square law; but he also knew that the assumption of concentric, circular planetary orbits is false. On more than one occasion he invoked the physical analogy between gravity and light, to justify an inverse-square law for gravity. But the passage just quoted suggests that Hooke when he wrote to Newton was relying on a different derivation.

From several of Hooke's writings we know that he had formulated for himself the following principle, which he believed to be generally true: "the comparative velocities of any body moved are in subduplicate proportion to [that is, as the square root of] the aggregates or sums of the powers by which it is moved." He had applied the principle successfully to a number of cases; for instance, to the speed of efflux of a liquid from a hole in the bottom of its containing vessel, where the speed varies as the square root of the depth of the liquid. In fact, Hooke's principle turns out to yield the right result whenever his "aggregates or sums of the powers" can be interpreted as work or potential energy. He seems to have derived the principle from Galileo's result that the square of the velocity of bodies descending without friction is proportional to the height of fall. From the passage quoted above, it would appear that Hooke is applying this

principle to planetary motion, assuming Kepler's inverse-distance law (see Chapter 5 above, p. 64), which in fact is not exact outside the apsidal line (it is the component of the planet's velocity at right-angles to the radius vector that, in all exactitude, is inversely as the distance from the central body). In his letter, Hooke gives no sign of realizing that the planet's speed taken simply is not inversely as the distance from the Sun; he appears rather to be assuming this inexact relation to be precisely true, and then on the basis of his general principle (which is not here applicable in the way he applies it) to be taking the inverse-square variation of the force as implying and implied by this relation. The derivation cannot stand: as previously stated, it is the transradial component of the planet's velocity that is inversely proportional to the distance from the attracting body, and this is true for all central forces, whether inverse-square or not.

We resume our quotation from Hooke's letter of 6 January:

And therefore (though in truth I agree with You that the Explicating the Curve in which a body Descending to the Centre of the Earth, would circumgyrate were a Speculation of noe Use yet) the finding out the proprietys of a Curve made by two such principles will be of great Concerne to Mankind. because the Invention of the Longitude by the Heavens [the finding by means of astronomical measurements of the longitude at sea] is a necessary Consequence of it: for the composition of two such motions I conceive will make out that of the Moon. What I mentioned in my last concerning the Descent within the body of the Earth was But upon the Supposall of such an attraction, not that I believe there really is such an attraction to the very Centre of the Earth, but on the Contrary I rather Conceive that the more the body approaches the Centre, the lesse will it be Urged by the attraction – possibly somewhat like the Gravitation on a pendulum or a body moved in a Concave Sphaere where the power Continually Decreases the neerer the body inclines to a horizontall motion, which it hath when perpendicular under the point of suspension, or in the Lowest point, and there the auges are almost opposite, and the nearest approach to the Centre is at about a quarter of a Revolution. But in the Celestiall Motions the Sun Earth or Centrall body are the cause of the Attraction, and though they cannot be supposed mathematicall points yet they may be Conceived as physicall and the attraction at a Consider-

able Distance may be computed according to the former proportion as from the very Centre. This Curve truly calculated will show the error of those many lame shifts made us[e] of by astronomers to approach the true motions of the planets with their tables. But of this more hereafter.

This passage is remarkable for the correctness of a number of its guesses. Hooke rightly supposes that the force of gravity beneath the Earth's surface decreases with approach to the centre, and in much the way that the force on a pendulum bob decreases with approach to the centre of its swing; from his experimentation with conical pendulums he also knows that the nearest approach to the centre follows about 90° after the furthest departure therefrom. His conception is that of an attraction resulting from the attractions of all the parts of which the attracting body is made up, and he recognizes the consequence that, close to the surface of the attracting body, the assumption that it may be thought of as attracting from its centre as from a point cannot be taken for granted. In imagining that the orbit deriving from an inverse-square central force will be something other than the ellipse used by practically all the astronomers of his day, and that the discovery of the true curve will lead, without consideration of perturbations, to so accurate a prediction of the Moon's motions as to result in "the invention of the longitude", Hooke is, of course, mistaken.

In the remainder of his letter Hooke informs Newton that he has performed the experiment of the falling body three times, and always found the body to fall to the southeast of the vertical; but as the amounts of deviation from the vertical were different in the three trials, he plans to repeat the experiment with further precautions until he obtains "a proof free from objections". (If Hooke found *any* appreciable deflection southward, then he found something that Gauss and other experimenters in the nineteenth century did not.)

To Hooke's letter of 6 January, Newton did not reply. On 17 January, Hooke wrote again, this time reporting that two trials of Newton's experiment carried out within doors had succeeded very well: "So that I am now persuaded the experiment is very certain, and that it will prove a demonstration of the diurnal motion of the earth as you have very happily intimated." And he concludes:

It now remains to know the proprietys of a curve Line (not circular nor concentricall) made by a centrall attractive power which makes the velocitys of Descent from the tangent Line or equall straight motion at all Distances in a Duplicate proportion to the Distances Reciprocally taken. I doubt not but that by your excellent method you will easily find out what the Curve must be, and its proprietys, and suggest a physicall Reason of this proportion. If you have had any time to consider of this matter, a word or two of your Thoughts of it will be very gratefull to the Society (where it has been debated) . . . .

To Hooke's letter of 17 January, Newton again did not reply; except that, nearly eleven months later, having to write Hooke about a different matter, he closed by remarking: "For the trials you made of an Experiment suggested by me about falling bodies, I am indebted to you thanks which I thought to have returned by word of mouth, but not having yet the Opportunity must be content to do it by Letter."

Nothing here about the calculation of orbits! Yet Newton had solved the problem that Hooke had set for him. He had satisfied himself that, on the assumption of an inverse-square force toward the central body, the planet or satellite will move in an ellipse with the central body at one focus.

We do not know for sure how Newton came to this result in 1679–80. But with high probability we can guess that it was the converse proposition – passing from the elliptical orbit with Sun at one focus to the inverse-square proportion – that Newton actually proved; and that he did so by assuming the areal rule, which made it possible for him to substitute area for time and so to reduce the problem to geometry. Props. 6 and 11 of Bk I of the *Principia* traverse the same logical path that Newton must have followed in 1679–80, but with refinements added at a later date.

And what about the converse that Newton in fact needed in order to reply to Hooke's question – the derivation of the elliptical orbit with centre of force at one focus from the inverse-square law? We would suggest that, as in the first edition of the *Principia*, Newton believed this to be a necessary consequence of the argument articulated in Prop. 17 of Bk I.

Thus, in the winter of 1679–80 Newton thought that the areal rule and stationary elliptical orbit were derivable from the assumption of a central force varying inversely as the square of the distance. We believe him also to have understood at this time that the third Keplerian law was similarly derivable – at least on the assumption of concentric circular orbits. The first two of these discoveries were quite unknown to anyone (Wren and Halley appear to have been on to the third). And yet Newton remained silent. Why?

No doubt psychology must be given its due here. Newton had made a firm resolve to withdraw from correspondence on philosophical subjects. And correspondence with Hooke, whose manner Newton felt to be condescending, was particularly disagreeable.

But the logical status of the theory thus far elaborated, in its relation to empirical facts, needs also to be considered. *If* there is an inverse-square attractive force toward the Sun, *then* the areal rule, elliptical orbit, and Kepler's third law follow; and the latter were known to be at least approximately true. By 1680, indeed, Flamsteed, following Horrocks, was using both the elliptical orbit and the areal rule in calculating tables of the Moon; and Streete's planetary tables of 1661 were already in accord with the third Keplerian law, and for this very reason were proving superior to other planetary tables (Halley would re-publish them in 1710 and 1716). It was not evident that the inverse-square law would lead to any *new* consequences, that could be tested against observations, or would prove useful to astronomers. Nor was it clear that the inverse-square law could be assumed to be exact: that might depend on what Hooke referred to as "the physical reason of this proportion", which he had invited Newton to speculate about. An inverse-square law of force operating on the planets towards the Sun was compatible with Newton's aethereal hypothesis of 1675; but how could one decide on the basis of that hypothesis that the inverse-square law would be exact?

As Newton probably knew, Hooke in his *Cometa* had suggested that a comet consists of matter that had lost its gravitating power, owing perhaps to a "jumbling" of its internal parts. If gravity depended on an internal arrangement of parts – indeed, if it depended on any kind of arrangement or motion of matter – then it was not universal. If it was universal, it had no such explanation.

But how in fact did comets move? The generally accepted notion in 1680 was that they moved in nearly rectilinear paths, slightly concave toward the Sun, accelerating as they came closer to the Sun, and decelerating afterward (see Chapters 10 and 12 above). Did the Sun act on them with precisely the same accelerative force that it would exercise on a planet at the same solar distance, as universal gravitation required, or only with a weakened force?

We cannot be sure that Newton entertained the foregoing thoughts in the winter of 1679–80. In the following winter a spectacular comet appeared, and Newton became avidly interested in the determination of its path. Had the idea of universal gravitation then occurred to him, and was he seeking to test it in the case of this comet?

Newton, Flamsteed, and the comet of 1680–81

The comet first appeared in November 1680, in the pre-dawn sky, heading in the direction of the Sun. The last observation of it (or its tail) in the morning sky appears to have been made on 8 December. On 10 December what was taken by most astronomers to be a different comet appeared in the evening sky, heading away from the Sun. It continued to be observed into March.

Was a single comet responsible for all the appearances from November through March, or were two comets involved?

The apparent paths of earlier comets across the celestial sphere had proved to be close to great circles, and these could be understood as the projections of rectilinear trajectories onto the stellar background. Wren and John Wallis had devised mathematical procedures for determining such trajectories on the basis of observation. If one assumed rectilinear or nearly rectilinear paths for comets, then the simplest hypothesis was that the comet appearing in November was different from the comet appearing from 10 December onwards, for the two apparent paths plotted on the celestial sphere did not fall on or near the same great-circle arc. Astronomers generally, including Cassini I and Huygens, took the two appearances to be two separate comets moving in opposite directions along nearly rectilinear paths.

Flamsteed, who had been appointed His Majesty's Astronomical Observator in 1675 and was thus the first full-time astronomer in England, in seeking to comprehend the cometary appearances

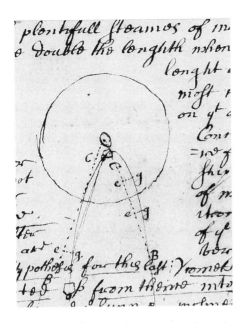

13.7. Diagram to illustrate the trajectory hypothesized by Flamsteed for the comet of 1680–81 (with his causal explanation thereof), in his letter to Halley of 17 February 1680/1.

13.8. The tail of the comet of 1680–81 as seen by Newton's colleague, Dr Humphrey Babington, over King's College Chapel, at 5 a.m. between 20 and 27 November 1679; the sketch is by Newton, who argues from it that the comet went farther south than Flamsteed supposes, and probably did not return to the north of the ecliptic.

of the winter of 1680–81, conceived a very different hypothesis. There was but one comet. As it came within the Earth's orbit:

. . . [the] attraction of the Sun would have drawne it neare him in a streight line, had not the laterall resistance of the Matter of the Vortex moved against it bent it into a Curve . . . the body of the Comet I conceave to have always the same part carried foremost in the line of its motion so that when [in Figure 13.7] it comes to C it moves contrary and crosse the motion of the Vortex till haveing the contrary End opposite to the Sun hee repells it as the North pole of the loadstone attracts the one end of the Magnetick needle but repells the other.

This act of repulsion would carry the Comet from the Sun in a streight line, were it not that the crosse motion of the vortex bendes it back. which yet the acquired velocity and strength of its motion may compensate and restore. so that the one countervaileing the other I see no reason why wee may not admitte it to have run from [the] Sun nearely in a Streight line.

Observational data led Flamsteed to believe that, *before* perihelion or closest approach to the Sun, the comet had twice pierced the plane of the ecliptic, first from north to south and then from south to north. This bend in the comet's path, Flamsteed

supposed, was due to the action of the Sun's "northern pole". Somehow the solar vortex then impinged on the comet in such a way as to keep it from falling into the Sun and to cause it subsequently to be oriented so as to be repelled by the Sun's magnetism. Thus, according to Flamsteed, the comet had been turned back before reaching the Sun, its hairpin turn being the result of the Sun's repulsion.

Flamsteed's observational data and an account of his hypothesis were conveyed to Newton through James Crompton, a Fellow of Jesus College in Cambridge. Newton was already familiar, through the writings of Johannes Hevelius and Hooke and the *Philosophical Transactions* of the Royal Society, with much of the thought and speculation about comets from Kepler's day to his own. In a response dated 28 February he pointed to a number of difficulties in Flamsteed's hypothesis: (1) If an attraction to the Sun caused the comet to bend from its original rectilinear path so that it was headed directly toward the Sun, then it would have continued thereafter in the same direction until it fell into the Sun. Flamsteed's invocation of the solar vortex is unhelpful, for the force of the vortex would have tended to turn the comet's path in the direction of the planetary motion (counterclockwise as seen from the ecliptic's north pole), not in the reverse direction assumed by Flamsteed. "The only way to releive this difficulty in my judgment is to suppose the Comet to have gone not between the Sun and Earth but to have fetched a compass about the Sun." (2) The Sun, being extremely hot, is not a magnet, for magnetic bodies when heated to redness lose their magnetic virtue. (3) Were the Sun a magnet, then as with terrestrial magnets its "directive virtue" would be more powerful than its "attractive virtue"; hence the cometary magnet would quickly be turned into such a position as to be attracted by the Sun, and would never thereafter be repelled. (4) The identity of the November and December comets is doubtful because (according to the data available to Newton) "if they were but one Comet, it's motion was thrice accelerated and retarded . . . . This frequent increas and decreas of motion is too paradoxical to be admitted in one and the same Comet without some proof that there was but one."

The fourth objection was clearly refutable. As Flamsteed pointed out in a letter to Crompton for Newton dated 7 March, Newton's conclusion of a triple acceleration and retardation resulted simply from a confusion of new-style and old-style dates. The comet had only once accelerated, and had thereafter slowed down.

Flamsteed then proceeded to present empirical evidence in support of his contention that the November and December comets were the same. The observations of M.A. Collio in Rome gave the first perigee or closest approach to the Earth as on 17 November (OS), and the motion of the comet from 11 to 17 November was 26° 30′ in six days. Flamsteed makes the second perigee to be on 30 December, and the motion from 30 December to 5 January 25° 36′ in six days. Again, the Roman observations make the motion from 17 to 26 November 41° 15′ and Flamsteed's observations of the motion in the symmetrically corresponding period before second perigee, from 21 to 30 December, give the motion as 42° 30′: "which considering the Courseness of the forreigne observations I look upon as a very good agreement and argument that

all the observations were of the same Comet."

Flamsteed also notes that the western elongation of the comet from the Sun on 26 November was 25° 18′, the eastern elongation of the comet on 21 December was 24°, and the perihelion almost exactly between these dates: he is once more arguing for the oneness of the comet from symmetry.

Another piece of evidence had to do with the latitudes of the comet. It went south of the ecliptic on 11 November, changing its latitude at the rate of about one degree per 10° of longitudinal motion. But from 18 to 27 November its latitude remained almost constant at about 1° south latitude. The December–March comet had northern latitude. "If Mr Newton cannot allow the Comet to have returned out of South into Northerne latitude hee will oblige me much if hee please to suggest some reason why without supposeing any attraction of the Sun the latitude should continue still nearly the Same so long as it was observed after the 18 to the 27 of Nov."

"Hence it should seeme", Flamsteed concludes from the observational evidence, "that the Comet was attracted and repelled by the Sun as I imagined and proposed in my last letter." As for Newton's objection against the Sun's magnetism, Flamsteed suggests that the attraction of the Sun may be of a different nature from that of terrestrial magnets, and the Sun although causing heat on the Earth may yet have a core which is not hot. He does not think Newton's claim that the "directive virtue" of a magnet exceeds its "attractive virtue" makes much against his hypothesis, "except hee can prove that a large fixed magnet would have that operation on a small one throwne violently by or about it".

Newton's answer came on 16 April, and began with an apology: "By some indisposition and other impediments I have deferred answering you longer than I intended." But as we now know, he had previously written two drafts of a reply, these drafts differing from each other as well as from the letter finally sent. In the first draft he acknowledges that, with the correction of his mistake about old-style and new-style dates, "the comets of November and December are less irreconcileable". On the other hand, the arguments from the symmetry of the observations about perihelion and the two perigees can be faulted, since Flamsteed writes only of geocentric longitudes and not of the positions of the comet in its path. Newton from an approximate calculation finds the motion of the November comet averaged 5° per day, that of the December comet about 4° 20′, but "I have not computed it exactly". (We note that the large inclination of the orbital plane of the comet to the ecliptic, and the inclination of the axis of the cometary orbit to the intersection of the two planes, along with the close approach of the Earth to the comet in its two perigees, made Earth-based observation an exceedingly tricky business.) Flamsteed's argument from the latitudes, Newton urged, is not compelling because the observations appear to have been insufficiently accurate: it is still possible to assume that the comet of November moved southward uniformly in accordance with a rectilinear trajectory.

Newton rejects the proposal that the Sun, although hot on its surface, might have a core that was not hot: "The whole body of the Sun therefore must be red hot and consequently voyd of magnetism unless we suppose its magnetism of another kind from any we know, which Mr Flamsteed seems inclinable to suppose." With respect to the comparative magnitudes of the directive and attractive virtues of a magnet, Newton argues that the former will always exceed the latter, however short a time the magnet is given in which to exert its force. Flamsteed's notion that the comet was repelled by the Sun after first being attracted has the difficulty that the comet would then have been accelerated in its recess from the Sun as well as in its approach and, moreover, would not have followed a path symmetrically placed with respect to the Sun. But

. . . all these difficulties may be avoyded by supposing the comet to be directed by the Sun's magnetism as well as attracted, and consequently to have been attracted all the time of its motion, as well in its recess from the Sun as in its' access towards him, and thereby to have been as much retarded in his recess as accelerated in his access. And by this continuall attraction to have been made to fetch a compass about the Sun . . ., the vis centrifuga at [perihelion] overpow'ring the attraction and forcing the Comet there notwithstanding the attraction, to begin to recede from the Sun.

In the final paragraph Newton begins by writing: "About the Comets path I have not yet made any computation though I think I have a direct method of doing it, whatever the line of its motion be." In

the remainder of the paragraph Newton gives his suppositions about the course of the comet as seen from the Sun, on the assumption that there was only one comet, and this comet "fetched a compass round the Sun".

What method of computation had Newton in mind? We do not know; no Newtonian manuscripts dealing with determination of cometary paths can be dated with certainty to March or April of 1680. And it is not at all clear how Newton expected to determine the comet's path "whatever the line of its motion may be"; the later successful methods made an assumption about what the line was, namely that it was a parabola, produced by the known accelerative force of the Sun's gravitational field.

Five and a half years later Newton had still not succeeded in devising a usable method for determining cometary paths from observations. In his letter to Halley of 20 June 1686, he remarked that Bk III of the *Principia*

. . . wants the Theory of Comets. In Autumn last I spent two months in calculations to no purpose for want of a good method, which made me afterwards return to the first Book and enlarge it with divers Propositions some relating to Comets others to other things found out last Winter.

The method Newton employed in the autumn of 1685 may have been an adaptation of the method of Wren, assuming that, for at least a part of its path, the comet can be taken to be moving rectilinearly. The assumption is a poor one, since comets approaching or receding from the Sun undergo very considerable changes of direction and speed. It must have been later than June 1686 that Newton discovered his method for fitting a parabolic trajectory to cometary observations. And then, finally, applying this method to the morning and evening comets of November–December and December–March of 1680–81, he claimed to find that their perihelia coincided, and that their speeds at equal distances from the Sun were equal, and so inferred that they were one and the same comet. (Encke in 1818, although obtaining elements close to Newton's, will find that the perihelion determined from the November appearances did not quite coincide with the perihelion determined from the December–March appearances – an effect presumably of non-gravitational accelerations induced by the solar wind.) We may conclude that Newton's hopefulness, in the first draft of his letter for Flamsteed, was premature.

The second draft, dated 12 April, differs from the letter finally sent in only one important respect, that it includes a paragraph omitted from the letter. The paragraph begins:

That the Comet went beyond the Sun I think myself pretty certain not only because it seems to me to be in your Hypothesis against the nature of the thing for the Comet to turn short of him but also because I think I have a way of determining the line of a Comets motion (what ever that line be) almost to as great exactness as the orbits of the Planets are determined, provided due observations made very exactly be had.

The faulty logic of the sentence may be one reason why Newton abandoned it. He appears to be thinking of the two comets as one, and we are tempted to think that he is entertaining the idea of universal gravitation, in which the acceleration toward the Sun is a function solely of distance from the Sun, whatever the body. But we cannot be sure.

What is certain is that in the rest of the second draft, and thus in the letter actually sent, Newton is arguing against the identity of the comet of November–December and that of December–March. One of his arguments is based on the misapprehension that the earlier comet, having crossed the ecliptic from north to south, then re-crossed it from south to north on 3 December. This meant that if it was bending round the Sun, it had already passed perihelion, since the nodes must be on opposite sides of the Sun. On the other hand, conjunction had occurred on 9 December; and the place of the comet at that time, on the opposite side of the Sun from the Earth, could not have been far from perihelion. The lapse of at least six days between perihelion and conjunction, in which the comet could have moved only a short distance at the very time it would be expected to move most rapidly of all, seemed to Newton too paradoxical to accept.

Newton's remaining argument was taken from the whole history of comets. If the comets of November and December were a single comet that moved in the bent trajectory Flamsteed imagines, then other comets would do the like, "and yet no such thing was ever observed in them but rather the contrary". Earlier comets that were seen both before and after their perihelia, for instance the comets of 1472, 1556, 1580, and 1664, began in one part of the heavens and ended in the opposite part, going through nearly a semicircle with motion first slow, then fast, then slow again, as if done in a straight line.

Let but the Comet of 1664 be considered where the observations were made by accurate men. This was seen long before its Perihelion and long after and all the while moved (by the consent of the best Astronomers) in a line almost straight. So neare was the line to a straight one that Monsieur Auzout on supposition that 'twas an arch of a great circle about the dog starr (as Cassini guessed and Auzout was afterward willing should be beleived) or rather a straight one (as the obviousness of the Hypothesis, easiness of the calculation, and number of observations on which 'twas founded makes me suspect) did from three observations predict the motion to the end without very considerable error.

Possibly Newton for some time in the spring of 1681 seriously entertained the idea that the comet of 1680–81 was subject to the same acceleration, at any given distance from the Sun, as a planet would be. But if so, he soon abandoned the idea, on the ground that comets had not previously been seen to "fetch a compass round the Sun".

By the autumn of 1684, he will have changed his mind. We do not know why.

The earliest version of the tract "De motu"

According to the well-known story (recorded in a Memorandum by John Conduitt now in Chicago University Library), Halley paid Newton a visit at Cambridge in August 1684 and asked him what would be the curve described by the planets supposing that the attraction towards the Sun diminished as the square of the distance; Newton immediately answered, *an ellipse*:

The Doctor struck with joy and amazement asked him how he knew it, why saith he I have calculated it, whereupon Dr Halley asked him for his calculation without any further delay, Sir Isaac looked among his papers but could not find it, but he promised him to renew it, and then to send it him. . . .

Whether or not the story is accurate in detail, it is clear that Halley's encouragement in 1684 was responsible for rekindling Newton's interest in the dynamics of planetary motion. By November 1684 he had composed a short tract *De motu corporum in*

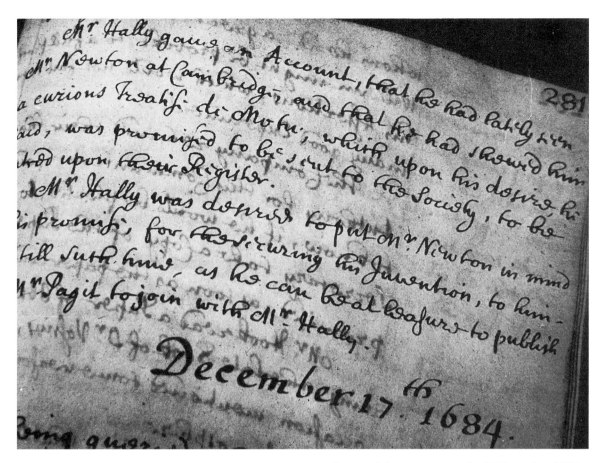

13.9. Registration of Newton's *De motu corporum in gyrum* in the Royal Society minutes for 17 December 1684.

gyrum (*On the Motion of Bodies in an Orbit*) which he sent to the Royal Society in London. It consisted of three definitions, four hypotheses, four theorems, seven solved problems, with added corollaries and scholia. It is an initial version of what, through a series of successive refinements and elaborations, was to become the *Principia*.

Problem 5 has to do with a body falling without resistance in an inverse-square force field; the final Problems 6 and 7 have to do with bodies moving through resisting media. The remainder of the tract is concerned with the resistance-free motion of bodies subject to the action of central forces.

Definition 1 is of "centripetal force". The term occurs here for the first time. "A centripetal force", Newton here says, is "that by which a body is impelled or attracted towards some point regarded as its centre." Newton uses the term throughout the tract, not only when referring to centripetal forces in general, but also when referring to the forces on the planets toward the Sun. In the case of the latter forces, as the manuscript shows, he had originally written 'gravitas', then replaced it by the more abstract term. The inference to draw would appear to be that Newton had hypothesized the centripetal force toward the Sun to be identical with gravity, then drew back from this identification, presumably as being premature. In a scholium to Problem 5, in referring to the motions of projectiles in our atmosphere, he remarks that "gravity is one species of centripetal force".

The explicitly formulated premises of the tract are four; Newton calls them "hypotheses". The first hypothesis originally postulated the absence of all resistance; when Problems 6 and 7 were added, Newton limited the statement to make it apply only to the earlier propositions, where the resistance is taken to be nil. According to Hypothesis 2: "Every body by its innate force alone proceeds uniformly into infinity following a straight line, unless it be impeded by something extrinsic." Hypothesis 3 is the parallelogram law for addition of forces. Hypothesis 4 states that in the beginning of a motion caused by a centripetal force the spaces traversed are as the squares of the times.

The tract is not, of course, hypothetical in the way in which the "Hypothesis of Light" of 1675 was so: it involves no hidden aethereal mechanism. Within a month Newton will have replaced the term "hypothesis" by "lex" (law). The search for a more adequate articulation of the premises will engage him again and again.

Theorem 1 is essentially the same as Prop. 1 of the *Principia*: "All orbiting bodies describe, by radii drawn to their centre, areas proportional to the times." Theorem 2, essentially the same as Prop. 4 of the *Principia*, shows that for bodies moving uniformly in circles the centripetal forces are as the squares of arcs simultaneously described, divided by the radii of their circles. In the fifth corollary following Newton applies this theorem to concentric circular orbits in which the squares of the periods are as the cubes of the radii, and concludes that the force varies inversely as the square of the distance from the centre. He then asserts that this case "does obtain in the major planets circling round the Sun and also in the minor ones orbiting round Jupiter and Saturn . . .".

The orbits of the planets, of course, though nearly circular are not concentric with the Sun, and Newton immediately passes on to Theorem 3, equivalent to Prop. 6 of the *Principia*, which permits the determination of the law of force in non-circular orbits with centre of force elsewhere than at the geometrical centre. Problems 1–3 apply this theorem to particular cases, and it is Problem 3 that gives us the planetary case: in elliptical orbits with centre of force at a focus, the force – positing that this is uniquely directed to the centre – varies inversely as the square of the distance. In an appended scholium Newton then concludes: "The major planets orbit, therefore, in ellipses having a focus at the centre of the Sun, and with their *radii* drawn to the Sun describe areas proportional to the times, exactly as Kepler supposed."

A question is in order concerning the logic of the inference here, but before attending to it, let us take note of Theorem 4 and Problem 4, the final proposi-

tions having to do with planets. Theorem 4 shows that, in an inverse-square field of force, Kepler's third law holds for elliptical orbits just as it does for circular orbits. And a scholium then points out in detail how this result can be used to determine the eccentricities and aphelia of the planetary orbits: the mean solar distances or semi-major axes are found from the periods by way of Kepler's third law (just as we have seen it done in Streete's *Astronomia Carolina*), then, using observations and geometrical procedures known earlier, the second focus or (in the case of the inner planets) the centre of the orbit is located with respect to the Sun.

Problem 4 is essentially the same as Prop. 17 of the *Principia*: it shows how to determine the conic section in which a body would move in an inverse-square field of force, given the absolute value of the force, and the initial direction and speed of the body. It is undoubtedly because he had obtained this theorem that Newton here and in the first edition of the *Principia* felt justified in asserting that in an inverse-square field the orbit would be a conic section.

Finally, in the scholium following Problem 4, Newton announces:

A bonus, indeed, of this problem, once it is solved, is that we are now allowed to define the orbits of comets, and thereby their periods of revolution, and then to ascertain from a comparison of their orbital magnitude, eccentricities, aphelia [! – Newton should have said perihelia], inclinations to the ecliptic plane and their nodes whether the same comet returns with some frequency to us.

Newton then proposes a method of determining the elliptical orbit of a comet. First he would use Wren's technique to find a straight-line path and speed for the comet from three observations. The ellipse is assumed to be tangent to this rectilinear trajectory at some point, and the ellipse is calculated as in Problem 4. The errors in position as derived from this elliptical orbit are then applied to the initial observations in order to compute a new rectilinear trajectory, from which a new elliptical orbit is derived. And so on, until (as Newton hopes) agreement between observation and prediction is reached. The truth is, the method will not work, as Newton himself discovered.

We have now to ask: what is the status of the results that Newton derives in the tract *De motu corporum in gyrum?* Are they certain, probable, or dubious?

Newton has shown that a parabolic, hyperbolic, or elliptical orbit traversed under the action of a central force implies an inverse-square law of force, and he believes he has proved the converse. Moreover, for elliptical orbits with centres of force at one focus, he has shown that if the squares of the periods are as the cubes of the mean solar distances, then the force varies inversely as the square of the distance, and he believes the converse to hold if the orbits are closed.

What is the relation of these results to empirical fact? Newton is aware that not all astronomers agree on the exactitude of Kepler's third law. He is aware that the elliptical orbit and areal rule have not been exactly verified: they were Kepler's conjectures. Why then should Newton assume the inverse-square law to hold exactly for the centripetal force toward the Sun?

The inverse-square field of force could be the consequence of an aethereal mechanism like that which Newton imagined in his "Hypothesis explaining the Properties of Light" of 1675. Other causes, such as pressure in an aethereal vortex, might also be operative, interfering with the exactitude of the inverse-square relation.

It is noteworthy that, in the *De motu corporum in gyrum*, there is not a single indication that Newton believes in universal gravitation, or has deduced any of the consequences of it that differ from those of the simple assumption of an inverse-square field of force. Kepler's third law, as enunciated by Kepler, and thus employed by Newton, is inexact and requires modification if Sun and planet attract each other. If the planets attract each other, the elliptical orbit and areal rule become inexact owing to perturbations.

Somehow, in this first tract, Newton has become 'embarked'. It is possible that the encouragement of Halley served a logical as well as a psychological purpose: Halley believed the comet of 1680–81 to be a single comet, and believed that it had been attracted by the Sun. The enterprise of the *De motu corporum in gyrum* was sustained not by assured foundations but by hope: hope that the inferences drawn by treating the Sun as an immobile centre of an inverse-square field of force were sufficiently exact, hope that comets were acted upon by this field in the same manner as planets, so that their orbits could be determined.

Newton will quickly deepen his analysis, as he discovers the possible implications of the course on which he is embarked.

The augmented tract "De motu corporum" (December 1684?)

In a revised and expanded version of the *De motu*, dated tentatively by D.T. Whiteside as from December 1684, the four "Hypotheses" of the original draft have become five "Laws". And here, in an addition to the scholium following Theor. 4, we find the first clear indication that Newton is thinking of gravitation as universal, and is deducing consequences from that assumption:

Moreover the whole space of the planetary heavens either rests (as is commonly believed) or moves uniformly in a straight line, and hence the common centre of gravity of the planets . . . either rests or moves along with it. In either case the motions of the planets among themselves are the same, and their common centre of gravity rests with respect to the whole space, and thus can be taken for the immobile centre of the whole planetary system. Hence in truth the Copernican System is proved a priori. For if in any position of the planets their common centre of gravity is computed, this either falls in the body of the Sun or will always be close to it. By reason of the deviation of the Sun from the centre of gravity, the centripetal force does not always tend to that immobile centre, and hence the planets neither move exactly in ellipses nor revolve twice in the same orbit. There are as many orbits of a planet as it has revolutions, as in the motion of the Moon, and the orbit of any one planet depends on the combined motion of all the planets, not to mention the action of all these on each other. But to consider simultaneously all these causes of motion and to define these motions by exact laws admitting of easy calculation exceeds, if I am not mistaken, the force of any human mind. Omit those minutiae, and the simple orbit and mean among all the deviations will be the ellipse of which I have already treated. If anyone tries to determine this ellipse by trigonometrical computation from three observations (as is customary), he will have proceeded with less caution. For those observations will share in the minute irregular motions here neglected and so make the ellipse to deviate a little from its just magnitude and position (which ought to be the mean among all the deviations), and so will yield as many ellipses differing from one another as there are trios of observations employed. Therefore there are to be joined together and compared with one another in a single operation a great number of observations, which temper each other mutually and yield the mean ellipse in both position and magnitude.

Here we find clearly affirmed the action of the planets on the Sun and on one another: in every case of action, two bodies are being said to *interact*. That such interaction actually occurs is not argued in the paragraph, but simply affirmed. The asserted interaction has the discouraging consequence that, in all probability, no exact predictive astronomy of the planetary motions will be possible; the best that can be done is to settle for mean orbits, and to console oneself with the thought that, because of the relatively immense mass of the Sun, the centre of gravity of the solar system is always within or close to the body of the Sun, so that the perturbations will be very small.

But how does Newton know where the centre of gravity of the solar system is? How has he "proved the Copernican system *a priori*"?

Only one way of comparing the masses of different celestial bodies was available to Newton, namely that which he employs in Cor. 3 of the first edition, or Cor. 2 of the later editions, of Prop. 8 of Bk III of the *Principia*. It requires that the body have a satellite. The central acceleration at the distance of the satellite is determinable from the period of the satellite. Then, on the assumption that this central acceleration varies inversely as the square of the distance, and is independent of the mass of the satellite, it is possible to compute the central acceleration of any satellite of the central body at any distance. Finally, the quantities of matter in different celestial bodies are taken to be as the centripetal accelerations at equal distances from their centres. In the *Principia* Newton is thus able to compute the relative masses of the Sun, Earth, Jupiter, and Saturn.

This computation presupposes two crucial premisses of Newton's final argument for universal gravitation. The first is that the acceleration of a body towards the Sun, Earth, Jupiter, or Saturn is independent of the mass of the body accelerated, and is thus a function solely of distance. Newton tacitly assumed the premiss in the first version of *De motu*, but in carrying out the comparison of the masses of different celestial bodies he had to articulate it explicitly. The second premiss is that the attractions exercised at a given distance from these several attracting bodies are as their masses. In the *Principia* this second premiss becomes, in Prop. 69 of Bk I, a deduction from the first premiss by way of the third law of motion. An empirical test of the first premiss makes its first appearance in the next document we have to examine.

But before leaving the initial revision we are examining we should take note of an addition Newton here makes to the scholium following Problem 5. After urging that the resistance of the aether in the interplanetary spaces must be nil, Newton proceeds as follows:

Motions in the heavens are ruled therefore by the laws demonstrated. But the motions of projectiles in our air, ignoring its resistance, are known by Problem 4, and also the motions of heavy bodies falling perpendicularly by Problem 5, given of course that gravity is inversely proportional to the square of the distance from the centre of the Earth. For gravity is one species of centripetal force; and by my calculations the centripetal force by which the Moon is held in her monthly motion about the Earth is very nearly to the force of gravity here at the surface of the Earth inversely as the square of the distances.

This is Newton's first explicit report of a "Moon test" of the inverse-square law.

Early revision of the definitions and laws
During the winter of 1684–85 and following spring Newton rapidly improved and deepened his treatment of the foundations of his argument. It is in an early expansion of the definitions that we find a report (in Latin) of an experiment to test the strict proportionality of what we now distinguish as mass and weight:

By Pondus *[the word can mean* either *mass or* weight*] I understand the quantity or amount of matter moved, abstracted from the consideration of gravity, whenever it is not a matter of gravitating bodies. To be sure, the* pondus *of gravitating bodies is proportional to their quantity of matter, and it is allowable to designate or explain analogues by each other. The analogy can actually be demonstrated as follows. The oscillations of two bodies of the same weight [*pondus*] with equal suspensions are counted, and the amount of mat-*

ter in them will be reciprocally as the number of oscillations made in the same time. But in experiments carefully made on gold, silver, lead, glass, sand, common salt, water, wood, and wheat I always found the same number of oscillations.

In a cancelled final sentence, Newton wrote: "On account of this analogy and for lack of a more convenient word I set forth and designate quantity of matter by *pondus*, even in bodies of which the gravitation is not considered." This awkward usage will be quickly abandoned in the revisions that follow, Newton maintaining henceforth the strict distinction between *pondus* and *quantitas materiae*, while demonstrating their proportionality.

The list of materials experimented on is the same as that given in Prop. 6 of Bk III of the *Principia*, where Newton describes the precautions he took in carrying out the experiment, and concludes that the proportion holds good to one part in a thousand.

The fundamental elements of Newton's argument for universal gravitation are now in place. In a careful experiment on nine different materials, carried out in the winter of 1684–85, Newton showed that the masses and weights of terrestrial bodies are proportional, so that they accelerate equally towards the Earth; it follows that the gravitational force to which each is subject is proportional to its mass. The Moon test, carried out for the first time during the same winter, showed that it is this same gravitational force towards the Earth that retains the Moon in her orbit. By inductive generalization the nature of gravity towards the planets and the Sun is then concluded to be the same as that toward the Earth, since the inverse-square law has been confirmed both for the circumsolar planets and for their satellites.

But then, by the third law of motion, or the equality of action and reaction, the attracting body must be in turn attracted; the force to which it is thus subject will be proportional to its own mass. The argument for this conclusion is as follows. The "absolute forces" or attractive powers of two bodies, A and B, may be compared by determining the accelerations towards the one and the other of a test body placed at a given distance from each. Or, since acceleration is here a function of distance only and not of the mass of the test body, A and B may be used as test bodies for each other, the

acceleration of each toward the other thus serving as a measure of the other's attractive power:

$$\frac{\text{the absolute force of } A}{\text{the absolute force of } B} = \frac{a_B}{a_A}.$$

By the third law of motion, $|m_A a_A| = |m_B a_B|$, and so

$$\frac{a_B}{a_A} = \frac{m_A}{m_B}.$$

Therefore,

$$\frac{\text{the absolute force of } A}{\text{the absolute force of } B} = \frac{m_A}{m_B}.$$

In all cases, then, the force of gravitation between two bodies will be proportional to the masses of both, and the distinction between "attracting" and "attracted" becomes only a matter of point of view. The argument no longer seems novel today; as Mach put it, an "uncommon incomprehensibility" has become a "common incomprehensibility". But if our reading of the documents is correct, it was very new in the winter of 1684–85, and a source first of wonder and then of satisfaction to Newton himself.

The firming of the argument

From the available evidence it appears that, during the winter of 1684–85, by an act of thought coupled with experiment and calculation, Newton emerged from uncertainty, and achieved the assurance he required in order to proceed with the composition of the *Principia*.

Two questions, inextricably related to one another, remained. What was the extent of the perturbations caused by the action of the planets on one another, the Sun on the satellites, the satellites on their primaries? And how could one be assured that there was no cause *other* than universal gravitation that had to be taken into account in the celestial motions?

In December 1684 Newton asked Flamsteed for his values for the radii and periods of Jupiter's satellites. Flamsteed replied on 27 December:

I find . . . that [their distances from Jupiter] are as exactly in sequialte proportion to theire periods [i.e., the periods are as exactly proportional to the $\frac{3}{2}$ power of the distances] as it is possible for our sences to determine . . . . I find I can answer all the Eclipses of the first [satellite] that have been carefully observed within *lesse than 2 minutes of time, the fourth has not faild me much more nor the 3d above thrice as much . . . . I use theire motions altogeather aequable onely allowing Roemers aequation of light, without which allowance the error of my tables would be above 10 minutes of time . . . .*

Newton replied on 30 December: "I thank you heartily for your kind information about those things I desired . . . . Your information about the Satellits of Jupiter gives me very much satisfaction. . . ." Newton's satisfaction stemmed not only from the exact verification of the inverse-square law, but also from the evidence that the satellites' motions were but minimally perturbed by the Sun. At the same time Newton was concerned about possible perturbations of Saturn by Jupiter, and in the letter from which we have just quoted wrote:

The orbit of Saturn is defined by Kepler too little for the sesquialterate proportion. This Planet so oft as he is in conjunction with Jupiter ought (by reason of Jupiter's action upon him) to run beyond his orbit about one or two of the Suns semidiameters or a little more and almost all the rest of his motion to run as much or more within it. Perhaps that might be the ground of Keplers defining it too little. But I would gladly know if you ever observed Saturn to err considerably from Keplers tables about the time of his conjunction with Jupiter.

Flamsteed's reply was dated 5 January:

As for the motion of Saturn I have found it about 27′ slower in the Antonicall appearances [= acronychal appearances or oppositions] since I came here, then Keplers numbers, and Jupiter's about 14 or 15′ swifter . . ., the error in Jupiter is not allwayes the same, by reason the place of his Aphelion is amisse in Kepler. Nor is the fault in Saturn allwayes the same . . . yet the differences in both are regular and may be easily answered by a small alteration in the Numbers . . . . I have corrected Jupiter my selfe so that hee has of late yeares answered my calculus in all places of his orbit but I have not beene strict enough to affirme that there is not such exorbitation as you suggest of Saturn . . . . I know Keplers distances of Saturn agree not with the sesquialter proportion and that Jupiter's too ought to be amended and both must be altered before wee set upon the enquiry whether Jupiter's motion had an influence on Saturn's . . . .

Flamsteed is here altogether too optimistic about the ease with which the theories of Jupiter and Saturn might be perfected, as he will be acknowledging in his final years; the difficulties will be resolved only in 1785, by Laplace. But Flamsteed's information was important for Newton, who replied as follows on 12 January:

Your information about the error of Keplers tables for Jupiter and Saturn has eased me of several scruples. I was apt to suspect there might be some cause or other unknown to me, which might disturb the sesquialtera proportion. For the influences of the Planets one upon another seemed not great enough though I imagined Jupiter's influence greater than Your numbers determin it. It would ad to my satisfaction if you would be pleased to let me know the long diameters of the orbits of Jupiter & Saturn assigned by your self and Mr Halley in your new tables, that I may see how the sesquiplicate proportion fills the heavens together with another small proportion which must be allowed for.

The "small proportion which must be allowed for" undoubtedly has to do with a modification of Kepler's third law that takes into account the masses of the Sun and planets. But what has most pleased Newton is Flamsteed's assurance that he need not trust Kepler's or any other astronomer's numbers; this assurance frees him to assume that the available data will not require any explanatory cause besides universal gravitation, which he now takes to be the sole cause of celestial phenomena.

There are, of course, many difficulties remaining that Newton must deal with. He must seek ways of determining perturbations more precisely (as in Prop. 66 of Bk I of the *Principia*); discover how spheres of gravitating matter attract one another, and whether any deviation from the inverse-square law is to be expected close to their surfaces; and solve the nagging problem of fitting orbits to the observations of comets. But he has now laid an experimental and observational foundation for his undertaking, and no fact known to him appears contrary to it. That foundation is the argument for universal gravitation. Its importance to him is shown by the care with which, in the first eight propositions of Bk III of the *Principia*, he chooses his way among purported empirical facts, distinguishing approximately confirmed ones from more precisely confirmed ones, and using each kind at its

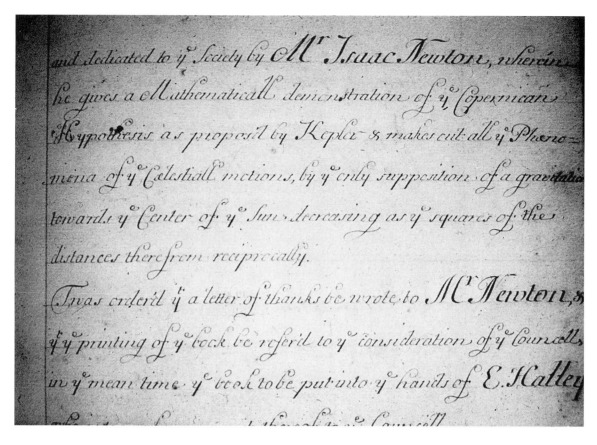

13.10. Report of the formal presentation of the manuscript of Newton's *Principia* (Book I only) to the Royal Society, in the Royal Society minutes for 28 April 1686. At this date Newton had already completed Book II, but Book III still lacked the Theory of Comets. The printing of the whole was finally completed on 5 July 1687.

warrantable value. The security possible in experimental philosophy – never more than temporary – is, at this point, his.

One feature of the argument, as presented in the first edition of the *Principia*, is specially relevant to our analysis; it indicates the extent to which universal gravitation emerged for Newton in contrast and opposition to aethereal hypotheses for gravity. The premises of Bk III in the first edition are there called "Hypotheses". The Hypotheses include the first two of the "Rules of Reasoning in Philosophy" found in the later editions, and most of what will be given in the later editions under the title "Phaenomena". Hypothesis III reads: "Every body may be transformed into a body of any other kind, and all intermediate degrees of qualities may be successively induced." This is a dialectical premiss: it is what the aethereal theorists (including, earlier,

Newton himself) have assumed. Hypothesis III is put to use in Cor. 2 of Prop. 6. The proposition states that: "all bodies gravitate toward each of the planets, and their weights toward any one planet, at equal distances from the centre of the planet, are proportional to the quantities of matter in the several bodies."

Corollary 1 states:

Hence the weights of bodies do not depend upon their forms and textures. For if the weights could be varied with the forms, then they would be greater or less, according to the variety of forms, in equal matter; altogether against experience.

Corollary 2 then tells us:

Therefore, all bodies which are about the Earth gravitate towards the Earth; and the weights of all that are

equally distant from the centre of the Earth are as the quantities of matter in them. For if aether or any other body were either deprived of its gravity altogether or were to gravitate less in proportion to its quantity of matter, then since this body does not differ from other bodies except in form of matter, it could by a change of form be gradually transmuted into a body of the same condition as those which gravitate most in proportion to their quantity of matter (by Hypothesis III), and reversely the heaviest bodies could gradually lose their gravity. And thus weights would depend on the forms of bodies, and could vary with the forms, against what was proved in the preceding Corollary.

In the second edition, Newton improved the argument by eliminating the dialectical assumption, Hypothesis III. Cor. 2 of Prop. 6 is changed so that the second of its three sentences just quoted is replaced by the following two sentences:

This is the quality of all bodies within the reach of our experiments; and therefore (by Rule III [of the Rules of Reasoning in Philosophy]) to be affirmed of all bodies whatsoever. If the aether, or any other body, were either altogether void of gravity, or were to gravitate less in proportion to its quantity of matter, then, because (according to Aristotle, Descartes, and others) there is no difference between that and other bodies but in mere form of matter, by a successive change from form to form, it might be changed at last into a body of the same condition with those which gravitate most in proportion to their quantity of matter; and, on the other hand, the heaviest bodies, acquiring the first form of that body, might by degrees quite lose their gravity.

Thus the assertion of the universality of gravity is now supported principally by a reference to Rule III, the rule of inductive generalization applying to the qualities of bodies:

The qualities of bodies, which admit neither intention nor remission of degrees, and which are found to belong to all bodies within the reach of our experiments, are to be esteemed the universal qualities of all bodies whatsoever.

This rule, according to Newton, is "the foundation of all philosophy".

Finally, in the third edition, to make the argument still more pointed, Newton adds Rule IV:

In experimental philosophy we are to look upon propositions inferred by general induction from phenomena

as accurately or very nearly true, notwithstanding any contrary hypotheses that may be imagined, till such time as other phenomena occur, by which they may either be made more accurate, or liable to exceptions.

Any aethereal hypothesis for gravity would presumably have to come in under the exception clause.

Newton, we conclude, was acutely conscious of the nature of his enterprise. He realized that it could not be freed from every possibility of doubt, that it remained risky. The rule of induction, more a procedure for justifying than a method of discovery, provided such security as was available.

The emergence for Newton of the argument that gravitation is universal and therefore mechanically inexplicable must have brought with it a sense of relief – the happiness of realizing that aethereal conjectures could be set aside as irrelevant. At the same time, to adopt the programme of accounting for celestial phenomena by means of universal gravitation was to assume an enormous task. To what Newton achieved and failed to achieve in this undertaking, we now turn.

II Newton's derivation of consequences from the doctrine of universal gravitation

A From the one-body to the two-body problem, and beyond

To the reader who wishes to understand the main argument of the *Principia*, Newton at the beginning of Bk III gives this advice: "read carefully the definitions, laws, and first three sections (that is, the first 17 propositions) of Bk I, then pass to Bk III "De mundi systemate", consulting the propositions there cited from the earlier books as the reading may require." In the first three sections of Bk I Newton considers the ideal case of a point-mass moving about a centre of force which is not itself attracted by the point-mass but may be at rest or moving in any way; universal gravitation is not here supposed. This was the sort of case that Hooke had provoked him into thinking about in 1679. By complicating the supposition – first replacing the centre of force by a point-mass and so arriving at the case of two mutually attracting bodies, then taking into account the presence of additional bodies that perturb the motions demonstrated in the two-body case, finally allowing that the bodies are not point-masses but extended – Newton in

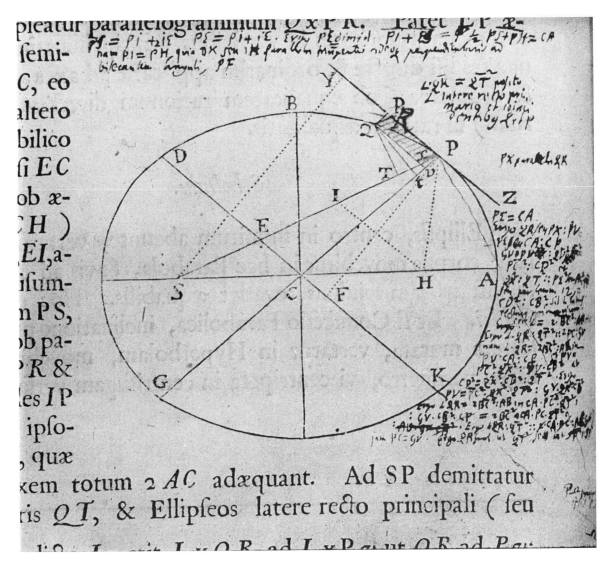

13.11. Leibniz's annotations to page 50 of the first edition of the *Principia*, where Newton shows that motion in an ellipse with centre of force in a focus *S* implies an inverse-square law of force. Shortly after the publication of the *Principia*, Leibniz, in his *Tentamen de motuum coelestium causis*, derived the elliptical orbit on the assumption of a set of 'harmonic' vortices, one for each planet, without invoking universal gravitation.

1684–85 brought his account by successive approximations into closer conformity with what universal gravitation implied. It is along this same course of successive approximations that he guides the reader of the *Principia*.

The logic of Newton's argument has sometimes – for instance by Newton's disciple Pemberton – been misconstrued. In Prop. 11 of Bk I Newton shows that if the orbit is elliptical, and the centre of force is a focus, then the force varies inversely as the square of the distance from the centre. But this proposition is never applied in Bk III: the ellipticity of the orbits is never taken to be an empirical law. In Bk III it is the converse of Props. 11–13, announced in Cor. 1 of Prop. 13, that is relevant: if the force varies inversely as the square of the distance from the centre, then the trajectory will be a conic section with focus at the centre.

The argument for this converse is not spelled out in the first edition of the *Principia*. Johann

Bernoulli, in a letter to Jacob Hermann publicized in the *Mémoires* of the Paris Academy of Sciences for 1710, complained of the logical lapse and supplied his own derivation – which was, however, no more than an analytical working through of the inverse-square case of Newton's general solution of the inverse problem of central forces in Prop. 41 of Bk I of the *Principia*. Newton himself had already in 1709 noted that the converse required explicit proof, and, preferring to argue for the exhaustiveness of the solution provided by the conic-sections, asked Roger Cotes, who was preparing the second edition for the press, to add the following two sentences to Cor. 1 of Prop. 13:

For, given a focus, the point of contact, and the position of the tangent, there can be described the conic which shall have a given curvature at that point. The curvature, however, is given from the given centripetal force and the velocity of the body; and two orbits which are tangent one to the other cannot be described by the same centripetal force and same velocity.

Newton's invocation of curvature here, in the absence of detailed explanation, appeared to some contemporaries to leave the conclusion doubtful. The final clause of the quotation presents a more serious difficulty: if it is intended as a purely mathematical proposition it needs to be proved, and if it is an assertion about physical possibility it is irrelevant, because mathematical and physical possibility do not imply each other.

Newton's attempted establishment of the elliptical orbit as the paradigm for planetary motion, as it emerges in Prop. 13 of Bk III of the *Principia*, assumes the inverse-square law as established earlier in the book on the basis of two independent sorts of empirical evidence: (1) data confirming Kepler's third law for the orbits of the circumsolar planets and of the satellites of Jupiter, taken in a first approximation as circles concentric with the central body; and (2) the fact – in truth more persuasive for the circumsolar planets although often passed over in silence by summarizers of Newton's argument – of the quiescence or near-quiescence of the orbital aphelia, the orbits here being recognized to be eccentric to the Sun as well as to one another (of this latter argument, more in Section B below).

Before the near-ellipticity of the orbits can be regarded as established, however, there are yet further steps in Newton's argument that must be taken into account. First, the abstract "centre of force" supposed in the first three sections of the *Principia* must be replaced by a body which not only attracts but is attracted. That Newton had derived the consequences of this step by 12 January 1685 may be inferred from his remarking, in a letter to Flamsteed of that date, that Kepler's third law requires to be modified by a small proportion.

In the *Principia* the consequences of the two-body problem are derived in Props. 57–60 of Bk I. After showing in Prop. 57 that two bodies attracting each other describe similar figures about their common centre of gravity and about each other mutually, Newton goes on in the following three propositions to derive the relations of the orbits, velocities, and periodic times in the one-body and two-body cases. With the same law of force, the orbits will be the same, and the areal rule will continue to hold, but the periodic time in the two-body case is shorter in a given proportion. From Prop. 60 the necessary modification of Kepler's third law follows directly: if $m_1, m_2, T_1, T_2, r_1, r_2$, are the masses, periodic times, and mean solar distances of two planets, and S is the mass of the Sun, then

$$\frac{T_1{}^2(S+m_1)}{T_2{}^2(S+m_2)} = \frac{r_1{}^3}{r_2{}^3}.$$

Although Newton expected to apply this relation to the determination of the mean solar distances of Jupiter and Saturn, the uncertainty in the periods of these planets before Laplace's discoveries in 1785–86 kept the application from yielding a significant result. Nevertheless, Newton's proof that under the assumptions of the two-body situation the elliptical orbit, areal rule, and a modified form of Kepler's third law follow from an inverse-square law of attraction is an essential stage in his "deduction of the motions of the heavens *a priori*", to use his phrase in Prop. 13 of Bk III.

The applicability of Kepler's laws as thus modified to the circumsolar planetary orbits presupposes in addition, of course, that the presence of more than one planet does not greatly disturb the consequences derived for the two-body case. In Prop. 66 of Bk I Newton argues that any one of several planets will move more nearly in an ellipse, and more nearly in accordance with the areal rule, if the central body (the Sun) is agitated by all and only the attractions that universal gravitation im-

Table 13.1. *Newton's values for the masses of Jupiter, Saturn, and the Earth, stated as fractions of the mass of the Sun, in the three editions of the "Principia"*

	First edn	Second edn	Third edn	Newcomb's values
Jupiter	1/1100	1/1033	1/1067	1/1047.38
Saturn	1/2360	1/2411	1/3021	1/3501.6
Earth	1/28700	1/227512	1/169282	1/329330

plies. But the masses of the planets, compared with the mass of the Sun, must also be so small that the centre of gravity of the system never departs far from the centre of the Sun, and that the mutual perturbations of the planets can be dismissed as negligible (Prop. 65 of Bk I).

In Cors. 1 and 2 of Prop. 8 of Bk III Newton undertakes to compute the relative masses of the planets with known satellites, using in effect the formula

$$m/S = (r/R)^3(T/t)^2,$$

where m is the mass of the planet and S that of the Sun, r is the mean distance of the satellite from the planet and R is the mean solar distance of the planet, T is the periodic time of the planet about the Sun and t that of the satellite about its primary. Newton's values for these mass-ratios in the first, second, and third editions of the *Principia*, as compared with Newcomb's values at the end of the nineteenth century, are given in Table 13.1. The manuscripts show Newton revising these results repeatedly. That his values for the Earth's mass err so much in excess is due to his faulty estimates of the Earth's solar parallax, 20″ in the first edition, 10″ in the second edition, and 10″.55 in the third edition, the correct value being 8″.8; the fraction r/R in the formula is a little more than sixty times the sine of this angle.

From these results Newton calculates (in Prop. 12 of Bk III) that the centre of gravity of the Sun and Jupiter falls a little outside the body of the Sun, the centre of gravity of the Sun and Saturn a little below its surface. He knows that the actions of these two bodies on each other, and particularly of Jupiter on Saturn, are not negligible; but imagines that the errors in prediction for Saturn can be largely avoided by placing the focus of its ellipse in the centre of gravity of Jupiter and the Sun (Prop. 13). In late 1694 Newton appears to have made an effort "to determine the orb of Saturn", introducing

a libratory motion into its line of apsides; but nothing came of the effort. Of the major, long-term perturbations that Jupiter and Saturn cause in each other's motions Newton had no inkling.

As for the masses of the remaining known planets, Mercury, Venus, and Mars, no way of determining them was available; but finding that the densities of the Earth, Jupiter, and Saturn diminish in the order of their solar distances, Newton in the final corollary to Prop. 8 conjectures that the relation holds for all the planets, "for the planets were to be placed at different distances from the Sun, [so] that, according to their degrees of density, they might enjoy a greater or less proportion of the Sun's heat". A monotonic relation between distance and density will again be conjectured by Leonhard Euler in 1754 and J.L. Lagrange in 1781, in order to obtain values for the masses of Mercury, Venus, and Mars, and thus of the perturbations they cause.

In any case, Newton believed he could conclude the masses of these planets to be very small. For according to the astronomers Streete and Mercator the apsides of the planets from Mercury to Mars are at rest with respect to the stars, while according to Kepler, Ismaël Boulliau, and Vincent Wing their motions are very small. Quiescence or near-quiescence of the aphelia Newton took to be an indication that the perturbations were negligible or very small. According to Prop. 14 of Bk III, the aphelia and nodes of the planets are fixed; in the text of the proposition, however, Newton allows that they can be subject to slow motions, owing to the mutual actions of planets and comets. In the second edition he adds a scholium in which he computes the motions of the apsides of Mercury, Venus, and the Earth, on the assumption that the motion of the Martian apse is $\frac{1}{3}′$ per year (an exaggerated value), and further that all these motions are caused solely by the action of Jupiter and Saturn – a questionable supposition, and the results are correspondingly poor. But the main fact remains: the near-quiescence of the planetary aphelia is a crucial element in Newton's argument.

B The motion of the apsides in general and of the lunar apse in particular

In his letter to Newton of 6 January 1679/80, Hooke – having no doubt learned a few things from Newton's letter of 13 December – mentions three different laws of attraction toward a centre, with the consequences they will have for orbital shape: (1) if the force is constant, the apses will fail to unite by about a third of a circle; (2) if the force varies inversely as the square of the distance from the centre, as Hooke supposes to be the case in the heavens, the successive apses will unite in the same part of the orbit, and be opposite the nearest approach to the centre; (3) if the force varies directly as the distance, as Hooke supposes to be the case in the motion of a simple pendulum and also for a body beneath the surface of the Earth, the successive apses will be 180° apart, with the nearest approaches to centre midway between. To these assertions among others Newton must be referring when in his letter to Halley of 20 June 1686 he complains that Hooke had "written in such a way as if he knew and had sufficiently hinted all but what remained to be determined by the drudgery of calculations and observations".

Hooke's assertions, derived in part from Newton's preceding letter, suggest thinking of the angular distance between consecutive apses of an orbiting body as a function of the law of centripetal force. For orbits approximating to circles, Newton reduces this idea to calculation in Props. 43–45 of Bk I of the *Principia*. Here he shows that a body may be made to move in a revolving eccentric orbit if to the centripetal force that would cause it to move in the same orbit at rest there is added a force varying inversely as the cube of the distance from the centre. (The motion of the body occurs under the limitation that its heliocentric longitude from the original apse always bears a constant ratio to the corresponding heliocentric longitude in the stationary orbit; thus the angular motion of the orbit is not uniform.) In Prop. 45 Newton then shows, for the case in which the orbit approximates closely to a circle, that if the centripetal force varies as the distance raised to the power $(n-3)$, the angle separating upper apse from the next succeeding lower apse will be $180°/n^{\frac{1}{2}}$, very nearly. Only for $n=1$, the

case of the inverse-square law, is this angle 180°. For $n>1$ the apse retrogrades; for $n<1$ it progresses. For $n=0$ (in which case the force varies inversely as the cube of the distance) the orbit ceases to have an apsidal line and becomes a Cotes spiral.

Newton uses this result in Cor. 1 of Prop. 45, and again in Prop. 3 of Bk III, to argue that if the upper apse advances about 3° per revolution, as in the case of the Moon, then the exponent of the distance in the law of force will be $(\frac{360}{363})^2 - 3 = -2\frac{4}{243}$ very nearly. "Therefore the centripetal force decreases in a ratio somewhat greater than the squared ratio, but which approaches $59\frac{3}{4}$ times nearer to the squared than the cubed." The departure from the inverse-square law, Newton asserts, is due to the action of the Sun.

Toward the end of Cor. 2 of Prop. 45 Newton enters on what appears to be the converse calculation: if from the attraction which the Earth exerts on the Moon is subtracted a force $\frac{1}{357.45}$ as great, and varying as the Moon's distance from the Earth, then it follows that the apse will advance 1° 31′ 28″ per revolution. To this calculation in the third edition Newton adds the statement: "The apse of the Moon is about twice as swift." Both calculation and appended statement puzzled later readers of the *Principia*. In Section C below we shall justify the number $\frac{1}{357.45}$, as representing the average value of the radial component of the Sun's perturbing force. What is left out in the calculation is the effect of the transradial perturbing force (at right angles to the radius vector).

Nowhere in the *Principia* does Newton explicitly undertake to calculate the motion of the Moon's apse. In the winter or early spring of 1687, however, it appears that he attempted such a derivation, intending to include it in Bk III as the culmination of Props. 25–35 on the perturbations of the Moon, though in the event the derivation remained in manuscript until modern times (the text of it is given by Whiteside in his edition of Newton's mathematical papers). The steps of the attempted derivation may be briefly characterized as follows.

First, assuming the lunar orbit to be an ellipse with the Earth at a focus, Newton obtains expressions for the instantaneous movement of the apse due to radial and transradial perturbing forces. Let V and W represent the radial and transradial perturbing forces, P the force of the Earth on the Moon, e the Moon's orbital eccentricity, r the radius vector, and v the Moon's true anomaly. The motion of the apse compared with the Moon's mean motion can then be expressed, in the case of the radial perturbation, as $(Vr/eP)\cos v$, and in the case of the transradial perturbation, as $(2Wr/eP)\sin v$.

It is in the second step of the argument that the crucial difficulty emerges. If the orbit, being circular or oval, had the Earth as its centre of symmetry, the perturbing forces would cause no apsidal motion; for in the course of a revolution each perturbative effect would be exactly cancelled by an opposing effect. But by how much, at each point, can the lunar orbit be said to depart from such symmetry? "The eccentric orbit in which the Moon revolves", Newton recognizes, "is not an ellipse but an oval of another kind." Nevertheless, "to lessen the difficulty of the computation", he assumes the orbit to be a Keplerian ellipse, with the Earth at one focus.

As the next step, we today might propose that the radius vector in the ellipse be expressed in terms of angle of anomaly, and the perturbative effects be added up (integrated) for a full synodic month. Whether Newton in fact ever carried through such a calculation may be a moot point, but Clairaut did so later and we know the result: it is only about half the observed motion of the lunar apse. The elliptical orbit is an insufficiently good approximation to the actual orbit to serve without refinement as a basis for determination of the apsidal motion.

Newton in fact proceeds as follows (we put r for the radius vector in the ellipse, and a for the semi-major axis). First, he multiplies the perturbative forces by (a/r), thus transferring the Moon from the eccentric ellipse into a concentric circle of radius a, in which no apsidal motion can occur. Next, he assumes that, in transferring the Moon back to the ellipse, the perturbative forces causing no motion of the apse will diminish, the radial forces as the square of the distance from the Earth, the transradial as the cube of that distance. Then he subtracts the perturbative forces thus calculated from the actual perturbative forces in the ellipse, to obtain the net perturbative forces causing motion of the apse.

Newton's decision here to multiply the transradial forces by $(a/r)^3$ lacks all logical basis; its sole justification is that it leads to a calculated value for

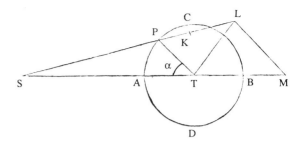

13.12. The orbit of the Moon (P) about the Earth (T), as perturbed by the Sun (S). The label α for angle PTS has been added to the diagram.

the apsidal motion in near agreement with observation.

Having thus 'fudged' the result, Newton in the final steps of the calculation obtains, first, the apsidal motion for a given position of the Moon in its orbit and given position of the line of syzygies; next, by integration, the average apsidal motion for a given elongation of the Sun from the apse; finally, by a further integration over a full cycle of the Sun's motion from the apse, the annual apsidal motion. In the MS the latter figure is given as 38° 51′ 51″, to be compared with the 40° 41′.5 found in Flamsteed's tables.

In a scholium to Prop. 35 of Bk III of the first edition of the *Principia*, Newton states that by computation he has found the motion of the lunar apse to be about 40° per year, and that his results differ somewhat from those in Flamsteed's tables; but "the computations, being too intricate and burdened with approximations, and moreover insufficiently accurate, it is not agreeable to append". In the later editions, the scholium is totally changed, and all reference to an attempted computation of the annual motion of the lunar apse deleted.

C. Other perturbations of the Moon

In corollaries to Prop. 66 of Bk I, Newton undertakes to determine how the motion of a body like the Moon, taken as pristinely circular and uniform, is altered by the action of a distant body such as the Sun. His analysis is based on the diagram in Figure 13.12. Here T is the Earth, S the Sun, P the Moon, $ADBC$ the circular orbit of the Moon, and $PTA = \alpha$ the angle of the Moon's elongation from the Sun. If TS is taken to represent the Sun's accelerative force on the Earth, and so also the Sun's *average* accelerative force on the Moon, then the Sun's accelerative force on the Moon at P will be given, in accordance with the inverse-square law, by

$$LS = TS(PS/TS)^{-2} = TS^3/PS^2.$$

The vectorial components of LS parallel to TS and PT are then (because of the similar triangles LMS and PTS)

and
$$MS = (TS/PS)LS = TS(TS/PS)^3$$
$$LM = (MS/TS)PT = PT(TS/PS)^3.$$

Now PS is $TS - PT \cos \alpha$ very nearly, and the fraction $(TS/PS)^3$ is equal in a first approximation to

$$1 + 3(PT/TS) \cos \alpha.$$

The net perturbing force parallel to TS is then

$$MS - TS = 3 \, PT \cos \alpha,$$

of which the radial component (directed outward along TP) is

$$3 \, PT \cos^2 \alpha = \tfrac{3}{2} PT \, (1 + \cos 2\alpha),$$

and the transradial component

$$-3 \, PT \sin \alpha \cdot \cos \alpha = -\tfrac{3}{2} PT \sin 2\alpha.$$

Meanwhile the component LM directed from P to T is $PT[1 + 3(PT/TS)\cos \alpha] \approx PT$ to within 1%, and thus the net radial perturbing force is

$$PT - \tfrac{3}{2} PT(1 + \cos 2\alpha) = -\tfrac{1}{2} PT \, (1 + 3 \cos 2\alpha).$$

But the average value of $\cos 2\alpha$ over a full synodic revolution is zero, and thus the average value of the radial perturbing force is $-\tfrac{1}{2}PT$, in comparison with TS taken as the acceleration of the Earth toward the Sun. From Cor. 2 of Prop. 4 of Bk I, with T_E and T_M as the periods of the Earth about the Sun and the Moon about the Earth, it then follows that

$$\frac{\text{radial perturbing acceleration}}{\text{acceleration of Moon to Earth}} = \frac{(PT/2TS)(TS/T_E^2)}{(PT/T_M^2)}$$

$$= \tfrac{1}{2}(T_M/T_E)^2$$
$$= \tfrac{1}{2}(27.3215/365.256)^2$$
$$= 1/357.45,$$

the very number which, as previously mentioned, Newton introduces without explanation in Cor. 2 of Prop. 45.

1. The variation. In Corollaries 2 and 3 of Prop. 66 of Bk I, Newton infers from the existence of the transradial perturbing force on a body such as the Moon that the rate of areal description increases as the body passes from quadrature to syzygy, and decreases as it passes from syzygy to quadrature;

thus, other things being equal, it moves more swiftly in the syzygies than in the quadratures. In Cors. 4 and 5 he points out that, from the combination of the transradial and radial perturbing forces, the orbit (other things being equal) will have greater curvature in the quadratures and less in the syzygies, and consequently the body will go further from the centre in the quadratures than in syzygies.

In Bk III Newton undertakes to derive quantitative results for these effects as they apply to the Moon. In Prop. 26, assuming a circular orbit, he develops by integration a numerical formula for the rate of description of area as it varies owing to the transradial perturbing force of the Sun. In Prop. 28, taking curvature to vary directly as the force of attraction and inversely as the square of the velocity, he finds the Moon's distances from the Earth in quadrature and syzygy to be as 70 to 69. From the inequalities in areal description and in orbital shape he then in Prop. 29 seeks to derive the effect on the Moon's longitude. This is the 'variation' which Tycho had discovered, and to which he had assigned the value $40'.5 \sin 2\alpha$; the coefficient is more nearly $39'.5$. Newton obtains the value $35'.16$.

In the first edition of the *Principia*, Newton after noting the discrepancy states that Halley has most recently found the maximum value of the variation to be $38'$ in the octants near the Moon's opposition with the Sun, and $32'$ in the octants near conjunction, so that the mean value is $35'$. "We have computed the mean value, neglecting the differences which can arise from the curvature of the Earth's orbit, and the Sun's greater action on the new and crescent Moon than on the full and gibbous Moon." By the time Newton had written his *Theory of the Moon's Motion*, first published by David Gregory in his *Astronomiae physicae & geometricae elementa* (1702), he had abandoned this explanation. But in agreement with Cor. 14 of Prop. 66 he observes that the variation is subject to an annual inequality, varying inversely as the cube of the Earth's distance from the Sun, so that its maximum values oscillate between $37' \, 25''$ and $33' \, 4''$.

The fact is that $35'$ is very nearly the optimum value for Newton to have arrived at. Flamsteed's value of $36' \, 45''$, obtained on the basis of observation, is similarly lower than the correct value. Newton, Halley, and Flamsteed all took as a basis for their determination of the variation the Horrocksian theory of the evection (see below, Section 3), which incorporated some 5'.25 of the variation.

2. The annual inequality. The average effect of the radial perturbing force, as stated above, is to diminish the centripetal force on the orbiting body P. In Cor. 6 of Prop. 66 of Bk I, Newton points out that if the perturbing force were to become alternately greater and smaller, so that the net centripetal force became alternately smaller and greater, the orbit would be alternately dilated and contracted; and the periodical time, varying jointly as the $\frac{3}{2}$ power of the radius directly and as the square root of the centripetal force inversely, must alternately decrease and increase. But the Earth is closer to the Sun at its perihelion in winter than at its aphelion in summer, and hence the perturbing force on the Moon is greater in winter than in summer. Thus Newton can conclude that "the mean motion of the Moon is slower in the perihelion of the Earth than in its aphelion" (Prop. 22 of Bk III). This is the annual inequality, which had been discovered first by Tycho and then by Kepler.

Kepler had incorporated the annual equation in the equation of time, in the belief that the Earth's relative proximity to the Sun in winter made it rotate faster during that half of the year than in the summer. Seventeenth-century astronomers generally followed this practice, although Huygens and Flamsteed objected to it. Newton's explanation of the annual equation, and his argument (in Cor. 22 of Prop. 66 of Bk I and Prop. 17 of Bk III) that the diurnal rotations of the planets must be uniform made the earlier supposition untenable.

In his *Theory of the Moon's Motion.* Newton gives 11' 49" as the maximum value of the annual inequality in the Moon's mean motion. In the scholium following Prop. 35 of Bk III of the second edition of the *Principia*, Newton states that he has computed this value from the theory of gravity, taking the eccentricity of the Earth's orbit to be $(16\frac{7}{8})/1000$. A first-order calculation of the effect, in accordance with the Newtonian rule (given above) for the variation of the lunar period, yields $3e'(T_M/T_E) \sin v'$ for the annual equation, where v' is the true anomaly of the Earth in its orbit, e' is the eccentricity of the Earth's orbit, and T_M, T_E are the mean periods of the Moon and the Earth; the maxi-

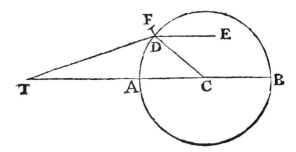

13.13. Newton's diagram for the Horrocksian theory of the lunar evection.

mum value, assuming Newton's values for the constants, is 13' 1". Just how Newton arrived at the value 11' 49" is unclear; it is larger than the correct value by about 37".

3. The evection and an oscillation of the lunar apse. As the Moon passes from apogee to perigee and back again, the varying of the net force to which it is subject is least in accord with the inverse-square law when the apsides are in the syzygies. It is at such times, Newton is thus able to argue (in Cor. 8 of Prop. 66 of Bk I), that the apse advances most rapidly; so arises what he called the semi-annual equation of the Moon's apse. The departure from the inverse-square law is also responsible for a semi-annual cycle in the value of the eccentricity of the lunar orbit: as Newton shows in Cor. 9 of Prop. 66, the eccentricity increases as the apsides are passing from quadratures to syzygies, and decreases as the apsides are returning from syzygies to quadratures.

These effects are qualitatively in agreement with Horrocks's lunar theory (see Chapter 10). In fitting the semi-annual inequalities of the apse and eccentricity with numbers in his *Theory of the Moon's Motion*, Newton in effect merely adjusts the parameters of Horrocks's theory. In the second and third editions of the *Principia*, in the scholium following Prop. 35 of Bk III, Newton gives a description of this theory. In Figure 13.13, let T be the Earth, TC the mean eccentricity of the lunar orbit, or $550\frac{1}{2}/10\,000$, and CB such that CB/TC is the sine of the greatest value of the semi-annual inequality in the position of the apogee (from observation Newton concludes this greatest value to be 12° 18'). According to the theory, the centre of the Moon's orbit rotates in a circle of radius CB about C;

the angle *BCD* being twice the Sun's angular distance from the mean position of the Moon's apogee. The angle *DTC* is the semi-annual equation of the lunar apogee, and *TD* is the corresponding value of the lunar eccentricity.

Neither the exact form of this theory nor its numerical parameters are derived by Newton from his theory of gravitation. The additional lines in the diagram are introduced to take account of the annual variation in the value of the solar perturbing force. The essential result is that the centre of the Moon's orbit is at *F* in the diagram, rotating annually in a small epicycle about *D*, which in turn rotates about *C* with twice the average angular speed of the Sun from the Moon's apogee.

4. Inequalities of the inclination and motion of the nodes. Of the two components into which Newton analyses the perturbing force in Prop. 66, namely *LM* and *MT* (see Figure 13.14), the first is in the plane of the orbit *CADB*, and so cannot affect the orientation of that plane. The second, however, is parallel to the line connecting *S* and *T*, and unless this line is also the line of nodes, the component $MT = 3PT \cos \alpha$ will be outside the plane of the orbit, and will act to change the orbital inclination and the position of the nodes. In Cors. 10 and 11 of Prop. 66, Newton describes what the effects will be: the inclination is greatest when the nodes are in the syzygies, and least when they are in the quadratures; the nodes recede most rapidly when in the quadratures, are quiescent when in the syzygies, and when between quadratures and syzygies alternately recede and progress, but mostly recede, so that the overall movement is one of regression.

The oscillation in the inclination of the lunar orbit had been discovered observationally in the late sixteenth century by Tycho Brahe, who then deduced the motions of the nodes from a theory embodying the changes in inclination. With Tycho's findings, the perturbational effects described in Cors. 10 and 11 of Prop. 66 agree qualitatively. In Props. 30–35 of Bk III Newton undertakes a quantitative derivation of these effects. For the mean annual motion of the nodes he obtains 19° 18′ 1″ as compared with the observational value 19° 21′ 21″; the difference he puts down to his not having taken account of the eccentricity of the lunar orbit. For the inequality in the

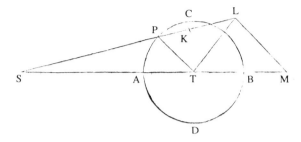

13.14. The orbit of the Moon (*P*) about the Earth (*T*), as perturbed by the Sun (*S*).

place of the nodes in the octants he finds 1° 30′, and for the maximum variation of the inclination from its mean value, 8′ 53″. The variations in the true place of the node and in true inclination being analogous to the evection in longitude, Newton supplies a procedure for calculating them analogous to the Horrocksian construction for the evection. These several results are among the most satisfactory of the computations of perturbational effects in the *Principia*.

5. Further refinements. In Prop. 22 of Bk III of the *Principia* (all editions), Newton points out that he has accounted for the variation, the annual equation, the evection, and the inequalities of the inclination and motion of the nodes that Tycho had discovered; and that "these are the principal [lunar] inequalities taken notice of by astronomers".

But, he adds, "there are yet other inequalities not observed by former astronomers, by which the motions of the Moon are so disturbed that to this day we have not been able to bring them under any certain rule". As we mentioned previously, both the variation and the evection undergo an annual oscillation in value. There are other such annual inequalities, Newton goes on to say, in the average rates of motion of the apse and node, and in the semi-annual inequality to which the motion of the node is subject. In all these cases the maximum quantity augments and diminishes inversely as the cube of the Sun's distance from the Earth, or (in Newton's formulation) directly as the cube of the Sun's apparent diameter.

In Newton's *Theory of the Moon's Motion*, the annual inequalities of the motions of the apse and node are assigned the maximum values 20′ and 9′ 30″; in the scholium following Prop. 35 of Bk III in the second edition of the *Principia*, these numbers are changed to 19′ 43″ and 9′ 24″, respectively.

Two further inequalities, unmentioned in the first edition of the *Principia* but given in both the *Theory of the Moon's Motion* and the later editions of the *Principia*, are according to Newton derivable from the theory of gravity. When the line of apsides is in the line of syzygies, and also when the line of nodes is in the line of syzygies, the action of the Sun on the Moon is greater than when the apses or nodes are in the quadratures. Whence arise oscillations in the Moon's mean motion, which may be expressed by

$$-3′\ 45″\ (1 - 3e'\cos v')\sin 2(L' - A_\mathrm{M}),$$

and

$$-47″\ (1 - 3e'\cos v')\sin 2(L' - N_\mathrm{M}),$$

where e' is the eccentricity of the Earth's orbit, v' is the Earth's (or Sun's) true anomaly, L' is the longitude of the Sun, A_M is the longitude of the aphelion of the Moon, and N_M is the ascending node of the Moon's orbit. The factor $(1 - 3e'\cos v')$ represents very nearly the inverse-cube variation with the Earth–Sun distance. Newton does not explain how these terms were derived.

Two further inequalities are described in the *Theory of the Moon's Motion*. No mention is made of them in any edition of the *Principia*; it may be that Newton had attempted to derive them, not from the theory of gravitation, but from the observations that Flamsteed transmitted to him in 1694–95. The first of them, which Newton calls "the sixth equation of the Moon's place", may be expressed by

$$-2′\ 10″\ \sin[(L - L') + (A_\mathrm{M} - A_\mathrm{S})],$$

where $(L - L')$ is the angular distance of the Moon from the Sun, and $(A_\mathrm{M} - A_\mathrm{S})$ is the angular distance between the apogees of the Moon and the Sun. What Newton labels "the seventh equation" is proportional to $-2′\ 20″\ \sin(L - L')$, but varies in magnitude according to the situation of the Moon's apogee. If this coincides with the Sun's apogee, the equation is 54″ greater; if with the Sun's perigee, 54″ less. And

. . . these things hold good when the lunar apogee is in the Sun's syzygies; but when it sticks in the Sun's quadrature, the aforesaid equation is to be diminished about 50″ or one whole minute, when the Moon's apogee and the Sun's are in conjunction; but if they are in opposition, I can't determine (for want of observations) whether it is to be increased or diminished.

Neither can I affirm anything certain concerning the aforesaid increase or decrease of the equation 2′ 20″, for want of observations exact enough.

How good was Newton's lunar theory as predictive astronomy? In introducing the theory when he published it in 1702, David Gregory said that Newton

. . . has compassed this extremely difficult matter, hitherto despaired of by astronomers; namely, by calculation to define the Moon's place even out of the syzygies, nay in the quadratures themselves so nicely agreeable to its place in the heavens (as he has experienced it by several of the Moon's places observed by the ingenious Mr Flamsteed) as to differ from it (when the difference is the greatest) scarce above two minutes in her syzygies, or above three in her quadratures; but commonly so little that it may well enough be reckoned only as a defect of the observation.

Gregory is quite mistaken here. When Flamsteed tested the theory against observations, he found errors in longitude as high as 8′ or 9′, and errors in latitude as high as 4′ (letter to John Caswell of 25 March 1703). Entirely similar errors were found years later by Halley. Of the four small inequalities or 'equations' added in the *Theory of the Moon's Motion* of 1702, only the term

$$3′\ 45″\ \sin(L' - A_\mathrm{M})$$

has the correct form and, very nearly, the correct coefficient. The task Newton set himself in the 1690s was to discover the small departures from truth of an essentially Horrocksian theory; to do this he used a combination of reasoning from his theory of perturbation as developed in Prop. 66 of Bk I, and analysis of a somewhat sparse series of lunar observations supplied by Flamsteed. Without adequate data, without a method of harmonic analysis, and without a method of developing the perturbations systematically out of a perturbing function, the difficulties proved too great. With hindsight we may say that Newton's effective adoption of Horrocks's theory, by interfering with on-going insight into perturbations not actually embraced in that theory, proved ultimately an insurmountable obstacle to him.

Newton's rules for calculating the place of the Moon were incorporated in the tables of Charles Leadbetter's *Uranoscopia* (1735); in the tables that Flamsteed constructed about 1702 and which, having been given by Halley to P.-C. Lemonnier, were published by the latter in his *Institutions astronomiques* of 1746; and in Halley's *Tabulae astronomicae* (1749). But they were not clearly better than a purely Horrocksian procedure for determining the place of the Moon; and the errors were far too large to permit longitude at sea to be determined to within 1° – the application of lunar theory most sought after during the early eighteenth century.

D. The attraction of spheres

In his letter to Newton of 6 January 1679/80, Hooke after urging that the forces between the celestial bodies obey an inverse-square law, added two qualifications:

Not that I believe there really is such an attraction to the very Centre of the Earth, but on the Contrary I rather Conceive that the more the body approaches the Centre, the lesse will it be Urged by the attraction . . . . But in the Celestiall Motions the Sun Earth or Centrell body are the cause of the Attraction, and though they cannot be supposed mathematicall points yet they may be Conceived as physicall and the attraction at a Considerable Distance may be computed according to the former [inverse-square] proportion as from the very Centre.

Six and a half years later, when Newton wrote to Halley on 20 June 1686, he had apparently forgotten these qualifying remarks, for he says that Hooke had told him

. . . that gravity in descent from hence to the centre of the Earth was reciprocally in a duplicate ratio of the altitude, that the figure described by projectiles in this region would be an Ellipsis and that all the motions of the heavens were thus to be accounted for . . . And yet . . . the first of those three things he told me is fals and very unphilosophical, the second is as fals and the third was more then he knew or could affirm me ignorant of by any thing that past between us in our letters.

Earlier in the letter Newton had stated that:

I never extended the duplicate proportion lower then to the superficies of the Earth and before a certain demonstration I found the last year have suspected it did not reach accurately enough down so low: and therefore in the doctrine of projectiles never used it nor considered the motions of the heavens. . . .

J.C. Adams in 1888 and again Florian Cajori in 1928 made this last-quoted remark the basis of an explanation of Newton's supposed "twenty-year delay" in publishing the theory of universal gravitation. The knowledge that spheres with radially symmetrical densities attract as though their masses were concentrated at their centres, and that this law extends down to the surface of the spheres, is indeed materially relevant to Newton's "Moontest" of the inverse-square law of universal gravitation. Pemberton's earlier explanation of the delay had been that Newton in comparing the centripetal acceleration of the Moon and a stone at the Earth's surface had used a faulty value (namely Galileo's) for the Earth's radius; but as Cajori points out, better values for the Earth's radius were available and known to Newton during the twenty-year "delay". The truth is, as we have argued in the first part of this essay, that Newton did not have the idea of universal gravitation in 1666, and thus there was no delay requiring explanation.

The available evidence suggests that Newton tackled the question of the gravitational attraction of spheres early in 1685, after his argument for universal gravitation had at length come to be formulated. The results of his investigation appear in Props. 70–76 of Bk I of the *Principia*, and include a proof that, on the assumption of an inverse-square force, a corpuscle within a uniform sphere of matter is attracted to the centre with a force varying as its distance from the centre, while a corpuscle exterior to the sphere is attracted to the centre with a force varying inversely as the square of the distance. Prop. 76 proves the theorem that is relevant astronomically: spheres within which the density varies solely as a function of distance from the centre attract one another with forces varying inversely as the square of the distance between their centres. (In the propositions following 76, Newton, ever the mathematician, goes on to consider how spheres attract under laws other than the inverse-square.)

There is no sign that the proofs of these propositions cost Newton exceptional travail. He proceeds in them by a combination of straightforward geometry with limiting procedures, in the manner to which his reader was assumed already to be accustomed.

E. The precession of the equinoxes

Before the appearance of Newton's *Principia*, nothing approaching a correct explanation of the precession of the equinoxes had ever been proposed. Newton managed to derive a figure for the mean rate of precession agreeing exactly with the empirical result obtained by astronomers in his day, namely 50″ per year. Of his derivation we may say that it postulates the true cause, but arrives at its conclusion by steps that are usually dubious when, upon narrow analysis, they are not plainly wrong.

As previously noted, in Cor. 11 of Prop. 66 of Bk I Newton shows that the nodes of the orbit of a body in the situation of our Moon will regress, owing to the perturbations caused by a third body such as the Sun. In Cor. 18 he asks us to imagine that many such moons, orbiting at equal distances from the central body, become fluid and unite so as to form a fluid ring. The ring, he asserts, will exhibit the same effects of perturbation as the orbit of the single moon: its inclination will be subject to an oscillation, and its nodes will regress. In Cor. 20 he then supposes the ring to solidify, and to be made to adhere to a much enlarged central body, so as to become an equatorial bulge. The central body will then have to participate in the motions of the solid annulus or bulge, and so the nodes or intersections of its equator with the plane defined by its centre and the orbit of the perturbing body will regress.

The idea is right, but Newton has not in fact demonstrated that the solid ring will show the same perturbational effects as the orbit of a single moon. In the first edition of the *Principia*, he reiterates the crucial assumption as Lem. III preceding Prop. 39 of Bk III: If all parts of the Earth were removed except the equatorial bulge, then, other things being unchanged, the motion of the equinoctial points would be the same whether the ring were fluid, or consisted of a hard and rigid matter. Newton appears to be explicitly acknowledging the unproved character of this proposition when in the second edition of the *Principia* he relabels it "Hypothesis II".

The derivation of the annual rate of precession in Bk III of the *Principia* involves further assumptions. In order to obtain the ratio of the equatorial to the polar radius of the Earth (and thence the thickness of the equatorial bulge), Newton in Prop. 19 supposed the Earth to be uniformly dense, and to have the figure of equilibrium that a rotating fluid body would assume under the action of the gravitational attractions of its parts. He concluded that the polar radius would be shorter than the equatorial radius by $\frac{1}{230}$ of the latter (the value of this fraction accepted today on the basis of observation is $\frac{1}{297}$).

The derivation also requires that we know the ratio of the forces of the Moon and Sun to move the seas or to act upon the Earth's equatorial bulge. For data, Newton used the heights of the tides in the Bristol Channel as measured by Samuel Sturmy, first when the Moon was in opposition or conjunction with the Sun, and then when it was in quadrature. The former of these figures, he stated, was proportional to the sum of the two forces, the latter to their difference. But the calculation, as Newton carried it out in Prop. 37, involved a number of refinements. The greatest tides occur not at the syzygies but somewhat later, when the Sun's force is no longer parallel to the Moon's; at quadrature the Moon's declination from the equator diminishes its force, as does its greater average distance (consideration of its orbital eccentricity being omitted) from the Earth. Adjusting for these factors, Newton reached the result that the Moon's and Sun's forces to move the seas were to one another as 4.4815 to 1.

The trustworthiness of this result was challenged in 1740 by Daniel Bernoulli, who by a calculation commencing with the periods of the tides rather than their heights arrived at a ratio of 2.5 to 1. But there is no trustworthy calculational path from data on the tides to the desired ratio. Today it is derived otherwise and is given a value of about 2.18 to 1.

Newton's calculation of the annual quantity of the precession is given in Prop. 39, and depends on the three lemmas immediately preceding. In the first two of these lemmas Newton attempts to compare the turning-moments of the Sun's force on the matter of the equatorial bulge (1) when this matter is imagined to be concentrated at a point on the Earth's equator farthest from the ecliptic; (2) when the same matter is distributed in a ring in the equatorial plane, concentric with the Earth's axis; (3) when it is distributed, as it is in fact, in a lunula of revolution. Although Newton correctly computes the comparative values of the turning-moments in these three cases, he assumes the resulting motions will be as the moments, without taking into account the inertial effect of the different distributions of mass. Because of this same omission, when he attempts in Lem. III to determine how the motion of precession in an equatorial ring will be shared with an underlying sphere to which the ring becomes attached, the result is wrong. Nowhere here does Newton take account of the angular momentum of the rotating Earth.

The calculation in Prop. 39 starts from the known regression of the nodes of the lunar orbit caused by the Sun, and on the assumption that the rates of regression for different moons would vary as their periods, determines the rate of regression for a moon circling once a day at the Earth's surface. This moon is then converted into an equatorial ring, and the various ratios derived in the lemmas are applied to it. The result is that the Sun acting on the Earth's equatorial bulge would cause an annual precession of $9'' 7''' 20^{iv}$. The Moon's action, being to the Sun's as 4.4815 to 1, then becomes $40'' 52''' 52^{iv}$. The sum of the two is $50'' 00''' 12^{iv}$, "the amount of which motion agrees with the phenomena, for the precession of the equinoxes, by astronomical observations, is about $50''$ yearly".

F. The determination of cometary orbits

As mentioned earlier, in the original tract *De motu* of November 1684, Newton proposed a method for determining the trajectories of comets that is quite impractical. On 19 September 1685 Newton wrote to Flamsteed that although he had "not yet computed the orbit of a comet", he was "now going about it: and taking that of 1680 into fresh consideration, it seems very probable that those of November and December were the same". Yet on 20 June 1686 he wrote to Halley that Bk III of the *Principia* "still wants the Theory of Comets. In Autumn last I spent two months in calculations to no purpose for want of a good method, which made me afterwards return to the first Book and enlarge it with divers Propositions some relating to Comets others to other things found out last Winter."

At some time between June 1686 and the end of March of the following year, when Newton sent his manuscript of Bk III to Halley for printing, he arrived at a new solution; it is presented in Lems. V–XI and the following Prop. 41 of that book. "This being a problem of very great difficulty", Newton

remarks at the beginning of Prop. 41: "I tried many methods of resolving it; and several of those problems, the composition whereof I have given in the first book, tended to this purpose. But afterwards I contrived the following solution, which is somewhat simpler."

If simpler than the procedures attempted earlier, Newton's method is nevertheless complex, and we shall not attempt a detailed description of it here. It is a way of fitting a parabolic path to three observations; the parabola being chosen because all parabolas have the same shape (that is, are similar in the strict geometrical sense), and so are determined by fewer conditions than ellipses or hyperbolas, while on the other hand the perihelial portion of an ellipse or hyperbola can always be approximated by a 'parabola of curvature'.

Crucial to the method is Lem. VIII, which shows how, given three positions A, B, C of a comet in a parabolic arc ABC about the Sun S (Figure 13.15), to find a point E dividing the chord AC, very nearly, in the ratio of the times of description of the arcs AB, BC. The construction involves extending the diameter of the segment μI (which bisects the chord AC and passes through the point of tangency of the tangent parallel to AC) to O so that $\mu O = \frac{1}{2}\mu I$, then extending OS to ξ so that $S\xi = 2SO$. The line ξB will then intersect the chord in the required point E, to a close approximation. The approximation improves as the point B approaches the vertex μ.

In determining a cometary orbit from observations, Newton applies the lemma in reverse. From three observations, separated by time intervals nearly equal (one of the observations, if necessary, can be 'constructed' out of other observations, using Newton's well-known method of interpolation by differencing), the lines of sight from Earth to comet, as projected onto the plane of the ecliptic, are obtained; in Figure 13.16, these are TA, tB, and τC. A point B is then chosen to represent the projection onto the ecliptic of the comet's position at the time of the middle observation; Newton shows how this may be done so that the placement is at least roughly right, and the remainder of the procedure is designed to correct for the error.

From B toward the Sun S the line BE is drawn of such a length as to represent the projection onto the ecliptic plane of the 'sagitta' of the parabolic arc, given that the 'sagitta' of the arc of the Earth's orbit during the same time is tV; the two sagittae

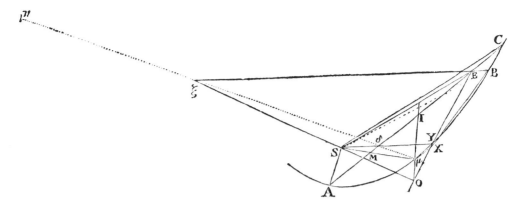

13.15. Newton's diagram for Lemma VIII, following Prop. 40 of Book III of the *Principia*: to divide the chord *AC* of a parabola into segments *AE* and *EC* very nearly proportional to the times required by a comet to traverse the arcs *AB* and *BC*.

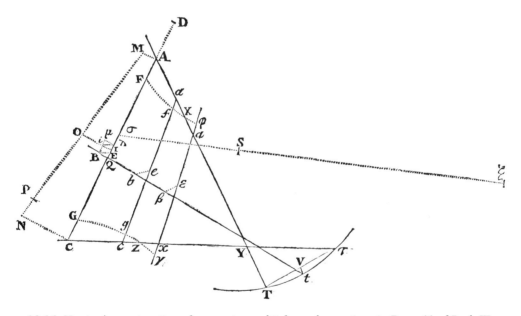

13.16. Newton's construction of a cometary orbit from observations in Prop. 41 of Book III.

must be to one another inversely as the squares of the solar distances of the comet and the Earth. Through *E* it is a matter of simple geometry to pass a straight line *AEC* in such a direction that the segments cut off on it by *TA* and *τC* will be in the required ratio of the time intervals; the line *AEC* is then a first approximation, in length and position, to the chord of the comet's parabolic arc.

Next, Newton shows how by finding the vertex *μ* of the parabolic segment to obtain a better approximation to the chord, in agreement with the proper-

ties of the parabola. Then, making use of the observed latitudes and the known relation between velocity in a parabolic orbit and solar distance, he finds by calculation the length the chord projected onto the plane of the ecliptic should have; this length may and most likely will differ from the length of the previously determined projection *AC*.

The entire process is repeated three times over, with different choices for the initial point *B*, and yielding each time a different placement of the projected chord and a different calculated value for

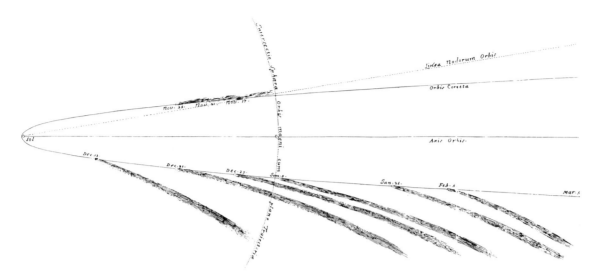

13.17. Newton's diagram of the trajectory of the comet of 1680–81, from the first edition of the *Principia*.

its proper length. A final position, in which both the orientation and length of the projected chord will be correct, is obtained by graphical interpolation.

Newton presents his method as partly arithmetical and partly graphical, but it is reducible to arithmetical calculations. In the particular case in which the lines of sight as projected onto the ecliptic prove to be parallel or nearly so, the method, taken in the strict sense of Newton's description, becomes unusable; but the difficulty can be overcome, A.N. Kriloff has shown, by making the same construction in a different plane, for instance in the plane of the cometary orbit. Another difficulty for which Kriloff has to make allowances arises from the fact that the projection of the cometary orbit onto the ecliptic, although parabolic, does not have the Sun at its focus. Kriloff applies the method to a number of comets, obtaining orbital elements in close agreement with those found by the methods, quite different from Newton's, that were commonly employed in the nineteenth and twentieth centuries.

Newton's own chief application, in the *Principia*, was to the comet of 1680–81, the trajectory of which he had struggled to determine six years before. This time (see Figure 13.17) he is more successful. In the first edition, after reviewing the observations, and from three of them computing the elements of the comet's path, he compares computed and observed positions of the comet for four dates from 12 December to 5 March, and finds that the errors in longitude never exceed 2'; one of the errors in latitude is as high as 10'.5. For three observations prior to perihelion, on November 17, 21, and 25, the errors are larger, going as high as 32' in longitude and 22' in latitude; they are to be explained, Newton suggests, both by the crassness of the observations and by the inaccuracies of the graphical method.

In the second edition of the *Principia*, we learn that Halley has recalculated the orbit by an arithmetical calculus, and comparing observed and computed positions of the comet for 16 dates from 12 December to 5 March finds that the errors in longitude rise no higher than 3' 5", and the errors in latitude no higher than 2' 14".

In the third edition, a further putative discovery is reported: Halley, taking the brilliant comets which had appeared in 44BC, AD531, and AD1106 to be previous apparitions, at a period of 575 years, calculated the "true" elliptical orbit of the comet so identified. From the elements found, he computed the positions of the comet for 25 observations from 3 November 1680 to 9 March 1680/1; the errors in longitude and latitude did not exceed about 2'.5. (However, J.F. Encke in 1818, using additional data for the comet of 1680–81, found Halley's supposition of a 575-year period unlikely; and J.R. Hind in 1852 finally disposed of it as altogether improbable.)

The comet, Newton remarks (in the second and third editions), while visible travelled over nine signs, from Capricorn to Gemini, and its motion as viewed from the Earth was very uneven, for:

. . . about the 20th of November it described about 5° a day; then with a retarded motion from Nov. 26 to Dec. 12, in the space of 15½ days it described only 40°; later with a motion once more accelerated, it described nearly 5° a day, until its motion again began to be retarded. And the theory that agrees excellently with a motion so unequable, through so large a part of the heavens, that observes the same laws as do the theories of the planets, and is accurately in agreement with accurate astronomical observations, cannot not be true.

The complete details of Newton's and Halley's determinations of the orbital elements of the comet of 1680–81 are not available to us. The method of Prop. 41 can have furnished no more than a first approximation, which was then refined by *ad hoc* adjustments. The method is not eminently practical, and has seldom been used since. Nevertheless, Newton's subsumption of cometary motion under the dynamics implied by universal gravitation marks an epoch, setting the stage for all future study of comets.

III The outlook after the *Principia*

The force of Newton's argumentation for universal gravitation, in the first edition of the *Principia*, was not universally persuasive. G.W. Leibniz, for instance, urged that the "mathematical attraction" to which Newton had recourse was no better than a miraculous and senselessly occult quality.

At noted earlier, Newton in the second edition revised his presentation of the opening assumptions of Bk III, introducing a special section entitled "Rules of Philosophizing", and here arguing that both the doctrine of atomism and that of universal gravitation were derived by a general induction from phenomena. And in the General Scholium appended at the end of the book he affirmed that "whatever is not deduced from the phenomena is to be called an hypothesis; and hypotheses, whether metaphysical or physical, whether of occult qualities or mechanical, have no place in experimental philosophy". So imperious an empiricism, at first glance justified by the enormous contrast between Newton's impressively confirmed theory and the conjectural hypotheses it replaced, fostered a distorted account of the history of Newton's great discovery, and for generations of natural philosophers after Newton exercised a dampening influence upon the freedom of hypothesizing.

Meanwhile, the Newtonian theory, considered as a program to account for the celestial motions, remained nearly at a standstill during the first half of the eighteenth century. The perturbations of the Moon were not yet accurately enough known to permit determination of the longitude at sea. The vagaries of Jupiter and Saturn remained unaccounted for.

At least some of the blame for this standstill must be placed upon Newton himself. In his later years he could not bear in any way to be seen to be wrong, nor to acknowledge that the solution of problems in the *Principia* might be improved upon.

When advances finally came to be made, they involved a new mode of attack. The perturbations of the Moon were adequately addressed after a new beginning was made from the equations of motion, without prior assumption of the theory of Horrocks. As for the perturbations of the planets, the only systematic method of calculation available before the middle of the eighteenth century was numerical integration, so laborious prior to the age of the electronic computer that it was applied to celestial motions but two or three times in the eighteenth century. The new start here needed was made when Euler in 1747 introduced the method of approximating by trigonometric series. Determination of perturbations now began with a force diagram, but geometry was then left behind, the problem being formulated in the symbolism of the Leibnizian calculus. The symbolic formulation by no means guaranteed the adequacy of the solution found, and it was some forty years before Laplace with an assist from Lagrange at last showed how the major perturbational terms could be identified. This symbolic working-out of the consequences of Newton's theory will be a major theme of Volume 2B.

Further reading

I. Bernard Cohen, Isaac Newton, *Dictionary of Scientific Biography*, vol. 10 (New York, 1974), 42–103

Alexandre Koyré and I. Bernard Cohen, *Isaac Newton's Philosophiae Naturalis Principia Mathematica: The Third Edition (1726) with Variant Readings* (Cambridge, 1971)

J.A. Lohne, Hooke *versus* Newton, *Centaurus*, vol. 7 (1960), 6–52

Richard S. Westfall, *Force in Newton's Physics* (London, 1971)

Richard S. Westfall, *Never at Rest: A Biography of Isaac Newton* (Cambridge, 1980), with extensive guide to further reading

D.T. Whiteside, Before the *Principia*: the maturing of Newton's thoughts on dynamical astronomy, 1664–1684, *Journal for the History of Astronomy*, vol. 1 (1970), 5–19

D.T. Whiteside, Kepler, Newton and Flamsteed on refraction through a 'regular aire': the mathematical and the practical, *Centaurus*, vol. 24 (1980), 288–315

D.T. Whiteside, Newton's early thoughts on planetary motion: a fresh look, *British Journal for the History of Science*, vol. 2 (1964–65), 117–37

D.T. Whiteside, Newton's lunar theory: from high hope to disenchantment, *Vistas in Astronomy*, vol. 19 (1975–76), 317–28

D.T. Whiteside, The mathematical principles underlying Newton's *Principia mathematica*, *Journal for the History of Astronomy*, vol. 1 (1970), 116–38 *acronychal:* happening in the evening or at nightfall [from Greek *akros*, tip or point + *nux*, night]. From Antiquity onwards, the acronychal risings of the superior planets Mars, Jupiter, and Saturn were important events for the astronomer, because when one of these planets rises as the Sun sets, it is directly opposite (or *in opposition to*) the Sun; and thus its geocentric longitude, or longitude as observed from the Earth, is the same as its heliocentric longitude, or longitude as it would be observed from the Sun. Whether the astronomer considered the Earth or the Sun as the centre of the planetary system, heliocentric longitudes of the planet were required in order to ascertain important parameters in its theory (eccentricity of the deferent or orbit, place or aphelion or apogee, etc.).

anomaly: unevenness or inequality of condition or motion [from Greek, *an*, privative + *homalos*, even]. The term has been used in astronomy for any deviation from regularity or uniformity of angular motion. Thus in the motion of the planets Ptolemy distinguishes between "the anomaly with respect to the Sun", which depends on the planet's elongation from the Sun and leads to stations and retrogradations, and "the zodiacal anomaly", which depends on the planet's position along the ecliptic; both kinds of anomaly are departures from the positions the planet would have if it moved with angular uniformity about the Earth. *Anomaly* is also used by Ptolemy to signify the mean motion of a planet on its epicycle, since this motion leads to the appearance of irregularity. The terms *mean anomaly, eccentric anomaly, co-equated* or *true anomaly* (listed separately in this glossary) were used from Kepler onwards in the analysis of elliptical planetary motion.

aphelion: the point of a planet's orbit that is farthest from the Sun [formed by Kepler from *aph = apo*, off, from + *hēlios*, Sun, in analogy to the term *apogaion* of Ptolemaic astronomy]. From Kepler's time until the end of the eighteenth century, the aphelion was taken as the zero of anomaly in planetary tables, and the apogee as the zero of anomaly in solar and lunar theory. The Paris Bureau des Longitudes shifted the zeroes of anomaly to perihelion and perigee in the first years of the nineteenth century, following a recommendation of Nicolas-Louis de Lacaille (1713–62), so that the rules of calculation used in determining planetary orbits could be applied without change to cometary paths.

apogee: the point in the orbit of the Moon, the Sun, or of any planet, that is farthest from the Earth [from Greek *apo*, off, from + *gē*, the Earth]. Compare *aphelion.*

apsides: plural of *apsis* [from Greek *apsis*, a fastening, the felloe of a wheel, hence a wheel, arch, or vault]; in astronomy, one of the two points in the orbit of a planetary body at which it is respectively at its greatest and least distance from the body about which it is conceived to revolve: the aphelion or perihelion, apogee or perigee. The straight line joining these two points is called *the line of apsides.*

areal rule: the rule introduced by Kepler (and now known as his second law) to determine the motion of a planet on its elliptical orbit. According to the rule, the areas swept out by the radius from Sun to planet are proportional to the times. Thus in Figure G.1, from Kepler's *Astronomia nova*, the times for the planet to move from aphelion at point *a* to points *m* and *b* are to one another in the ratio of the area *amn* to the area *abn*, where *n* is the position of the Sun. Since the areas *amn* and *abn* in the ellipse are proportional to the areas *akn* and *aen* in the circumscribing circle, Kepler nor-

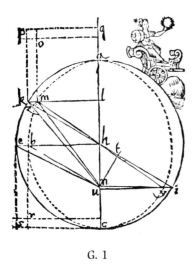

G. 1

mally substitutes the latter for the former in applying his rule.

bisection of the eccentricity: in Ptolemy's theories for Venus and the outer planets, Mars, Jupiter, and Saturn, the eccentricity of the deferent circle (or distance from the Earth to the centre of this circle) is just half the eccentricity of the equant (or distance from the Earth to the point about which the motion on the deferent is assumed to be angularly uniform). Kepler refers to such an arrangement as "the bisection of the eccentricity". In his studies of the motions of Mars, Kepler found that the midpoint of the line of apsides fell half way between the Sun and the point round which the planet's motion came nearest to being angularly uniform. Similarly, in the case of the motion of the Sun relative to the Earth, he found that the midpoint of the line of apsides could not be identical with an equant point, or centre of uniform angular motion, as in all earlier solar theories, but must be placed at only half the distance of the putative equant point from the Earth. These confirmations of the "bisection of the eccentricity" proved to be important steps preparing the way for Kepler's discovery of the elliptical orbit.

coequated (or *true*) *anomaly:* in seventeenth- and eighteenth-century astronomy, the angle between the highest apse (aphelion or apogee) and the true position of the planet or satellite on its orbit, as seen from the central body. The anomaly is said to be 'coequated' because it has been calculated from the *mean anomaly* by the ad-

dition or subtraction of an *equation* such as to yield a result equal to or coequal with the true position of the planet. In Kepler's planetary theory the passage from mean anomaly to true anomaly involves two stages: first the *eccentric anomaly* is obtained from the *mean anomaly*, either by interpolation in Kepler's tables or by approximative solution of the equation $M = E + e$ sin E, where M is mean anomaly, E eccentric anomaly, and e the eccentricity; then the *true anomaly* (v) is obtained from the *eccentric anomaly* either by interpolation in Kepler's tables or by solution of the equation

$$\tan(v/2) = \tan(E/2)[(1 - e)/(1 + e)]^{\frac{1}{2}}.$$

In Figure G.1, the coequated anomaly when the planet is at m is $\angle$ *anm*.

deferent: in pre-Keplerian astronomy, a circle bearing an epicycle [from the Latin *deferre*, to carry down or away]; thus the centre of the epicycle was conceived to move on the circular deferent, while the planet moved on the epicycle. In the Ptolemaic system, the deferent circle of an outer planet corresponds approximately to the modern orbit of the planet, and the epicycle to the orbit of the Sun or Earth; for an inner planet the correspondence is reversed.

dichotomy, lunar: see *lunar dichotomy.*

eccentric: an eccentric circle, that is, a circle or orb not having the Earth (or Sun) precisely in its centre [from Greek *ek*, out of + *kentron*, centre]. Thus Ptolemy employs an eccentric circle in accounting for the anomaly of the Sun. Kepler refers to the circle circumscribing an elliptical orbit as an eccentric; whence his term 'eccentric anomaly' (which see).

eccentric anomaly: in Keplerian planetary theory, an angle with vertex at the centre of the circle circumscribing the elliptical orbit, and one side extending to the upper apse (aphelion or apogee), the other to a point obtained by projecting the planet's position onto the circumscribing circle along a line perpendicular to the line of apsides. See Figure G.1, where if the planet is at m on the ellipse, the eccentric anomaly is $\angle$ *ahk.*

eccentricity: the distance between the centre of an eccentric circle or orbit and the body (Sun or planet) about which the epicycle's centre or planet or satellite moves, considered as a fraction of the semi-diameter of the eccentric circle or semi-transverse axis of the orbit. In Figure G.1 the eccentricity is given by *hn/ha*.

elongation: the difference in longitude between a planetary body and the Sun. From Antiquity onward, the observations most necessary to construction of the theories of the inner planets were those of maximum eastern and western elongations, in which the line of sight can be taken to be tangent to the planet's path. In the case of the outer planets, the elongations take on all possible values, and there are no maxima; but determinations of the longitude and latitude of the planet when its elongation from the Sun is 180° (so that its rising is *acronychal* (which see)) were of special importance in constructing the theory.

ephemeris, plur. *ephemerides:* a table showing the predicted positions of a heavenly body for every day (or multiple of days) during a given period [from Greek, *eph* = *epi*, on + *hēmera*, day: lasting only for a day].

epicycle: a small circle whose centre moves on the circumference of a greater circle [from Greek *epi*, on + *kyklos*, circle]. In the medieval period the greater circle came to be called the *deferent* (which see).

equant: let *E* be the Earth or Sun, *P* a planet that moves on the eccentric circle with fixed centre *C*, and *Q* a point on the line of apsides drawn through *E* and *C* (Figure G.2). In the Ptolemaic theories for Venus and the outer planets, *CQ* is made equal to *EC*, and the motion of the planet on the eccentric circle with centre *C* is made to be angularly uniform about *Q*, so that ∟ *PQA* increases uniformly with time. The point *Q* received from medieval astronomers the name *punctum aequans*, shortened later in English to *equant* [from Latin *aequans*, present participle of *aequare*, to make equal].

equation: a term introduced by medieval astronomers for the angle to be added to or subtracted from a mean motion in order to 'correct' it, that is, in order to obtain a theoretical position in agreement with the position observed. Thus the 'equation of centre' for a planet at point *P* on the eccentric circle (Figure G.2) is the angle *QPE*, because this angle when added to the mean anomaly (∟ *AQP*) yields the coequated anomaly (∟ *AEP*). The word 'equation', in the special sense of a quantity to be added or subtracted in order to 'correct' a mean motion, has continued

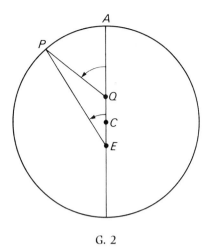

G. 2

in use among astronomers down to the present; as in the term 'equation of time'.

equation of time: the difference between mean and apparent solar time. To understand the definition, we must know that *hour circles* are great circles of the celestial sphere passing through the celestial poles; and that the *hour angle* of a star is the arc of the celestial equator included between the meridian and the star's hour circle. *True* or *apparent solar time* may then be defined as the hour angle of the Sun, measured in units of time, with twenty-four hours corresponding to the whole circuit of the equator. The apparent motion of the Sun as projected onto the celestial equator by hour circles is not uniform; first because the Sun's apparent motion along the ecliptic is not uniform, and second because the ecliptic is inclined to the celestial equator, so that equal arcs on the ecliptic do not correspond to equal arcs on the equator. If we imagine a fictitious sun (called the Mean Sun) that moves uniformly on the celestial equator with the mean speed of the apparent Sun, its hour angle will be *mean solar time*. The equation of time, or difference between apparent and mean solar time, varies through the year and rises at maximum to about sixteen minutes in November.

evection: the name given by Ismaël Boulliau (in his *Astronomia Philolaica* (Paris, 1645)) to the Moon's second or synodic inequality [from Latin *e-*, out + *vehere*, to carry], on the assumption that the Moon's elliptical orbit was simply displaced or 'evected' in the course of the synodic month, without change of shape. The theory is wrong, but the name has persisted. From its discovery by Ptolemy to the seventeenth century the second inequality was regarded as menstrual, so as to be completed in a synodic month; Kepler transformed it into a semestral inequality with a period of 6.75 months, in which time the line of syzygies returns to the line of apsides; and Horrocks gave it the form in which Newton encountered it, that of an ellipse whose line of apsides and eccentricity both undergo sinusoidal variations with the period of 6.75 months.

geoheliocentric system: the planetary system developed by Tycho Brahe and others in which the planets Mercury, Venus, Mars, Jupiter, and Saturn circle the Sun, while the Sun goes round the Earth at rest in the centre. The history of this idea is discussed in Chapters 1 and 3 of the text.

lunar dichotomy: the situation in which the surface of the Moon facing Earth is divided into two precisely equal halves by the terminator – the sunrise or sunset line between the dark and sunlit parts of the Moon. In this situation the lines connecting the centres of the Earth, Moon, and Sun form a right triangle with the right angle at the Moon. In the third century BC Aristarchus of Samos first proposed the use of lunar dichotomy to find the ratio of the distances of the Sun and the Moon from the Earth: it is only necessary to measure the elongation of the Moon from the Sun when the Moon is dichotomized in order to determine this ratio by trigonometry. Unfortunately, to ascertain the moment of exact dichotomy is well-nigh impossible, given the unevenness of the lunar surface. Aristarchus's value of the ratio of Earth–Sun to Earth–Moon distance is 19:1, too small by a factor of 20. In the seventeenth century, attempts were made to use the telescope in determining the moment of dichotomy; in this way Harriot found 140:1 for the distance ratio, and Riccioli found 123:1 (the correct value is 389:1).

mean anomaly: an angle proportional to the time that has elapsed since the planet was last at the upper apse of its orbit, and such that 360° corresponds to a complete period. It is a primary aim of any predictive planetary theory to provide a way, starting from the mean anomaly as datum, to calculate the true or coequated anomaly (which see). In the elliptical astronomy introduced by Kepler, the mean anomaly becomes proportional to area (see the entry areal rule).

Mean Sun: a fictitious sun that is imagined to move eastward along the celestial equator with the mean speed of the true or apparent Sun along the ecliptic, so as to return to the vernal equinox in 365.2422 days. Up to Kepler's time the observationally determined elongations of the planets used in constructing predictive theories of their motions were reckoned from the Mean Sun. Kepler, believing the real Sun to have a causal role in moving the planets, introduced the practice of taking the true Sun, rather than the Mean Sun, as the point of reference.

nodes of an orbit: the points in which a planetary orbit intersects the plane of the ecliptic, or plane defined by the apparent annual motion of the Sun. The node in which the planet passes from the southern to the northern side of the ecliptic is the ascending node; the opposing node is the descending node.

octant: in astronomical theory, a point in the course of a planet or satellite in which it is 45° (an eighth of a circle) from a particular point such as apogee or perigee, aphelion or perihelion, conjunction or opposition. Thus the Moon is said to be in the octants when it is 45° either way from conjunction with or opposition to the Sun; a planet is said to be in the octants of anomaly when it is 45° either way from aphelion or perihelion.

parallax: in the most general sense of the term, the difference between the directions of an object as seen from two different points. In astronomy the two different points are commonly (1) a point of reference such as the centre of the Earth or Sun, and (2) some point of observation on the surface of the Earth. The necessity of taking account of the Moon's parallax in deriving its apparent position from theory was recognized in Antiquity; as Ptolemy puts it (Almagest IV, 1), "the straight line drawn from the centre of the Earth . . . through the centre of the Moon to a point on the ecliptic, which determines the true position . . ., does not in this case always coincide, even sensibly, with a line drawn from some point on the Earth's surface, that is, the observer's point of view, to the Moon's centre, which determines its apparent position." The question of the magnitude of solar parallax, that is, the difference between the position of the Sun as seen from the Earth's surface and the position it would have if observed from the Earth's centre, became a pressing one for the first time in the seventeenth century; see Chapter 7. Tycho, adopting the ancient and exaggerated value of 3' for the solar parallax when the Sun is on the horizon, and consequently 'correcting' observed meridian altitudes of the Sun in accordance with this mistaken value, arrived at a solar theory with exaggerated eccentricity; and the error necessarily propagated itself into all the planetary theories of Kepler's *Rudolphine Tables*. Improvements in the predictive theories of the planets during the seventeenth century were thus dependent on reductions in solar parallax (in addition, they depended on the introduction of better tables of astronomical refraction). Claims to have detected *stellar parallax* – the difference in direction of a star as seen from different points of the Earth's orbit – were made by Hooke in 1674 and by Flamsteed in 1698; these claims were mistaken, but form the background to the observations of Bradley that led to the discovery of the aberration of light and nutation of the Earth's axis.

perigee: the point in the orbit of the Moon, the Sun, or of any planet, that is closest to the Earth [from Greek *peri*, around + *gē*, the Earth]; opposed to *apogee.*

perihelion: the point of a planet's orbit that is closest to the Sun [formed by Kepler from *peri*, around + *hēlios*, the Sun, on the analogy of *perigee*]; opposed to *aphelion.*

precession of the equinoxes: the earlier occurrence of the equinoxes in each successive sidereal year, due to the retrogradation of the equinoctial points along the ecliptic, produced by the slow rotation of the Earth's axis round the poles of the ecliptic. The quantity of the retrogradation is about 50.3″ per year. Thus the longitudes of the stars, which are measured from the vernal equinox eastward along the ecliptic, are augmented by this quantity annually.

prosthaphaeresis: the general sense of the term is of a quantity to be added or subtracted, or of a procedure involving addition or subtraction of a quantity [apparently a contraction from Greek *prosthesis*, addition + *aphairesis*, subtraction]. Ptolemy uses the word to denote what was later called the *equation of centre* (which see): the angle to be added to or subtracted from the mean place of a planet in order to find its true or actual, apparent place. The term in this astronomical sense was still occasionally used in the sev-

teenth century. During the sixteenth century it was also used to designate various processes in algebra and trigonometry, of which the most important for astronomy were the reductions of products of sines and cosines to formulae involving only addition or subtraction:

$$\sin x \cdot \sin y = \tfrac{1}{2}[\cos(x-y) - \cos(x+y)],$$
$$\cos x \cdot \cos y = \tfrac{1}{2}[\cos(x-y) + \cos(x+y)].$$

Products of sines and cosines occur in the solution of spherical triangles, which are required for the reduction of observations to celestial coordinates; since logarithms were not yet available, the multiplications were time-consuming and for that reason subject to calculational error. The above formulae permitted replacement of the multiplications by swiftly performed and easily checked additions and subtractions. Johann Werner (1468–1522) has been credited with the discovery of the first of these prosthaphaeretic formulae, but his work on the subject was never published, and has been lost. By what route or routes the two formulae came to be included in the manual of trigonometry used by Tycho on Hven was a disputed question in the sixteenth century, and has remained vexed.

quadrant: in astronomical theory, a point in the course of a planet or satellite in which it is 90° or a quarter of a circle from a particular point such as apogee or perigee, aphelion or perihelion, conjunction or opposition. Thus a planet is said to be in the quadrants of anomaly when it is 90° either way from aphelion or perihelion.

quadrature: the position of one heavenly body relative to another when they are 90° apart. In particular, the Moon is said to be in quadrature when it is 90° distant from the Sun, or midway between the points of conjunction and opposition.

retrogradation: the portion of the courses of the five planets Mercury, Venus, Mars, Jupiter, and Saturn, in which they reverse their normal direction of motion from west to east against the background of the stars, and move instead from east to west.

syzygy: the conjunction or opposition of the Moon with the Sun [from *syn*, with + *zygos*, yoke].

Ṭūsī couple: a device invented by the Sufi philosoher and mathematician Naṣīr al-Dīn al-Ṭūsī (1201–74) to produce harmonic motion along a straight line or along a great arc of a sphere. The device consists of an epicycle mounted on a deferent of equal size, the motion of the point or body on the epicycle being double that of the epicycle's centre on the deferent. This 'couple' and various generalizations of it were used in the thirteenth and fourteenth centuries by Persian astronomers of the so-called 'Maragha School', in order to avoid such violations of the principle of uniform circular motion as Ptolemy's use of the equant. Knowledge of some of this work made its way into Italy in the fifteenth century; and it is probable that Copernicus became aware of it during his sojourn in Padua, 1501–03. Copernicus's use of the Ṭūsī couple made it well-known and indeed popular among European astronomers of the late sixteenth century; Boulliau was still using it in the mid-seventeenth century.

variation: a lunar anomaly discovered by Tycho Brahe, in accordance with which the Moon speeds up in the syzygies and slows down in the quadratures. The resulting displacements of the Moon from its mean positions are maximal in the octants, and given by the formula $39.5' \sin 2D$, where D is the difference between the mean longitudes of the Moon and the Sun. (Tycho's value for the coefficient was $40.5'$.)

vicarious hypothesis: Kepler's name for an equant-style theory of Mars that he developed in the early years of his "war on Mars", and which has the merit of predicting with considerable accuracy the heliocentric longitudes of the planet (the largest deviation of the theory from observation, he found, was a little over $2'$). But, as Kepler next found, the theory is false in its predictions of Mars–Sun distances. The hypothesis is 'vicarious' because, in all Kepler's further study of Mars, it serves as an interpolating device, yielding nearly correct values for the heliocentric longitude of Mars at any time: in this one respect, it can 'stand in' for the true theory (see Chapter 5).

ILLUSTRATIONS: ACKNOWLEDGEMENTS AND SOURCES

1.1 Tycho Brahe, *Astronomiae instauratae mechanica* (Wandsbeck, 1598), f. H^v.

1.2 Tycho Brahe, *De stella nova* (Copenhagen, 1573), f, B^v.

1.3 Vienna MS 10689.17, f. 8a. By permission of Osterreichische Nationalbibliothek, Vienna, and courtesy of Owen Gingerich.

1.4 Tycho Brahe, *De mundi aetherei recentioribus phaenomenis* (Uraniburg, 1588), p. 189.

1.5 *Ibid.*, p. 191.

1.6 Vienna MS 10658, f. 106v–107. By permission of Osterreichische Nationalbibliothek, Vienna, and courtesy of Owen Gingerich.

1.7 Vienna MS 10686.84, f. 21. By permission of Osterreichische Nationalbibliothek, Vienna, and courtesy of Owen Gingerich.

1.8 Tycho Brahe, *Astronomiae instauratae mechanica* (Wandsbeck, 1598), f. E4v.

1.11 Courtesy of Owen Gingerich.

2.1 Václav Dasypodius, *De miserando rerum statu* (Prague, 1579), courtesy of Zdeněk Kopal.

2.2 Thaddaeus Hagecius, *Dialexis de novae . . . stellae . . . apparitione* (Frankfurt, 1574), p. 20.

2.3 Courtesy of Uppsala Observatory Library and Owen Gingerich.

2.4 From the copy of *De revolutionibus*, Stadtbibliothek Schweinfurt 6796.

2.5 Courtesy of Hessisches Landesmuseum, Kassel.

2.6 Vatican Ottoboniana 1902, f. 210v, courtesy of Owen Gingerich.

2.7 Thomas Digges, *A Perfit Description of the Caelestiall Orbes*, an addition to Leonard Digges, *A Prognostication Everlastinge* (London, 1576).

3.1 Nicolaus Reymers Ursus, *Fundamentum astronomicum* (Strassburg, 1588). Permission of the Houghton Library, Harvard University, photograph by Owen Gingerich.

3.2 Nicolaus Reymers Ursus, *De hypothesibus astronomicis* (Prague, 1597), collection of Owen Gingerich.

3.3 Courtesy of the Master and Fellows of Trinity College, Cambridge.

3.4 P. Gassendi, *Institutio astronomica* (Paris, 1647), Lib. III, Cap. XII, p. 212.

3.5 G.B. Riccioli, *Almagestum novum* (Bologna, 1651), frontispiece. Collection of Owen Gingerich.

3.6 Christoph Rothmann, MS astron., 11, 9, courtesy of Landesbibliothek und Murhardsche Bibliothek der Stadt Kassel.

4.1 W. Gilbert, *De magnete*, 2nd (pirated) edn (Stettin, 1628), frontispiece.

4.2 J. Grandami, *Demonstratio immobilitatis terrae* (Le Flèche, 1645), frontispiece.

5.1 Adapted from Johannes Kepler, *Mysterium cosmographicum* (Tübingen, 1596), p. 9, courtesy of Owen Gingerich.

5.2 Adapted from Kepler, *Mysterium cosmographicum*, Tab. III, courtesy of Owen Gingerich.

5.4 From Kepler's Mars workbook of 1600; archives of the Academy of Sciences of the USSR, Leningrad Kepler MSS, XIV, f. 95v, courtesy of Owen Gingerich.

5.5 J. Kepler, *Astronomia nova* (Prague, 1609), p. 132.

5.6 *Ibid.*, p. 108.

5.7 From Kepler's Mars workbook of 1600; archives of the Academy of Sciences of the USSR, Leningrad Kepler MSS, XIV, f. 71v, courtesy of Owen Gingerich.

5.8 J. Kepler, *Astronomia nova* (Prague, 1609), p. 149.

5.11 J. Kepler, *Harmonice mundi* (Linz, 1619), p. 207.

5.12 J. Kepler, *Tabulae Rudolphinae* (Ulm, 1627), frontispiece, courtesy of Owen Gingerich.

6.1 Galileo, *Istoria e Dimonstrazioni intorno alle Macchie Solari* (Rome, 1613), frontispiece.

6.2 Woodcut sketches of the Moon, first published in Galileo's *Sidereus nuncius* (Venice, 1610), and here taken from the reprint of the woodcuts included in *De phoenomenis in orbe Lunae* by G.C. La Galla (Venice, 1612), courtesy of Bibliotheca Nazionale, Rome, and the Rev. O. Van der Vyer, SJ.

6.3 Reproduced from *Le opere di Galileo Galilei*, ed. by A. Favaro (Florence, 1890–1909), vol. 3, pt 2, p. 427.

6.4 Galileo, *Sidereus nuncius* (Venice, 1610), frontispiece.

6.5 *Ibid.*, f. 23r.

6.6 *Ibid.*, f. 16r.

6.7 (*a*) Letter of Galileo to Belisario Vinta, in *Le Opere di Galileo Galilei*, ed. by A. Favaro (Florence, 1890–1909), x, p. 410.
(*b*) Munich Bayrlsche Staatsbibliothek, M3 Clm 12425, f. 66.
(*c*) Drawing in Galileo's papers, reproduced in *Le Opere di Galileo Galilei*, xii, p. 276.
(*d*) P. Gassendi, Commentarii de rebus caelestibus, in *Opera omnia* (Lyons, 1658), iv, p. 362.
(*e*) F. Fontana, in Galileo MS 95, 81^r, Biblioteca Nazionale Centrale, Florence, reproduced in *Physis*, vi (1964), p. 441.
(*f*) J. Hevelius, *Selenographia* (Danzig, 1647), facing p. 42.
(*g*) E. Divini . . . reproduced from *Bulletino di Bibliografia e di Storia délle Scienze Matematiche e Fisiche*, xx (1887), facing p. 614.
(*h*) G. Odierna, *Protei caelestis vertigines seu Saturni systema* (Palermo, 1657), frontispiece.
(*i*) C. Wren, De corpore Saturni, in C. Huygens,

Oeuvres complètes (The Hague, 1888–1950), iii, facing p. 424.

(*j*) C. Huygens, *Systema Saturnium* (The Hague, 1659), p. 55.

(*k*) G. Campani, *Ragguaglio di Due Nuove Osservazioni* (Rome, 1664), facing p. 8.

(*l*) G.D. Cassini, from letter to Oldenburg, *Philosophical Transactions*, no. 128 (1676).

6.8 Vienna MS 10702, f. 74. By permission of Osterreichische Nationalbibliothek, Vienna, and courtesy of Owen Gingerich.

6.9 J. Kepler, *Epitome astronomiae Copernicanae* (Linz/Frankfurt, 1618–21), p. 833.

6.10 Galileo, *Istoria e Dimonstrazioni Intorno alle Macchie Solari* (Rome, 1613), pp. 66–7.

6.11 Courtesy of the Biblioteca Nazionale Centrale, Florence, shelf-mark Palat C.10.6.26.

6.12 Galileo, *Dialogo Sopra i Due Massimi Sistemi del Mondo* (Florence, 1632), frontispiece.

6.13 *Ibid.*, title-page.

6.14 *Ibid.*, p. 320.

7.2 P. Gassendi, *Mercurius in Sole visus*, appended to revised edition of his *Institutio astronomica* (The Hague, 1656), p. 184.

7.3 Courtesy of the British Library.

7.4 G.B. Riccioli, *Almagestum novum* (Bologna, 1651), i, p. 712, collection of Owen Gingerich.

7.5 Christiaan Huygens, *Cosmotheoros* (The Hague, 1698), p. 122.

8.1 Codex Atlanticus, ff. 310r and 674v, courtesy of Giunti Barbéra, Florence.

8.2 From the original in the British Library.

8.3 Based on the original MS, courtesy of the Earl of Egremont and Leconfield.

8.4 Courtesy of the Earl of Egremont and Leconfield.

8.5 (*a*) Galileo, *Sidereus nuncius* (Venice, 1610), p. 10; (*b*) drawing accompanying the first draft of Galileo's *Sidereus nuncius*, courtesy of Biblioteca Nazionale Centrale, Florence.

8.6 (*a*) C. Scheiner, *Disquisitiones mathematicae* (Ingolstadt, 1614), p. 58; (*b*) C. Malapert, *Oratio habita* (Douai, 1619); (*c*) G. Biancani, *Sphaera mundi* (Bologna, 1620); (*d*) C. Borri, *Collecta astronomica* (Lisbon, 1631), 137; and (*e*) A. Argoli, *Pandosion sphaericum*, 1st edn (Padua, 1644), p. 228.

8.7 First quarter and full Moon, courtesy of Cabinet des Estampes, Bibliothèque Nationale, Brussels; last quarter, courtesy of Observatorio de Marina, San Fernando.

8.8 Crawford Library, Royal Observatory, Edinburgh, courtesy of the Astronomer Royal for Scotland.

8.9 J. Hevelius, *Selenographia* (Danzig, 1647), following p. 316.

8.10 *Ibid.*, Fig. R.

8.11 *Ibid.*, Fig. Q.

8.12 Courtesy of Observatorio de Marina, San Fernando.

8.13 G.B. Riccioli, *Almagestum novum* (Bologna, 1651), i, following p. 204.

8.14 *Ibid.*

8.15 R. Hooke, *Micrographia* (London, 1665), Schema 38.

8.16 Courtesy of the Sid W. Richardson History of Science Collection, University of Texas at Austin.

8.17 (*a*) Courtesy of the Paris Observatory.

8.18 P. de la Hire, *Tabulae astronomicae* (Paris, 1702), Plate 4.

9.2 G.D. Cassini, *Ephemerides Bononienses Mediceorum syderum* (Bologna, 1668), pp. 2–3, courtesy of Paris Observatory.

9.3 Courtesy of Paris Observatory.

10.2 J. Horrocks, *Philosophicall Exercises* (unpublished), RGO MS 1/68B, f. 75v (p. 45 of Horrocks's text), courtesy Royal Greenwich Observatory.

10.3 I. Boulliau, *Astronomia Philolaïca* (Paris, 1645), p. 32, collection of Owen Gingerich.

10.10 T. Streete, *Astronomia Carolina* (London, 1661), p. 16, collection of Owen Gingerich.

10.11 N. Mercator, *Institutionum astronomicarum libri duo* (London, 1676), p. 163, collection of Owen Gingerich.

10.14 Collection of Owen Gingerich.

10.15 T. Streete, *A Compleat Ephemeris for . . . 1682* (London, 1682), entries for November.

Graphs 1–5 Owen Gingerich and Barbara Welther, *Planetary, Lunar, and Solar Positions, New and Full Moons, A.D. 1650–1806* (American Philosophical Society, Philadelphia, 1983), pp. xiii–xvii.

10.16 I. Boulliau, *Astronomia Philolaïca* (Paris, 1645), p. 103, collection of Owen Gingerich.

10.18 J. Horrocks, *Philosophicall Exercises* (unpublished), RGO MS 1/68B, f. 82v, courtesy of Royal Greenwich Observatory.

10.19 J. Horrocks, *Opera posthuma* (London, 1673), p. 471.

11.1 R. Descartes, *Principia philosophiae* (Amsterdam, 1644), p. 110

11.2 *Ibid.*, p. 92.

11.3 *Ibid.*, p. 185.

11.4 *Ibid.*, p. 220.

12.1 W. Gilbert, *De magnete* (London, 1600), p. 191.

12.2 J. Wilkins, *A Discourse Concerning a New World and Another Planet* (3rd impression, corrected and enlarged, London, 1640), Bk I, p. 212.

12.3 R. Hooke, *Cometa*, in *Lectures and Collections* (London, 1678), Tab. V.

13.2 Trinity College Cambridge Library, MS R.4.48[1], by permission of the Master and Fellows.

13.3 Hooke's letter to Newton, 9 December 1679; reproduced by permission of Yale University Library.

13.4 Hooke's letter to Newton, 9 December 1679; reproduced by permission of Yale University Library.

13.5 Newton's letter to Hooke, 13 December 1679; reproduced by permission of the British Library.

13.7 Library of Corpus Christi College, Oxford, MS 361, f. 3r, courtesy of the President and Fellows.

13.8 From Newton's letter to Crompton for Flamsteed of 28 February 1680/81; reproduced by permission of the President and Fellows of Corpus Christi College, Oxford.

13.9 Photograph by Charles Eames, by permission of the Royal Society of London.

13.10 Photograph by Charles Eames, by permission of the Royal Society of London.

13.11 By permission of Bodmeriana Library, Geneva, photograph by Owen Gingerich.

13.12 I. Newton, *Principia*, 3rd edn (London, 1726), p. 428.

13.13 *Ibid.*, p. 462.

13.14 *Ibid.*, p. 428.

13.15 *Ibid.*, p. 488.

13.16 *Ibid.*, p. 492.

13.17 I. Newton, *Principia*, 1st edn (London, 1687), facing p. 496.

INDEX